Earth
Science

Interactive
Textbook

HOLT, RINEHART AND WINSTON
A Harcourt Education Company

Orlando • Austin • New York • San Diego • London

ISBN-13: 978-0-03-099442-5
ISBN-10: 0-03-099442-X

12 0868 15 14 13 12
4500376197

Contents

CHAPTER 18 Studying Space

CHAPTER 19 Stars, Galaxies, and the Universe

CHAPTER 20 Formation of the Solar System

CHAPTER 21 A Family of Planets

CHAPTER 22 Exploring Space

SECTION 1 | Branches of Earth Science

BEFORE YOU READ

After you read this section, you should be able to answer these questions:

• What are the four major branches of Earth science?

• What are some special branches of Earth science?

What Is Geology?

Earth is a large and complicated place. How do scientists study it? The answer is that no one scientist studies all parts of Earth. Instead, different scientists study different parts of the planet. The study of different parts of Earth is called *Earth science*. There are many different *branches*, or types, of Earth science.

Geology is one branch of Earth science. **Geology** is the study of the origin, history, and structure of Earth. It also includes the study of the processes that shape Earth. A scientist who studies geology is called a *geologist*.

In most cases, a geologist studies one specific part of the Earth. For example, *volcanologists* study volcanoes. *Seismologists* study earthquakes. *Paleontologists* study the history of life on Earth. ☑

What Is Oceanography?

Another branch of Earth science is oceanography. **Oceanography** is the study of the sea. Scientists who study oceanography are called *oceanographers*.

Like geologists, oceanographers may focus on certain areas of oceanography. For example, *biological oceanographers* study the living things in the oceans. *Chemical oceanographers* study the amounts of different chemicals in ocean water.

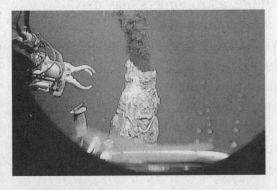

Many oceanographers use special tools, such as this submarine, to study the oceans.

STUDY TIP

Describe As you read this section, make a chart describing the four main branches of Earth science. In your chart, define each branch and give the term used to describe a scientist who studies it.

READING CHECK

1. Describe What do seismologists study?

_____

_____

TAKE A LOOK

2. Identify What do oceanographers study?

_____

_____

What Is Meteorology?

Meteorology is a branch of Earth science that deals with Earth's atmosphere, especially weather and climate. Scientists who study meteorology are called *meteorologists.* ☑

Many meteorologists try to *forecast*, or predict, the weather. In most cases, weather forecasts help to make our lives more comfortable. Sometimes, meteorologists can help save people's lives by predicting severe weather, such as hurricanes and tornadoes. These predictions can warn people to leave an area before severe weather strikes.

☑ **READING CHECK**

3. Define What is meteorology?

_____

_____

_____

TAKE A LOOK

4. Explain How can meteorology save people's lives?

_____

_____

_____

_____

These meteorologists are risking their lives to gather information about tornadoes. This information may help other meteorologists better predict where and when tornadoes will strike.

What Is Astronomy?

Astronomy is the study of the universe. *Astronomers* are scientists who study stars, asteroids, planets, and other objects in space.

Most objects in space are very far away. Therefore, astronomers depend on technology to help them study these objects. For example, astronomers may use telescopes to study distant stars and planets.

You may wonder why astronomy is a branch of Earth science if astronomers study objects far from the Earth. The reason is that many astronomers use information about other planets and stars to learn more about the Earth. For example, some astronomers study ancient stars in the universe. The information they gather can help them to predict how changes in our sun may affect the Earth.

Critical Thinking

5. Compare What is the main difference between astronomy and other branches of Earth science?

_____

_____

_____

_____

_____

_____

SECTION 1 Branches of Earth Science *continued*

What Are Some Other Branches of Earth Science?

Geology, oceanography, meteorology, and astronomy are the four main branches of Earth science. However, there are many other branches of Earth science.

ENVIRONMENTAL SCIENCE

Environmental science is the study of how humans interact with the environment. Environmental scientists help people learn ways to preserve the environment and to use resources wisely. ☑

READING CHECK

6. Identify Give two things that environmental scientists can help people learn to do.

_____

_____

ECOLOGY

Ecology is the study of relationships between living things and their environments. Ecologists study communities of organisms and their environments to better understand how organisms behave. Ecologists work in many fields, including agriculture and forestry.

GEOCHEMISTRY

Geochemistry combines the studies of geology and chemistry. Geochemists study the chemicals that make up Earth materials such as rocks, minerals, and soil. They can use this information to learn how the Earth materials formed. Geochemists may also study the effects of human-made chemicals on the environment.

Say It

Discuss In a small group, talk about different jobs that Earth scientists can have. Who may Earth scientists work for? What kinds of work could they do every day?

Geochemists may take rock samples and analyze them in a laboratory.

TAKE A LOOK

7. Describe What do geochemists study?

_____

_____

GEOGRAPHY AND CARTOGRAPHY

Geography is the study of the surface features of the Earth, such as continents, rivers, and mountains. Many geographers work in *cartography*, or map-making. Cartographers use information from photographs and computers to make maps. They may also study the ways that areas change with time.

Section 1 Review

SECTION VOCABULARY

astronomy the scientific study of the universe **geology** the scientific study of the origin, history, and structure of Earth and the processes that shape Earth	**meteorology** the scientific study of Earth's atmosphere, especially in relation to weather and climate **oceanography** the scientific study of the ocean, including the properties and movements of ocean water, the characteristics of the ocean floor, and the organisms that live in the ocean

1. List What are the four major branches of Earth science?

_____

_____

_____

_____

2. Infer What kind of Earth scientist would most likely study thunderstorms? Explain your answer.

_____

_____

3. Explain Why is astronomy a branch of Earth science?

_____

_____

_____

4. Compare How is environmental science different from ecology?

_____

_____

_____

_____

5. Explain Why do astronomers depend on technology?

_____

_____

6. Define What is geography?

_____

_____

CHAPTER 1 | The World of Earth Science

SECTION 2

Scientific Methods in Earth Science

After you read this section, you should be able to answer these questions:

• What are the steps used in scientific methods?

• How is a hypothesis tested?

• Why do scientists share their findings with others?

How Do Scientists Learn About the World?

Imagine you are standing in a thick forest. Suddenly, you hear a booming noise, and you feel the ground shake. You notice a creature's head looming over the treetops.

The creature's head is so high that its neck must be 20 m long! Then, the whole animal comes into view. Now you know why the ground is shaking. The giant animal is *Seismosaurus hallorum*, the "earthquake lizard."

This description of *Seismosaurus hallorum* is not just from someone's imagination. Since the 1800s, scientists have gathered information about dinosaurs and their environment. Using this knowledge, scientists can infer what dinosaurs may have been like hundreds of millions of years ago.

How do scientists piece all the information together? How do they know if they have discovered a new species of dinosaur? Asking these questions is the first step in using scientific methods to learn more about the world.

STUDY TIP

Outline As you read this section, make a chart showing the ways that David Gillette used the steps in scientific methods to learn more about the dinosaur bones he studied.

Seismosaurus hallorum was one of the largest dinosaurs that ever lived.

Math Focus

1. Make Comparisons
When a *Seismosaurus* held its head up as high as it could, it could have been 25 m tall. What fraction of Seismosaurus's height is your height? Give your answer as a decimal.

What Are Scientific Methods?

Scientific methods are a series of steps that scientists use to answer questions and to solve problems. Although each question is different, scientists can use the same methods to find answers. ☑

Scientific methods have several steps. Scientists may use all of the steps or just some of them. They may even repeat some of the steps.

The goal of scientific methods is to come up with reliable answers and solutions. These answers and solutions must be able to stand up to the testing of other scientists.

2. Define What are scientific methods?

_____

_____

_____

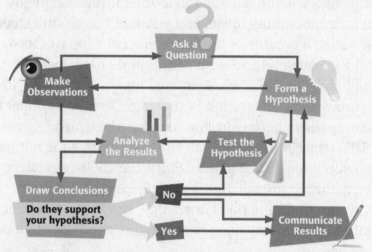

There are many steps in scientific methods. Notice that there are many ways to move through the different steps.

TAKE A LOOK
3. Use Models Starting with "Ask a question," trace two different paths through the figure to "Communicate results." Use a colored pen or marker to trace your paths.

Why Is It Important to Ask a Question?

Asking a question helps scientists focus on the most important things they want to learn. The question helps to guide the research that the scientist does.

David D. Gillette is a scientist who studies fossils. In 1979, he began to study some fossil bones from New Mexico. He knew they came from a dinosaur, but he did not know which kind.

Gillette began his study by asking, "What kind of dinosaur did these bones come from?" We will use a table to follow Gillette as he tried to answer his question using scientific methods.

Step in scientific methods	How did David Gillette apply this step?
Asking questions	He wondered what kind of dinosaur the bones came from.

SECTION 2 Scientific Methods in Earth Science *continued*

How Do Scientists Form a Hypothesis?

When scientists want to investigate a question, they form a hypothesis. A **hypothesis** (plural, *hypotheses*) is a possible answer to a question. It is sometimes called an educated guess. ☑

The hypothesis is a scientist's best answer to the question. However, a scientist can't just assume that a hypothesis is the correct answer. The scientists must test the hypothesis to see if it is true.

From his observations and knowledge about dinosaurs, Gillette formed a hypothesis about the bones. He thought that the bones came from a kind of dinosaur that had not been discovered yet. To test his hypothesis, Gillette had to do a lot of research.

READING CHECK
4. Define What is a hypothesis?
_____
_____

Step in scientific methods	How did David Gillette apply this step?
Forming a hypothesis	

TAKE A LOOK
5. Identify Fill in the blank spot in the table with David Gillette's hypothesis.

How Do Scientists Test a Hypothesis?

To see if an idea can be proven scientifically, scientists must test the hypothesis. They do this by gathering data. *Data* (singular, *datum*) are pieces of information gathered through observation or experimentation. Scientists use data to learn if their hypotheses are correct. ☑

READING CHECK
6. Define What are data?
_____
_____
_____

TESTING WITH EXPERIMENTS

To test a hypothesis, a scientist may perform a controlled experiment. A *controlled experiment* tests only one factor, or *variable*, at a time. No other variables change. By changing only one variable, scientists can see the results of changing just one thing.

For example, suppose a scientist does an experiment to learn the temperature that a rock melts at. The scientist uses several samples of the same kind of rock. She heats each sample to a different temperature and records whether the rock melts. The type of rock does not change, but the temperature does. Therefore, temperature is the variable.

SECTION 2 Scientific Methods in Earth Science *continued*

KEEPING ACCURATE RECORDS

During experiments, scientists must keep accurate records of everything that they do and observe. This includes failed attempts, too. Keeping detailed records helps scientists to show that their results are accurate. Accurate records can also help other scientists to repeat an experiment.

TESTING WITHOUT EXPERIMENTS

Sometimes, it is not possible to do a controlled experiment. In such cases, scientists depend on observation to test their hypotheses. By observing nature, scientists can collect large amounts of data. If the data support a hypothesis, the hypothesis is probably correct.

To test his hypothesis, Gillette took hundreds of measurements of bones. He also visited museums and talked with other scientists.

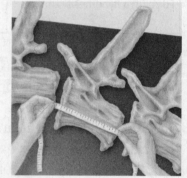

To test his hypothesis, Gillette took hundreds of measurements of the sizes and shapes of the bones.

Critical Thinking

7. Compare and Contrast What is the difference between a controlled experiment and observation?

_____

_____

_____

_____

TAKE A LOOK
8. Identify In the table, fill in the way that David Gillette tested his hypothesis.

Step in scientific methods	How did David Gillette apply this step?
Testing hypotheses	

How Do Scientists Analyze Results?

When scientists finish collecting data, they must analyze the results. Analyzing results helps scientists explain their observations. Their explanations are based on the evidence they collect. ☑

To arrange their data, scientists often make tables and graphs. Gillette organized his data in a table that compared the sizes and shapes of his dinosaur bones. He compared his measurements to measurements of bones from known dinosaurs. When he analyzed his results, he found that the mystery dinosaur's bones did not match the bones of any known dinosaur.

✓ READING CHECK
9. Explain Why do scientists analyze their data?

_____

_____

_____

SECTION 2 Scientific Methods in Earth Science *continued*

Step in scientific methods	How did David Gillette apply this step?
Analyzing results	

TAKE A LOOK
10. Identify In the table, fill in the way that David Gillette analyzed his results.

What Are Conclusions?

After analyzing results, a scientist must decide if the results agree with, or support, the hypotheses. This is called *drawing conclusions*. Finding out that a hypothesis is not true can be as valuable as finding out that a hypothesis is true.

Sometimes, the results do not support the hypothesis. When this happens, scientists may repeat the investigation to check for mistakes. Scientists may repeat experiments hundreds of times. Another option is to ask another question and make a new hypothesis.

From all his work, Gillette concluded that the bones found in New Mexico were from an unknown dinosaur. From his data, he learned that the new dinosaur was about 35 m long and had a mass of 30 to 70 metric tons. The dinosaur definitely fits the name Gillette gave it— *Seismosaurus hallorum*, or the "earthquake lizard."

Critical Thinking
11. Infer How can finding out that a hypothesis is not true be useful for a scientist?

_____

_____

_____

_____

Step in scientific methods	How did David Gillette apply this step?
Drawing Conclusions	

TAKE A LOOK
12. Identify In the table, fill in David Gillette's conclusions about his dinosaur bones.

Why Do Scientists Share Their Findings?

After finishing a study, scientists share their results with others. They write reports and give presentations. They can also put their results on the Internet.

Sharing information gives other scientists the chance to repeat the experiments for themselves. If other scientists get different results, more studies must be done to find out if the differences are important. ☑

In many cases, the results of an investigation are reviewed year after year as new evidence is found. In the case of *Seismosaurus*, the debate still continues. Some scientists think that *Seismosaurus* is a new genus of dinosaur. Other scientists think that it belongs to the genus *Diplodocus*. As scientists gather more data on the fossil bones, they may change their conclusions.

READING CHECK
13. Describe Why is it important for scientists to share their results?

_____

_____

_____

_____

Section 2 Review

SECTION VOCABULARY

hypothesis a testable idea or explanation that leads to scientific investigation	**scientific methods** a series of steps followed to solve problems

1. Describe How can a scientist test a hypothesis if it is not possible to do a controlled experiment?

_____

_____

_____

2. Explain Why is it important for scientists to ask questions?

_____

_____

_____

3. Apply Procedures You observe that your tongue sticks to a very cold ice pop when you lick it. You ask yourself, "Why does my tongue stick to the ice pop?" Make a hypothesis about why your tongue sticks to the ice pop.

_____

_____

_____

4. Identify How could you share the results of an experiment with the rest of your class? Give three ways.

_____

_____

_____

5. Infer Why might a scientist need to repeat a step in scientific methods?

_____

_____

_____

_____

_____

CHAPTER 1 | The World of Earth Science

SECTION 3 | # Scientific Models

After you read this section, you should be able to answer these questions:

• How do scientists use models?
• What are three kinds of scientific models?

What Are Models?

Why do scientists use crash-test dummies to learn how safe cars are? By using crash-test dummies, scientists can learn how to make cars safer without putting real people in danger. A crash-test dummy is a model of a person. A **model** is something scientists use to represent an object or event in order to make it easier to study.

Scientists use models to study things that are very small, like atoms, or things that are very large, like Earth. Some scientists use models to predict things that haven't happened yet, or to study events that happened long ago. Some models, like crash-test dummies, allow scientists to study events without affecting or harming the things they are studying. ☑

Models are very useful for scientists. However, you cannot learn everything by studying a model, because models are not exactly like the objects they represent.

PHYSICAL MODELS

Physical models are models that you can see or touch. Many physical models look like the things they represent. Other physical models may look different from the things they represent. For example, a map is a physical model of Earth. However, a flat map looks very different from the round Earth! ☑

STUDY TIP

Compare As you read, make a table to show the features of physical models and mathematical models.

READING CHECK

1. Identify Give two reasons scientists use models.

_____
_____

READING CHECK

2. Define What is a physical model?

_____
_____

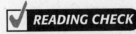

A globe is a physical model of the Earth.

SECTION 3 Scientific Models *continued*

MATHEMATICAL MODELS

A *mathematical model* is made up of data and mathematical equations. A *mathematical equation* shows how data are related to each other. Some mathematical models are simple. They can help you calculate things such as how far a car will travel in an hour. Other models are more complicated. These models can contain a lot of data related by complicated equations. ☑

Meteorologists often use mathematical models called *climate models* to help them study the Earth's climate. Most climate models include large amounts of data. The data may be measurements of temperatures or amounts of rainfall.

Climate models use equations to represent different parts of Earth's climate. For example, some equations represent the way that ocean water moves. Others represent the way that the amount of carbon dioxide in the air changes with time.

You may wonder how scientists can use models that contain so many data and equations. Scientists use computers to help them process these complicated models. Because computers can deal with large amounts of data, they can solve many mathematical problems at once. Computers can do complicated calculations more quickly and accurately than people can. ☑

Climate models, like most mathematical models, do not make exact predictions. Instead, they estimate what may happen. Scientists and lawmakers can use the estimates to help them plan for the future.

READING CHECK

3. Define What is a mathematical model?

_____

_____

READING CHECK

4. Explain Why do scientists use computers to process many mathematical models?

_____

_____

_____

TAKE A LOOK

5. Identify What are two kinds of data that may be part of a climate model?

_____

_____

The climate model in this picture was produced by a computer. The computer combined huge amounts of data and equations into the climate model. Without computers, scientists would not be able to use complicated models like this.

SECTION 3 Scientific Models *continued*

CONCEPTUAL MODELS

A *conceptual model* is a diagram, drawing, or spoken description of how something works or is put together. Conceptual models may be made of many different hypotheses. Each hypothesis is supported by scientific methods. For example, the conceptual model below shows how mercury moves through the environment. Scientists have used scientific methods to learn how mercury from coal burning can affect humans.

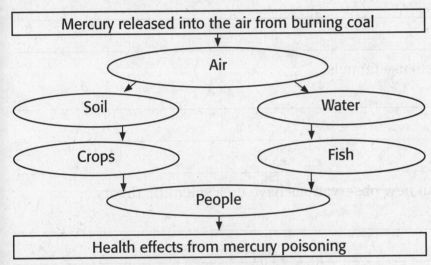

Why Do Scientists Use Models?

Scientists often use models to help explain or support scientific laws and theories. A scientific *law* is a statement or equation that can predict what will happen in certain situations. A scientific **theory** is an explanation that connects and explains many observations.

Name	What it is
Scientific theory	an explanation that connects and explains evidence and observations
Scientific law	a statement or equation that predicts what will happen in a certain situation

Scientific theories are based on observations. They explain all of the observations about a topic that scientists have at a certain time. However, scientists are always discovering new information. This new information may show that a theory is incorrect. When this happens, the theory must be changed so that it explains the new information. Sometimes, scientists have to develop a totally new theory to explain the new and old information.

TAKE A LOOK
6. Use a Model Use a colored pen or marker to trace two different ways that mercury in the air can affect people.

Critical Thinking
7. Compare How is a scientific theory different from a scientific law?

_____
_____
_____
_____
_____

Section 3 Review

SECTION VOCABULARY

model a pattern, plan, representation, or description designed to show the structure or workings of an object, system, or concept	**theory** a system of ideas that explains many related observations and is supported by a large body of evidence acquired through scientific investigation

1. Identify How are scientific theories related to observations and evidence?

_____

_____

_____

2. Explain Why do scientists use models?

_____

_____

_____

3. Describe What effect can new observations have on a scientific theory?

_____

_____

_____

4. List Give one example of a physical model and one example of a mathematical model.

_____

_____

5. Explain Why do scientists use computers to process climate models?

_____

_____

6. Infer A globe is a model of the Earth. Give two ways a globe is like the Earth and two ways a globe is not like the Earth.

_____

_____

_____

_____

CHAPTER 1 | The World of Earth Science

SECTION 4 | # Measurement and Safety

After you read this section, you should be able to answer these questions:

• What is the SI?

• How do scientists measure length, area, mass, volume, and temperature?

How Do Scientists Measure Objects?

Scientists make many measurements as they collect data. It is important for scientists to be able to share their data with other scientists. Therefore, scientists use units of measurement that are known to all other scientists. One system of measurement that most scientists use is called the International System of Units.

THE INTERNATIONAL SYSTEM OF UNITS

The *International System of Units*, or *SI*, is a system of measurement that scientists use when they collect data. This system of measurement has two benefits. First, scientists around the world can easily share and compare their data because all measurements are made in the same units. Second, SI units are based on the number 10. This makes it easy to change from one unit to another.

It is important to learn the SI units that are used for different types of measurements. You will use SI units when you make measurements in the science lab.

LENGTH

Length is a measure of how long an object is. The SI unit for length is the **meter** (m). Centimeters (cm) and millimeters (mm) are used to measure small distances. There are 100 cm in 1 m. There are 1,000 mm in 1 m. Kilometers (km) are used to measure long distances. There are 1,000 m in 1 km.

Length	SI Unit: meter (m)	
	kilometer (km)	1 km = 1,000 m
	centimeter (cm)	1 cm = 0.01 m
	millimeter (mm)	1 mm = 0.001 m

STUDY TIP

Compare As you read this section, make a table comparing how scientists measure length, area, mass, volume, and temperature. Include the units of measurement that scientists use.

Critical Thinking

1. Predict Consequences What could happen if all scientists used different systems of measurement to record their data?

_____

_____

_____

TAKE A LOOK
2. Identify What is the SI unit for length?

_____

AREA

Area is the measure of how much surface an object has. For most objects, area is calculated by multiplying two lengths together. For example, you can find the area of a rectangle by multiplying its length by its width. Area is measured in square units, like square meters (m^2) or square centimeters (cm^2). There are 10,000 cm^2 in 1 m^2. ☑

✓ READING CHECK

3. Explain How can you find the area of a rectangle?

Area	square meter (m^2)	
	square centimeter (cm^2)	1 cm^2 = 0.0001 m^2

VOLUME

Volume is the amount of space an object takes up. There are two main ways to find the volume of an object. You can find the volume of a box-shaped object by multiplying its length, width, and height together. To find the volume of other objects, measure the volume of water that they push out of a container. ☑

The volume of a solid object is often measured in cubic units. For example, very large objects can be measured in cubic meters (m^3). Smaller objects can be measured in cubic centimeters (cm^3). There are 1 million cm^3 in 1 m^3.

The volume of a liquid is usually given in units of liters (L) or milliliters (mL). One mL has the same volume as one cm^3. There are 1,000 mL in 1 L. There are 1,000 L in one m^3.

✓ READING CHECK

4. Define What is volume?

Math Focus
5. Calculate How many milliliters are there in 100 L?

Volume	cubic meter (m^3)	1 cm^3 = 0.000001 m^3
	cubic centimeter (cm^3)	1 L = 0.001 m^3
	liter (L)	1 mL = 1 cm^3
	milliliter (mL)	1 mL = 0.001 L

MASS

Mass is a measure of the amount of matter in an object. The SI unit for mass is the kilogram (kg). The masses of large objects, such as people, are measured using kilograms. The masses of smaller objects, such as apples, are measured in grams (g) or milligrams (mg). There are 1,000 g in 1 kg. There are 1 million mg in 1 kg. ☑

✓ READING CHECK

6. Identify What is the SI unit for mass?

Mass	SI Unit: kilogram (kg)	
	gram (g)	1 g = 0.001 kg
	milligram (mg)	1 mg = 0.000001 kg

TEMPERATURE

Temperature is a measure of how hot or cold an object is. The SI unit for temperature is the Kelvin (K). However, most people are more familiar with other units of temperature. For example, most people in the United States measure temperatures using degrees Fahrenheit (°F). Scientists often measure temperatures using degrees Celsius (°C). ☑

Temperature	SI Unit: Kelvin (K)	0°C = 273 K
	degrees Celsius (°C)	100°C = 373 K

READING CHECK

7. Define What is temperature?

_____

_____

DENSITY

Density is a measure of how closely packed the particles in a substance are. You can calculate an object's density by dividing the object's mass by its volume. There is no SI unit for density. Scientists usually use the units grams per milliliter (g/mL) or grams per cubic centimeter (g/cm^3) to measure density.

Density	grams per milliliter (g/mL)	1 g/mL = 1 g/cm^3
	grams per cubic centimeter (g/cm^3)	

How Can You Stay Safe in Science Class?

Science can be exciting, but it can also be dangerous. In order to stay safe while you are doing a science activity, you should always follow your teacher's directions. Read and follow the lab directions carefully, and do not take "shortcuts." Pay attention to safety symbols, such as the ones in the figure below. If you do not understand something that you see in a science activity, ask your teacher for help.

Safety Symbols

Eye protection

Clothing protection

Hand safety

Heating safety

Electrical safety

Chemical safety

Animal safety

Sharp object

Plant safety

TAKE A LOOK

8. Investigate Look around your classroom for safety symbols like the ones in the figure. Give two examples of places where safety symbols are found in your classroom.

_____

_____

Section 4 Review

SECTION VOCABULARY

area a measure of the size of a surface or a region	**meter** the basic unit of length in the SI (symbol, m)
density the ratio of the mass of a substance to the volume of the substance	**temperature** a measure of how hot (or cold) something is; specifically, a measure of the average kinetic energy of the particles in an object
mass a measure of the amount of matter in an object	**volume** a measure of the size of a body or region in three-dimensional space

1. Identify What are two units that scientists use to measure temperature?

_____

_____

2. List What are two benefits of using the SI?

_____

_____

3. Describe How can you find the volume of a shoebox?

_____

_____

4. Identify Fill in the blank spaces in the table below. Give at least two examples for each measurement

Type of measurement	Examples of units used for this measurement
length	
area	
mass	
volume	

5. Apply Ideas An object has a mass of 10 g and a volume of 5 cm³. What is its density?

_____

4. Explain Give three ways that you can stay safe while doing a science activity.

_____

_____

_____

| CHAPTER 2 | Maps as Models of the Earth |

SECTION 1 | You Are Here

BEFORE YOU READ

After you read this section, you should be able to answer these questions:

• What is a map?

• What are latitude and longitude?

• How can you find locations on Earth?

What Is a Map?

A **map** is a model that shows the features of an object. Most maps that people use show the features of Earth's surface. Some maps show all of the Earth's surface. Other maps show only part of it. Maps can show natural features, such as rivers. They can also show features made by people, such as roads.

FINDING DIRECTIONS ON EARTH

The Earth's shape is similar to a sphere, but the Earth is not a true sphere. A *true sphere* has no top, bottom, or sides—it looks the same from all directions. In addition, a true sphere has no reference points. *Reference points* are certain locations that never change. They can be used to define directions. ☑

Unlike a true sphere, the Earth has two reference points. They are located where the Earth's axis of rotation passes through the Earth's surface. The reference points are called the North Pole and the South Pole. The North and South Poles are known as *geographic poles*. Since these poles never move, they are used as reference points to define directions on Earth.

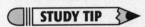

STUDY TIP

Discuss Read this section quietly to yourself. In a small group, try to figure out anything you didn't understand.

READING CHECK

1. Define What are reference points?

_____

_____

_____

The North Pole and the South Pole can be used as reference points to define directions on Earth.

North Pole

Earth's axis of rotation

South Pole

TAKE A LOOK

2. Identify What are two reference points on Earth?

_____

_____

CARDINAL DIRECTIONS

A reference point alone will not help you give good directions. You also need to describe how to get from one place to another. To do this, you can use the cardinal directions. The *cardinal directions* are north, south, east, and west. To use cardinal directions properly, you will need to use a compass. ☑

USING A COMPASS TO FIND DIRECTIONS

The Earth is like a giant magnet. It has two *magnetic poles*, which are located near the geographic poles. You can use the magnetic poles to help you find the cardinal directions.

A *compass* is a tool that uses the Earth's natural magnetism to show direction. The needle on a compass points to the magnetic pole that is near the geographic North Pole. Therefore, you can use a compass to learn which direction is north. ☑

TRUE NORTH AND MAGNETIC DECLINATION

There is a difference between the location of the geographic North Pole and the magnetic pole. Therefore, a compass needle cannot show you exactly where the geographic North Pole is. When you use a compass, you have to correct for this difference. **True north** is the direction of the geographic North Pole. The angle between true north and the direction a compass needle points is called **magnetic declination**.

READING CHECK

3. Identify What are the four cardinal directions?

_____

_____

_____

_____

READING CHECK

4. Describe In which general cardinal direction does a compass needle point?

_____

_____

TAKE A LOOK

5. Define What is true north?

_____

_____

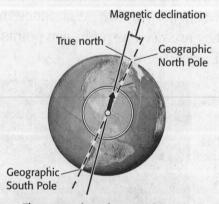

The magnetic and geographic poles are not in the same place. Therefore, compass readings must be corrected to learn the direction of true north.

USING MAGNETIC DECLINATION

Magnetic declination is measured in degrees east or west of true north. The magnetic declination is different for different points on the Earth's surface. Once you know the declination for your area, you can use a compass to determine the direction of true north. For example, suppose the magnetic declination in your area is 10°W. This means that true north is 10°E of the direction a compass needle points.

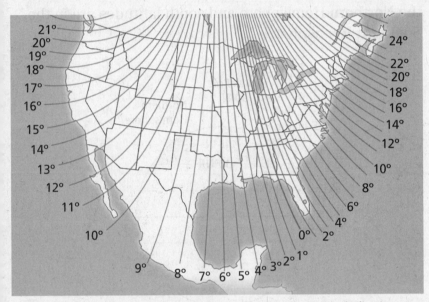

The lines on this map connect points that have the same magnetic declination.

Critical Thinking

6. Apply Concepts The magnetic declination for an area is 8°E. Compared to the direction a compass needle points, in which direction does true north lie?

_____

How Do People Find Specific Locations on Earth?

All of the houses and buildings in your neighborhood have addresses that give their locations. These addresses may include a street name and a number. You can tell someone exactly where you live by giving them your address. In a similar way, you can use latitude and longitude to give an "address" for any place on Earth.

TAKE A LOOK

7. Read a Map Roughly what is the magnetic declination in the area in which you live?

_____

LATITUDE

The **equator** is a circle halfway between the North and South Poles. It divides Earth into two *hemispheres*, or halves—the Northern Hemisphere and the Southern Hemisphere. *Lines of latitude*, or *parallels*, are imaginary lines on Earth's surface that are parallel to the equator. ☑

Latitude is the distance north or south from the equator. Latitude is measured in degrees. The equator represents 0° latitude. The North Pole is 90° north latitude and the South Pole is 90° south latitude. North latitudes are in the Northern Hemisphere and south latitudes are in the Southern Hemisphere.

READING CHECK

8. Define What are parallels?

_____

_____

_____

TAKE A LOOK
9. Describe Which parallel is farther from the equator: 10° N latitude or 10° S latitude?

_____

_____

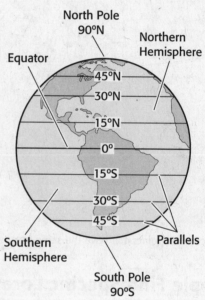

North Pole
90°N

Northern Hemisphere

Equator

45°N

30°N

15°N

0°

15°S

30°S

45°S

Parallels

Southern Hemisphere

South Pole
90°S

LONGITUDE

Lines of longitude, or *meridians*, are imaginary lines that link the geographic poles. Lines of longitude are similar to the lines on a basketball. The lines all touch at the poles. They are farthest apart at the equator. The **prime meridian** is the line that represents 0° longitude. **Longitude** is the distance east or west of the prime meridian. Like latitude, longitude is measured in degrees.

The prime meridian does not circle the whole globe. It runs from the North Pole, through Greenwich, England, to the South Pole. On the other side of the globe, the 180° meridian runs from the North to the South Pole. Together, the prime meridian and the 180° meridian divide the Earth into Western and Eastern Hemispheres.

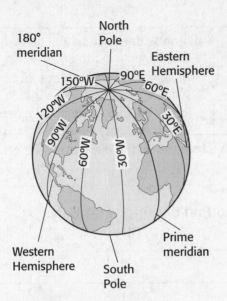

Critical Thinking

10. Apply Concepts A friend asks you what the distance is between 80°E longitude and 90°E longitude. What else do you need to know in order to answer your friend's question? Explain why you need this piece of information.

_____

_____

_____

_____

_____

_____

USING LATITUDE AND LONGITUDE

Lines of latitude and longitude cross to form a grid. This grid is shown on maps and globes. You can use the lines of latitude and longitude to tell someone the location of any point on the Earth's surface. First, find the point on a map like the one below. Then, estimate the latitude and longitude of the point, using the lines closest to it.

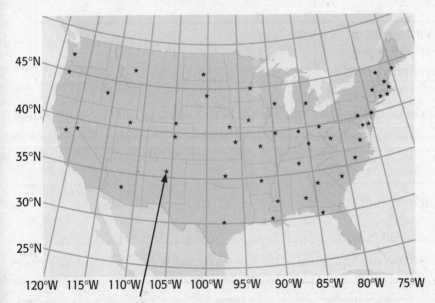

You can use latitude and longitude to locate points on a map. For example, this city is located very close to latitude 35°N, longitude 105°W.

TAKE A LOOK

11. Read a Map Circle the city on the map that is closest to latitude 45°N, longitude 100°W.

Section 1 Review

SECTION VOCABULARY

equator the imaginary circle halfway between the poles that divides the Earth into the Northern and Southern Hemispheres	**magnetic declination** the difference between the magnetic north and the true north
latitude the distance north or south from the equator; expressed in degrees	**map** a representation of the features of a physical body such as Earth
longitude the angular distance east or west from the prime meridian; expressed in degrees	**prime meridian** the meridian, or line of longitude, that is designated as 0° longitude
	true north the direction to the geographic North Pole

1. Explain How can you use a compass to find directions on Earth?

_____

_____

_____

_____

2. Compare Fill in the table below to compare latitude and longitude.

Name	What it is	Measured in	Distance apart
latitude		degrees	always the same
longitude	distance east or west of the prime meridian		

3. Identify What is the latitude of the North Pole? What is the latitude of the South Pole?

_____

_____

4. Identify Where do all lines of longitude meet?

_____

5. Describe What two lines divide the Earth into the Eastern and Western Hemispheres?

_____

_____

CHAPTER 2 | Maps as Models of the Earth

SECTION 2 **Mapping the Earth's Surface**

After you read this section, you should be able to answer these questions:

- What is a map projection?
- What are four types of map projections?
- What information should every map have?
- What are four tools used by modern mapmakers?

What Happens When a Globe Becomes a Map?

Because a globe is a sphere, a globe is the most accurate model of Earth. However, a globe is too small to show many details of the Earth's surface, such as roads and rivers. In order to show these details, we use maps.

A map is a flat model of Earth's curved surface. Maps can show many details that globes cannot. However, moving information from a curved to a flat surface causes errors in shapes, sizes, and distances. These errors are called *distortions*.

Below is a picture of an orange. It has a black imprint of Earth's land masses on it. If you peel the orange in one piece, you can try to make a flat map from the curved surface. Notice how the distances and shapes are distorted in the peeled orange.

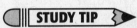
STUDY TIP

Compare As you read, make a chart comparing the features of cylindrical, conic, and azimuthal map projections. Describe the way that each projection is made and what features it distorts.

The continents have been drawn on the peel of this orange.

This is what the peel of the orange looks like when it is flattened out. Notice how shapes and distances between points on the orange are distorted when the peel is flattened.

TAKE A LOOK

1. Explain Why do all maps have distortions?

_____

_____

_____

SECTION 2 Mapping the Earth's Surface *continued*

What Are Map Projections?

A *map projection* is a certain way of transferring the curved surface of the Earth to a flat map. Mapmakers use map projections to move the image of the Earth's curved surface onto a flat map. ☑

To understand how map projections are made, think of a clear globe with a light inside it. If you wrap a piece of paper around the globe, the continents will make shadows on the paper. If you wrap the paper in different ways, the shadows will look different. Each way of wrapping the paper produces a different map projection.

There are four main kinds of map projections that mapmakers use—cylindrical, conic, azimuthal, and equal-area projections.

CYLINDRICAL PROJECTION

A cylinder is a shape like a an empty paper-towel tube. If you wrap the paper in a cylinder around the globe, the resulting map projection is called a **cylindrical projection**. The most common map projection is a cylindrical projection called a *Mercator projection*. In a Mercator projection, latitude and longitude lines are straight.

The Mercator projection has two main distortions. First, the space between longitude lines is equal everywhere, instead of being wider at the equator. Second, latitude lines are farther apart in the far north and south, instead of being evenly spaced everywhere. ☑

The distortions in a cylindrical projection make areas near the poles look wider and longer than they really are. For example, the Mercator projection makes Greenland look larger than South America!

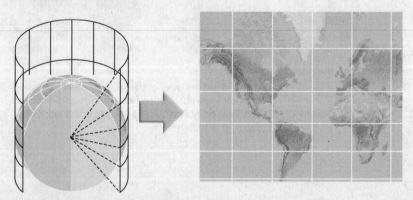

A cylindrical projection causes areas near the poles to look larger than they actually are.

CONIC PROJECTION

A cone is shaped like a party hat. If you wrap the paper in a cone around one of the Earth's hemispheres, the resulting map projection is called a **conic projection**.

The cone touches the globe at each line of longitude. Therefore, a conic projection shows each longitude line accurately. However, the cone only touches the globe at one line of latitude. Areas on this line of latitude are not distorted. However, areas far from this latitude look distorted in a conic projection. ☑

The distortions in a conic projection are worst from north to south. Therefore, conic projections are best for mapping land masses with more area east and west. They are good for mapping continents that are wide, such as the United States.

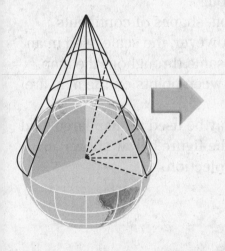 

A conic projection causes more distortion from north to south than from east to west.

READING CHECK

4. Explain Why do conic projections show longitude accurately?

_____

_____

_____

TAKE A LOOK
5. Identify Which parts of a conic projection are most distorted?

_____

_____

AZIMUTHAL PROJECTION

If you place a flat piece of paper on top of the globe, the resulting map projection is called an **azimuthal projection**.

On an azimuthal projection, the plane touches the globe at only one point. In most azimuthal projections, this point is one of the geographic poles. The distortion near this point is very small. However, distortions of direction, distance, and shape increase as you move away from this point.

Azimuthal projections are most often used to map areas that are near the North and South Poles. The figure at the top of the next page shows an azimuthal projection.

Critical Thinking

6. Infer Why are azimuthal projections most often used to map areas that are near the poles?

_____

_____

_____

_____

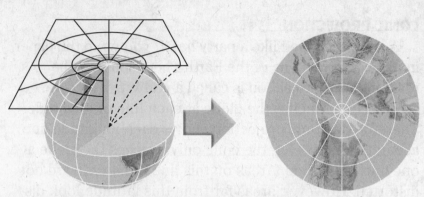

TAKE A LOOK
7. Identify Where is the smallest distortion in an azimuthal projection?

_____

_____

An azimuthal projection produces very little distortion at the point of contact. However, distortion increases as you move away from this point.

EQUAL-AREA PROJECTION

In an *equal-area projection*, the area between the latitude and longitude lines is the same as on a globe. An equal-area projection can be made from a cylindrical, conic, or azimuthal projection.

In an equal-area projection, shapes of continents and oceans are distorted. However, the scale used in an equal-area projection is the same throughout the map. Therefore, the distances between points on the map are accurate.

Equal-area projections may be used to map large land areas, such as continents. The figure below shows an example of an equal-area projection.

Critical Thinking
8. Infer Why are many road maps drawn as equal-area projections?

_____

_____

_____

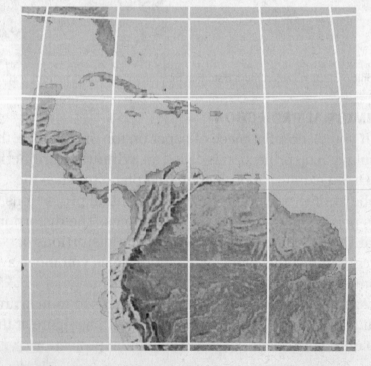

TAKE A LOOK
9. Identify What is distorted on an equal-area projection?

_____

_____

Equal-area projections are useful for determining distance on a map.

SECTION 2 Mapping the Earth's Surface *continued*

How Is Information Shown on Maps?

The information on a map is in the form of symbols. To read a map, you must understand the symbols on the map. Maps can show many different kinds of information. Almost all maps contain five pieces of information: a title, an indicator of direction, a scale, a legend, and a date. ☑

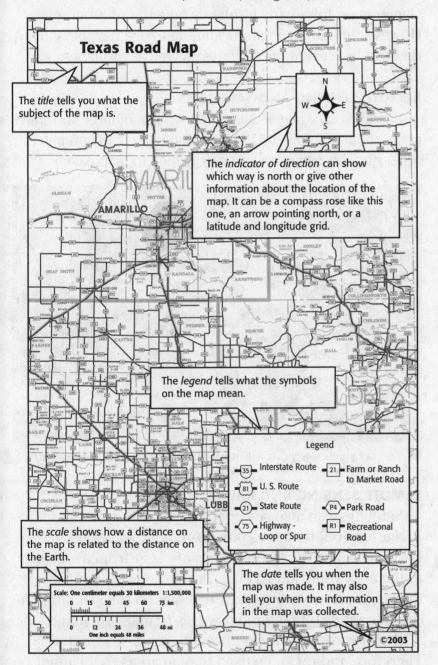

Texas Road Map

The *title* tells you what the subject of the map is.

The *indicator of direction* can show which way is north or give other information about the location of the map. It can be a compass rose like this one, an arrow pointing north, or a latitude and longitude grid.

The *legend* tells what the symbols on the map mean.

Legend

35 Interstate Route 21 Farm or Ranch to Market Road
81 U. S. Route
21 State Route P4 Park Road
75 Highway - Loop or Spur R1 Recreational Road

The *scale* shows how a distance on the map is related to the distance on the Earth.

Scale: One centimeter equals 30 kilometers 1:1,500,000
0   15   30   45   60   75 km
0   12   24   36   48 mi
One inch equals 48 miles

The *date* tells you when the map was made. It may also tell you when the information in the map was collected.

©2003

READING CHECK

10. Describe How is information shown on a map?

_____

_____

TAKE A LOOK

11. Identify What is the subject of the map in the figure?

_____

_____

Critical Thinking

12. Reason Why is it important for a map to have a legend?

_____

_____

_____

_____

SECTION 2 Mapping the Earth's Surface *continued*

How Are Maps Made?

Much of the information used to make maps today comes from remote sensing. **Remote sensing** is a way to gather information about an object without actually touching the object. ☑

Today, most maps are made from photographs. Cameras on low-flying airplanes take the photographs. However, mapmakers are beginning to use new equipment that can be carried on satellites as well as airplanes.

REMOTE SENSING AND SATELLITES

Satellites can take detailed photographs of large and small areas of land. They send the information back to computers on Earth. These pictures are very helpful to mapmakers.

Satellites can produce very detailed images of the Earth's surface. The satellite that took this picture was 423 mi above the Earth's surface!

REMOTE SENSING AND RADAR

Satellites can also see energy that your eyes cannot. Radar is a tool that uses powerful radio waves to map Earth's surface. First, the waves are sent from a satellite to the area you want to map. Then, the waves reflect off the land and travel back to a receiver on the satellite. The satellite interprets the radio waves to make an image.

Radio waves can move through clouds and water. Therefore, radar can be used to map areas that cannot be easily seen. For example, the planet Venus has a very thick and cloudy atmosphere. However, scientists have been able to map the surface of Venus using radar from satellites.

✓ **READING CHECK**

13. Define What is remote sensing?

_____

_____

_____

Math Focus

14. Calculate How far above the Earth's surface, in kilometers, was the satellite that took this picture?

1 km = 0.62 mi

_____

_____

SECTION 2 Mapping the Earth's Surface *continued*

GLOBAL POSITIONING SYSTEM

Satellites can keep you from getting lost. The *global positioning system* (GPS) can help you find where you are on Earth. GPS is a system of satellites that travel around the Earth. The satellites send radio waves to receivers on Earth. The receivers calculate the latitude, longitude, and elevation of a certain place. ☑

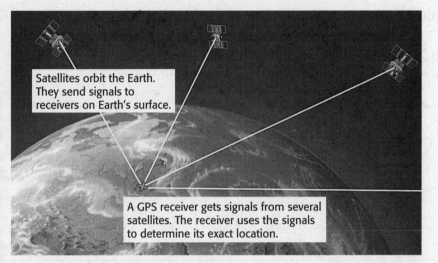

Satellites orbit the Earth. They send signals to receivers on Earth's surface.

A GPS receiver gets signals from several satellites. The receiver uses the signals to determine its exact location.

GPS is very common in people's lives today. Mapmakers use GPS to check the location of boundary lines between countries. Airplane and boat pilots use GPS for navigating. GPS receivers are even put into cars and watches!

GEOGRAPHIC INFORMATION SYSTEMS

A *geographic information system* (GIS) is a computerized system that shows information about an area. Mapmakers can use a GIS to store and view many pieces of information about an area. Mapmakers can enter different types of information about an area into the GIS. Each piece of information is stored in a separate layer. Then, the mapmakers can combine many layers into a single map. ☑

A GIS can be a very useful tool for a mapmaker. Using a GIS, a mapmaker can produce many maps of the same area. Each map shows different information about the area. For example, a GIS can be used to make a map that shows only the roads and buildings in an area. The mapmaker can then add another layer that shows the rivers in the area. Using a computer to do this is much faster and more accurate than doing it by hand.

☑ **READING CHECK**

15. Identify What is GPS?

_____

_____

TAKE A LOOK

16. Describe What do GPS satellites do?

_____

_____

☑ **READING CHECK**

17. Define What is a GIS?

_____

_____

_____

Section 2 Review

SECTION VOCABULARY

azimuthal projection a map projection that is made by moving the surface features of the globe onto a plane	**cylindrical projection** a map projection that is made by moving the surface features of the globe onto a cylinder
conic projection a map projection that is made by moving the surface features of the globe onto a cone	**remote sensing** the process of gathering and analyzing information about an object without physically being in touch with the object

1. Compare How is a conic projection different from a cylindrical projection? How are they similar?

_____

_____

_____

2. List What are five pieces of information that are found on all maps?

_____

_____

_____

_____

3. Identify Which type of map projection is most useful for showing the distances between two points on the Earth?

_____

4. Explain Which type of map projection is often used to map the United States?

_____

5. Describe Why is radar useful for mapping areas that are covered by clouds?

_____

_____

6. Identify Give three types of technology that mapmakers use to make maps.

_____

_____

_____

CHAPTER 2 Maps as Models of the Earth

SECTION 3
Topographic Maps

After you read this section, you should be able to answer these questions:

• What is a topographic map?

• How do topographic maps show the features of the Earth's surface?

What Is a Topographic Map?

If you were going hiking in the wilderness, you would want to take a compass and a map. Because there are no roads in the wilderness, you would not take a road map. Instead, you would take a topographic map.

A **topographic map** is a map that shows the surface features, or *topography*, of an area. Topographic maps show natural features, such as rivers and lakes. They show some features made by people, such as bridges. Topographic maps also show elevation. **Elevation** is the height of an object above the surface of the sea. The elevation at sea level is 0 m. ☑

CONTOUR LINES

How can a flat map show elevations? Contour lines are used to show elevation on a topographic map. **Contour lines** are lines on a map that connect points with the same elevation. Each contour line on a map shows a different elevation. Here are some rules for using contour lines:

• Contour lines never cross. All points on a contour line are at the same elevation.

• The space between contour lines depends on the slope of the ground. Contour lines that are close together show a steep slope. Contour lines that are far apart show a gentle slope.

• Contour lines that cross a valley or stream are V-shaped. The V points toward the area of higher elevation, or upstream.

• The tops of hills, mountains, and depressions (dips) are shown by closed circles. Depressions are marked with short, straight lines inside the circle. The lines point toward the center of the depression.

STUDY TIP

Learn New Words As you read, underline words that you don't know. When you figure out what the words mean, write the words and their definitions in your notebook.

READING CHECK

1. Define What is elevation?

_____

_____

Critical Thinking

2. Explain Why can two contour lines never cross?

_____

_____

_____

_____

_____

_____

SECTION 3 Topographic Maps *continued*

CONTOUR INTERVALS AND RELIEF

Each contour line represents a certain elevation. The difference in elevation between one contour line and the next is called the **contour interval**. For example, a map with a contour interval of 20 m has contour lines drawn at 0 m, 20 m, 40 m, and so on. The contour interval of a map is usually given in or near the map's legend. The contour interval on a map is based on the relief in the area. ☑

Relief is the difference in elevation between the highest and lowest points in the area on the map. Mountains have high relief. They are usually mapped with large contour intervals. Plains have low relief. They are usually mapped with small contour intervals.

INDEX CONTOURS

The many contour lines on a map can make it hard to read. An index contour is used to make reading the map easier. An **index contour** is a darker, heavier contour line that is labeled with an elevation. In most maps, every fifth contour line is an index contour. For example, a map with a contour interval of 20 m may have index contours at 0 m, 100 m, 200 m, and so on.

COLORS

Topographic maps use colors and symbols to show different features of Earth's surface. Buildings, bridges, and railroads are shown by special symbols drawn in black. Contour lines are brown. Major roads are red. Bodies of water are blue. Wooded areas are shaded in green. Cities are shaded in gray or red. ☑

Topographic maps contain a lot of information. This information can be confusing at first. However, if you practice, you will be able to read topographic maps more easily. When you look at a topographic map, ask yourself these questions to help you read the map:

• What area does the map show?

• What is the contour interval of the map?

• What is the relief of the area in the map?

• What kinds of features are shown on the map?

The map on the next page is an example of a topographic map. Try to answer the four questions above for the map on the next page.

READING CHECK

3. Define What is a contour interval?

_____

_____

Math Focus

4. Calculate A map has index contours at 250 m, 500 m, and 750 m. What is the contour interval?

_____

_____

READING CHECK

5. Identify How do topographic maps show information?

_____

_____

SECTION 3 Topographic Maps *continued*

Topographic Map of El Capitan

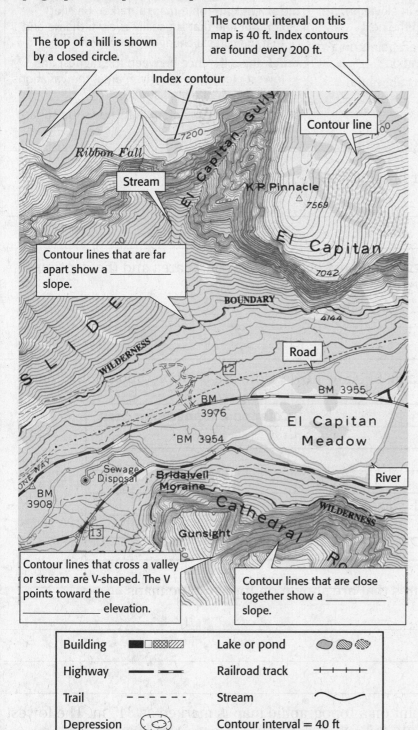

The top of a hill is shown by a closed circle.

The contour interval on this map is 40 ft. Index contours are found every 200 ft.

Index contour

Contour line

Stream

Contour lines that are far apart show a _____ slope.

Contour lines that cross a valley or stream are V-shaped. The V points toward the _____ elevation.

Contour lines that are close together show a _____ slope.

Road

River

Building	▬ ▭ ▨	Lake or pond	◯ ◯ ◯
Highway	——— -·-	Railroad track	+++++
Trail	- - - -	Stream	∿
Depression	◌	Contour interval = 40 ft	

 **Say It**

Compare How is this topographic map different from other maps that you have seen? How is it similar? In a small group, talk about the similarities and differences between topographic maps and other kinds of maps.

TAKE A LOOK
6. Identify Fill in the blanks in the figure to explain how to use contour lines.

Section 3 Review

SECTION VOCABULARY

contour interval the difference in elevation between one contour line and the next **contour lines** lines which connect points on a map that are at the same elevation **elevation** the height of an object above the surface of the sea	**index contour** on a map, a darker, heavier contour line that is usually every fifth line and that indicates a change in elevation **relief** the difference between the highest and lowest elevations in a given area; the variations in elevation of a land surface **topographic map** a map that shows the surface features of Earth

1. Define What are contour lines?

_____

_____

2. Explain What is the relationship between the relief of an area and the contour interval on a map of the area?

_____

_____

3. Describe Complete the table to show how colors are used on topographic maps.

Feature	Color on a topographic map
Contour lines	
	blue
Major roads	
Buildings and bridges	
	green
Cities	

4. Identify Give three features that are shown on topographic maps.

_____

_____

_____

5. Calculate The highest point on a topographic map is marked as 345 m. The lowest contour line is at 200 m. What is the relief of the area in the map?

_____

6. Describe How is the top of a mountain shown on a topographic map?

_____

_____

CHAPTER 3 | Minerals of the Earth's Crust

SECTION 1 | What Is a Mineral?

BEFORE YOU READ

After you read this section, you should be able to answer these questions:

• What are minerals?

• What determines the shape of a mineral?

• What are two main groups of minerals?

What Are Minerals?

A **mineral** is a naturally formed, inorganic solid that forms crystals and is always made of the same elements. The figure below shows four questions that you can ask in order to learn whether something is a mineral.

Is it nonliving? Minerals are inorganic. This means that they are not made of living things or their remains.

Is it a solid? Minerals are not gases or liquids.

Does it have a crystalline structure? Minerals are crystals. Each mineral has a certain crystal structure that is always the same.

Does it form naturally? Minerals are not made by people.

All minerals have four features, as described in the figure.

TAKE A LOOK

1. Explain Why are diamonds that are made by people not considered minerals?

_____

_____

Critical Thinking

2. Apply Concepts Coal is made from the remains of dead plants. Is coal a mineral? Explain your answer.

_____

_____

_____

_____

You might not be familiar with the term "crystalline structure." To understand what crystalline structure is, you need to know a little about how elements form minerals. **Elements** are pure substances that cannot be broken down into simpler substances. Oxygen, chlorine, carbon, and iron are examples of elements. Elements can come together in certain ways to form new substances, such as minerals. All minerals are made of one or more elements.

COMPOUNDS AND ATOMS

Most minerals are made of compounds of several different elements. A **compound** is a substance made of two or more elements that are chemically bonded. For example, the mineral halite is a compound of sodium, Na, and chlorine, Cl. A few minerals, such as gold and silver, are made of only one element. A mineral that is made of only one element is called a *native element*. ☑

Each element is made of only one kind of atom. An *atom* is the smallest part of an element that has the properties of that element. Like other compounds, minerals are made up of atoms of one or more elements.

CRYSTALS

Remember that minerals have a definite crystalline structure. This means that the atoms in the mineral line up in a regular pattern. The regular pattern of the atoms in a mineral causes the mineral to form crystals. **Crystals** are solid, geometric forms of minerals that are formed by repeating a pattern of atoms. ☑

The shape of a crystal depends on how the atoms in it are arranged. The atoms that make up each mineral are different. However, there are only a few ways that atoms can be arranged. Therefore, the crystals of different minerals can have similar shapes.

Although different minerals may form similar shapes, each mineral forms only one shape of crystal. Therefore, geologists say that a mineral has a definite crystalline structure. This means that crystals of a certain mineral always form the same shape.

READING CHECK

3. Define What is a compound?

_____

_____

_____

READING CHECK

4. Explain What causes minerals to form crystals?

_____

_____

TAKE A LOOK

5. Identify What shape are gold crystals?

_____

_____

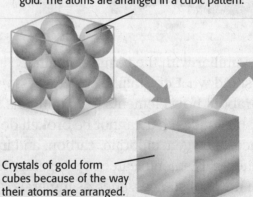

The mineral gold is made of atoms of the element gold. The atoms are arranged in a cubic pattern.

Crystals of gold form cubes because of the way their atoms are arranged.

Real crystals of gold may not be perfect cubes because the crystals may be damaged or not form completely. However, the atoms are still arranged in a cubic pattern.

SECTION 1 What Is a Mineral? *continued*

How Do Geologists Classify Minerals?

Geologists classify minerals based on the elements or compounds in the minerals. Two main groups of minerals are silicate minerals and nonsilicate minerals.

SILICATE MINERALS

Silicon and oxygen are two of the most common elements in the Earth's crust. Minerals that contain compounds of silicon and oxygen are called **silicate minerals**. Silicate minerals make up more than 90% of the Earth's crust. Most silicate minerals also contain elements other than silicon and oxygen, such as aluminum, iron, or magnesium.

Common Silicate Minerals

Quartz is a mineral that is found in many rocks of the Earth's crust.

Mica breaks into sheets easily.

Feldspar is also common in the rocks of the Earth's crust. Feldspar can contain many elements other than silicon and oxygen, such as potassium or sodium.

TAKE A LOOK
6. Identify What two elements are found in all of the minerals in the figure? Explain your answer.

_____

_____

_____

NONSILICATE MINERALS

Minerals that do not contain compounds of silicon and oxygen are called **nonsilicate minerals**. Some of these minerals are made of elements such as carbon, oxygen, fluorine, and sulfur.

Types of Nonsilicate Minerals

Native elements are minerals that are made of only one element. Copper, gold, silver, and diamonds are native elements.

Copper

Oxides are minerals that contain compounds of oxygen and another element, such as iron or aluminum. Rubies and sapphires are forms of the mineral corundum, which is an oxide mineral.

Corundum

Carbonates are minerals that contain compounds of carbon and oxygen. Calcite is a carbonate mineral.

Calcite

Sulfates are minerals that contain compounds of oxygen and sulfur. Gypsum is a sulfate mineral.

Gypsum

Halides are minerals that contain the elements fluorine, chlorine, iodine, or bromine. Fluorite and halite are halide minerals.

Fluorite

Sulfides are minerals that contain compounds of sulfur and an element other than oxygen, such as lead, iron, or nickel. Galena and pyrite ("fool's gold") are sulfide minerals.

Galena

TAKE A LOOK
7. Compare How are sulfate minerals different from sulfide minerals?

_____

_____

_____

_____

Name _____ Class _____ Date _____

Section 1 Review

SECTION VOCABULARY

compound a substance made up of atoms of two or more different elements joined by chemical bonds	**mineral** a naturally formed, inorganic solid that has a definite chemical structure
crystal a solid whose atoms, ions, or molecules are arranged in a regular, repeating pattern	**nonsilicate mineral** a mineral that does not contain compounds of silicon and oxygen
element a substance that cannot be separated or broken down into simpler substances by chemical means	**silicate mineral** a mineral that contains a combination of silicon and oxygen and that may also contain one or more metals

1. Identify What are four features of a mineral?

_____

_____

_____

_____

2. Compare What is the difference between an atom and an element?

_____

_____

_____

3. Infer What determines the shape of a crystal?

_____

_____

4. Apply Concepts Why is the ice in a glacier considered a mineral, but the water in a river is not considered a mineral?

_____

_____

5. Describe What are the features of the two major groups of minerals?

_____

_____

_____

6. List Give four types of nonsilicate minerals.

_____

_____

_____

CHAPTER 3 | Minerals of the Earth's Crust

SECTION 2 | # Identifying Minerals

BEFORE YOU READ

After you read this section, you should be able to answer these questions:

• What seven properties can be used to identify a mineral?

• What are some special properties of minerals?

How Can You Identify Minerals?

If you close your eyes and taste different foods, you can usually figure out what the foods are. You can identify foods by noting their properties, such as texture and flavor. Minerals also have properties that you can use to identify them.

COLOR

The same mineral can have many different colors. For example, the mineral quartz can be clear, white, pink, or purple. Minerals can also change colors when they react with air or water. For example, pyrite ("fool's gold") has a golden color. If pyrite is exposed to air and water, it can turn brown or black. Because the color of a mineral can vary a lot, color is not the best way to identify a mineral. ☑

LUSTER

The way a surface reflects light is called **luster**. When you say that something looks shiny, you are describing its luster. A mineral can have a metallic, submetallic, or nonmetallic luster. The table below gives some examples of different kinds of luster.

Luster	Description	Examples
Metallic	bright and shiny, like metal	gold, copper wire
Submetallic	dull, but reflective	graphite (pencil "lead")
Nonmetallic		
Vitreous	glassy, brilliant	glass, quartz
Waxy	greasy, oily	wax, halite
Silky	looks like light is reflecting off long fibers	satin fabric, asbestos
Pearly	creamy	pearls, talc
Resinous	looks like plastic	plastic, sulfur
Earthy	rough, dull	concrete, clay

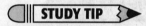

STUDY TIP

Reading Organizer As you read this section, create an outline of the section. Use the properties of minerals to form the headings of your outline.

READING CHECK

1. Explain How can the color of a mineral change?

_____

_____

_____

📣 **Say It**

Apply Ideas In a small group, think of a list of 10 to 15 everyday materials. Together, try to describe the luster of each material using the terms in the table.

SECTION 2 Identifying Minerals *continued*

STREAK

The color of a mineral in powdered form is called its streak. You can find a mineral's streak by rubbing the mineral against a piece of unglazed porcelain. The piece of unglazed porcelain is called a *streak plate*. The mark left on the streak plate is the streak.

Streak is a more useful property than color for identifying minerals. This is because the color of a mineral's streak is always the same. For example, the color of the mineral hematite may vary, but its streak will always be red-brown. ☑

CLEAVAGE AND FRACTURE

Different minerals break in different ways. The way that a mineral breaks depends on how its atoms are arranged. When some minerals break, the surfaces that form are smooth and flat. These minerals show the property of cleavage. Other minerals break unevenly, along curved or rough surfaces. These minerals show the property of fracture.

The mineral biotite, a type of mica, shows the property of cleavage. It breaks easily into thin, flat sheets. ▶

The mineral halite also shows the property of cleavage. Its crystals break into cubes. ▼

The mineral quartz shows the property of fracture. It breaks along a curved surface. This kind of fracture is called *conchoidal* fracture.

DENSITY

Density is a measure of how much matter is in a given amount of space. Density is usually measured in grams per cubic centimeter (g/cm^3). For example, the density of water is 1 g/cm^3.

Geologists often use specific gravity to describe the density of a mineral. A mineral's *specific gravity* is the density of the mineral divided by the density of water. For example, gold has a density of 19 g/cm^3. Its specific gravity is 19 $g/cm^3 \div 1 \ g/cm^3 = 19$.

READING CHECK

2. Explain Why is streak more useful than color in identifying a mineral?

_____

_____

_____

TAKE A LOOK

3. Identify What kind of fracture does quartz show?

_____

_____

Math Focus

4. Calculate How many times denser is gold than water?

_____

_____

SECTION 2 Identifying Minerals *continued*

HARDNESS

A mineral's resistance to being scratched is its **hardness**. Scientists use the *Mohs hardness scale* to describe the hardness of minerals. The harder a mineral is to scratch, the higher its rating on the Mohs scale. Talc, one of the softest minerals, has a rating of 1. Diamond, the hardest mineral, has a rating of 10. ☑

Scientists use reference minerals to find the hardness of unknown minerals. They try to scratch the surface of the unknown mineral with the edge of a reference mineral. If the reference mineral scratches the unknown mineral, the reference mineral is harder than the unknown mineral.

You probably don't have pieces of these reference minerals. However, you can find the hardness of a mineral using common objects. For example, your fingernail has a hardness of about 2 on the Mohs scale. A piece of window glass has a hardness of about 5.5.

Hardness	Mineral
1	Talc
2	Gypsum
3	Calcite
4	Fluorite
5	Apatite

Hardness	Mineral
6	Orthoclase
7	Quartz
8	Topaz
9	Corundum
10	Diamond

SPECIAL PROPERTIES

Some minerals have special properties. These properties can be useful in identifying the minerals.

Special Properties of Some Minerals

 Calcite and fluorite show the property of *fluorescence*. This means that they glow under ultraviolet light.

 Calcite produces a *chemical reaction* when a drop of weak acid is placed on it. It fizzes and produces gas bubbles.

 Some minerals, such as this calcite, show *optical properties*. Images look doubled when they are viewed through calcite.

 Magnetite shows the property of *magnetism*. It is a natural magnet.

 Halite has a salty *taste*. You should not taste a mineral unless your teacher tells you to.

 Minerals that contain radioactive elements may show the property of *radioactivity*. The radiation they give off can be detected by a Geiger counter.

READING CHECK

5. Define What is hardness?

_____

_____

Critical Thinking

6. Apply Concepts A scientist tries to scratch a sample of orthoclase with a sample of apatite. Will he be able to scratch the orthoclase? Explain your answer.

_____

_____

_____

_____

TAKE A LOOK

7. Describe Under ultraviolet light, what happens to minerals that show the property of fluorescence?

_____

Section 2 Review

SECTION VOCABULARY

cleavage in geology, the tendency of a mineral to split along specific planes of weakness to form smooth, flat surfaces	**hardness** a measure of the ability of a mineral to resist scratching
density the ratio of the mass of a substance to the volume of the substance	**luster** the way in which a mineral reflects light
fracture the manner in which a mineral breaks along either curved or irregular surfaces	**streak** the color of a mineral in powdered form

1. Compare How are cleavage and fracture different?

_____

_____

_____

2. Explain Why is color not the best property to use to identify a mineral?

_____

_____

3. Identify Give five properties that you can use to identify a mineral.

_____

_____

_____

_____

_____

4. Apply Concepts A geologist has found an unknown mineral. She finds that a sample of calcite will not scratch the unknown mineral. She also finds that a sample of apatite will scratch the unknown mineral. About what is the unknown mineral's hardness? Explain your answer.

_____

_____

_____

_____

_____

5. Calculate The density of a mineral is 2.6 g/cm³. What is its specific gravity?

_____

_____

SECTION 3 The Formation, Mining, and Use of Minerals

After you read this section, you should be able to answer these questions:

• How do minerals form?

• How are mineral resources used?

How Do Minerals Form?

Different minerals form in different environments. The table below shows five ways that minerals can form.

Process	Description	Minerals that form this way
Evaporation	When a body of salt water dries up, minerals are left behind. As the water evaporates, the minerals crystallize.	gypsum, halite
Metamorphism	High temperatures and pressures deep below the ground can cause the minerals in rock to change into different minerals.	garnet, graphite, magnetite, talc
Deposition	Surface water and ground water carry dissolved minerals into lakes or seas. The minerals can crystallize on the bottom of the lake or sea.	calcite, dolomite
Reaction	Water underground can be heated by hot rock. The hot water can dissolve some minerals and deposit other minerals in their place.	gold, copper, sulfur, pyrite, galena
Cooling	Melted rock can cool slowly under Earth's surface. As the melted rock cools, minerals form.	mica, feldspar, quartz

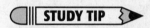

STUDY TIP

Describe As you read this section, make a chart showing the uses of different rock and mineral resources.

TAKE A LOOK

1. Identify Give three minerals that form by metamorphism and three minerals that form by reaction.

Metamorphism:

_____

_____

_____

Reaction:

_____

_____

_____

How Are Minerals Removed from the Earth?

People mine many kinds of minerals from the ground and make them into objects we need. Some minerals have more useful materials in them than others. An **ore** is a rock or mineral that contains enough useful materials for it to be mined at a profit.

There are two ways of removing ores from Earth: surface mining and subsurface mining. The type of mining used depends on how close the ore is to the surface.

SURFACE MINING

People use surface mining to remove ores that are near Earth's surface. Three types of surface mines include open pits, surface coal mines, and quarries.

Open-pit mining is used to remove large, near-surface deposits of gold and copper. Explosives break up the rock layers above the ore. Then, trucks haul the ore from the mine to a processing plant. ☑

Quarries are open mines that are used to remove sand, gravel, and crushed rock. The layers of rock near the surface are removed and used to make buildings and roads.

Strip mines are often used to mine coal. The coal is removed in large pieces. These pieces are called *strips*. The strips of coal may be up to 50 m wide and 1 km long.

SUBSURFACE MINING

People use subsurface mining to remove ores that are deep underground. Iron, coal, and salt can be mined in subsurface mines. ☑

✔ **READING CHECK**

2. Identify Give two minerals that are mined using open-pit mining.

_____

_____

✔ **READING CHECK**

3. List Give three resources that can be mined using subsurface mining.

_____

_____

_____

TAKE A LOOK

4. Identify What are three kinds of tunnels used in subsurface mining?

_____

_____

_____

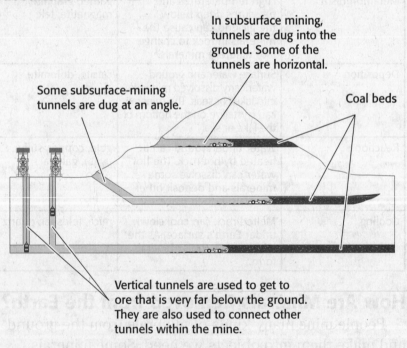

In subsurface mining, tunnels are dug into the ground. Some of the tunnels are horizontal.

Some subsurface-mining tunnels are dug at an angle.

Coal beds

Vertical tunnels are used to get to ore that is very far below the ground. They are also used to connect other tunnels within the mine.

SECTION 3 The Formation, Mining, and Use of Minerals *continued*

RESPONSIBLE MINING

Mining can help us get the resources we need, but it can also create problems. Mining may destroy or harm the places where plants and animals live. The wastes from mining can be poisonous. They can pollute water and air. ☑

One way to reduce these problems is to return the land to nearly its original state after mining is finished. This is called **reclamation**. Since the mid-1970s, laws have required the reclamation of land used for mining.

Another way to reduce the problems with mining is to reduce our need for minerals that are mined. For example, when you recycle materials made from minerals, you reduce the need for further mining. If you recycle the aluminum in your soda can, less aluminum has to be removed from the Earth. ☑

How Are Minerals Used?

We can use some minerals just as they are. However, most minerals must be processed before they can be used. The table below shows how some common minerals are used. The figure on the next page shows some of the processed minerals that are used in a bicycle.

Mineral	Uses
Bauxite (aluminum ore)	source of aluminum for cans, foil, appliances, and utensils
Copper	electrical wire, plumbing, coins
Diamond	jewelry, cutting tools, drill bits
Galena (lead ore)	source of lead for batteries and ammunition
Gold	jewelry, computers, spacecraft
Gypsum	plaster, cement, wallboard
Halite	table salt, road salt, water softener
Quartz	glass, source of silicon for computer chips
Silver	photography, electronic products, jewelry
Sphalerite (zinc ore)	jet aircraft, spacecraft, paint

✓ **READING CHECK**

5. Describe What are two problems with mining?

_____

_____

_____

✓ **READING CHECK**

6. Explain How can recycling help reduce the problems with mining?

_____

_____

_____

TAKE A LOOK

7. Identify Give two uses for the mineral silver and two uses for the mineral bauxite.

Silver:

_____

_____

Bauxite:

_____

_____

SECTION 3 The Formation, Mining, and Use of Minerals *continued*

TAKE A LOOK

8. Identify Name four minerals that are used in the parts of a bicycle.

_____

_____

_____

_____

Critical Thinking

9. Infer Electricity can pass through metals easily. How does this make metals useful in computers and other electronic appliances?

_____

_____

_____

_____

Minerals Used in the Parts of a Bicycle

Handlebars
titanium from ilmenite

Frame
aluminum from bauxite

Spokes
iron from magnetite

Pedals
beryllium from beryl

METALLIC MINERALS

Many minerals contain metals. Many of the features of metals make them useful in aircraft, automobiles, computer parts, and spacecraft. All metals have the features given below:

• Metals have shiny surfaces.

• Light cannot pass through metals.

• Heat and electricity can pass through metals easily.

• Metals can be rolled into sheets or stretched into wires.

Some metals react easily with air and water. For example, iron can react with oxygen in the air to produce rust. However, many of these metallic minerals can be processed into materials that do not react with air and water. For example, iron can be used to make stainless steel, which does not rust. Other metals do not react very easily. For example, gold is used in parts of aircraft because it does not react with many chemicals.

Many metals are strong. Their strength makes them useful in making ships, automobiles, airplanes, and buildings. For example, tall buildings are too heavy to be supported by a wooden frame. However, steel frames can support skyscrapers that are hundreds of meters tall.

NONMETALLIC MINERALS

Many minerals also contain nonmetals. Some important features of nonmetals are given below:

• Nonmetals can have shiny or dull surfaces.

• Light can pass through some kinds of nonmetals.

• Heat and electricity cannot pass through nonmetals easily.

Nonmetallic minerals are some of the most widely used minerals in industry. For example, the mineral calcite is used to make concrete. The mineral quartz is used to make glass. Quartz can also be processed to produce the element silicon, which is used in computer chips. ☑

GEMSTONES

Some nonmetallic minerals are considered valuable because of their beauty or rarity. These minerals are called *gemstones*. Important gemstones include diamond, ruby, sapphire, emerald, topaz, and tourmaline.

Color is the feature that determines the value of a gemstone. The more attractive the color, the more valuable the gemstone is. The colors of many gemstones are caused by impurities. An *impurity* is a small amount of an element not usually found in the mineral. For example, rubies and sapphires are both forms of the mineral corundum. Rubies look red because they have chromium impurities. Sapphires look blue because they have iron impurities. ☑

Most gemstones are very hard. This allows them to be cut and polished easily. For example, corundum (rubies and sapphires) and diamond are the two hardest minerals. They are also some of the most valuable gemstones.

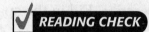

READING CHECK

10. Identify Give two nonmetallic minerals that are used in industry.

_____

_____

READING CHECK

11. Explain What gives many gemstones their color?

_____

_____

Diamonds are some of the most valuable gemstones. They are used in jewelry and in other items, such as this scepter.

Section 3 Review

SECTION VOCABULARY

ore a natural material whose concentration of economically valuable minerals is high enough for the material to be mined profitably	**reclamation** the process of returning land to its original condition after mining is completed

1. Define Write your own definition for ore.

_____

_____

_____

2. Describe Fill in the spaces in table to describe metals and nonmetals.

Type of material	Main features	Common objects made from it

metal	has shiny surfaces
	does not transmit light
	transmits heat and
	   electricity easily
	can be rolled into sheets
	   or stretched into wires	

nonmetal		

3. List What are three ways minerals can form?

_____

_____

_____

4. Identify Give three types of surface mines and an example of the kind of material that each is used to mine.

_____

_____

_____

CHAPTER 4 | Rocks: Mineral Mixtures
SECTION 1 **The Rock Cycle**

BEFORE YOU READ

After you read this section, you should be able to answer these questions:

• What is a rock?

• How are rocks classified?

• What does the texture of a rock reveal about how it was formed?

Why Are Rocks Important?

You know that you can recycle paper, aluminum, and plastic. Did you know that the Earth also recycles? One thing the Earth recycles is rock. A **rock** is a naturally occurring solid mixture of one or more minerals. Some rocks also contain the remains of living things.

Rock is an important resource for human beings. Early humans used rocks as hammers and other tools. They shaped rocks like chert and obsidian into spear points, knives, and scrapers. Rock is also used in buildings, monuments, and roads. The figure below shows how rock has been used as a building material in ancient and modern civilizations.

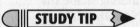

The ancient Egyptians used a rock called **limestone** to build the pyramids at Giza (left-hand figure). The Texas state capitol building in Austin is constructed of a rock called **granite** (right-hand figure).

It may seem like rocks never change, but this is not true. In fact, rocks are changing all the time. Most of these changes are slow, which is why it seems like rocks do not change. The processes by which new rocks form from older rock material is called the **rock cycle**.

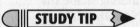
STUDY TIP

Describe As you read this section, make a chart describing the processes of weathering, erosion, and deposition.

TAKE A LOOK

1. Identify What are two kinds of rocks that people have used for constructing buildings?

_____

_____

SECTION 1 The Rock Cycle *continued*

What Processes Shape the Earth's Surface?

Many different processes are part of the rock cycle. These processes shape the features of our planet. They form the mountains and valleys that we see around us. They also affect the types of rock found on the Earth's surface.

WEATHERING, EROSION, AND DEPOSITION

Weathering happens when water, wind, ice, and heat break down rock into smaller fragments. These fragments are called *sediment*. Sediment can move over the Earth's surface through erosion and deposition.

Erosion happens when water, wind, ice, or gravity move sediment over the Earth's surface. Over time, sediment that has been eroded stops moving and is deposited. When sediment stops moving, it is called **deposition**. Sediment can be deposited in bodies of water and other low-lying areas.

STANDARDS CHECK

ES 1c Land forms are the result of a combination of constructive and destructive forces. Constructive forces include crustal deformation, volcanic eruption, and deposition of sediment, while destructive forces include weathering and erosion.

2. Explain How does weathering shape the Earth's surface?

_____

_____

_____

_____

TAKE A LOOK
3. Identify Give two things that may have caused the weathering and erosion in Bryce Canyon.

_____

_____

The rocks in Bryce Canyon, Utah, have been shaped by weathering and erosion. Although these processes can be slow, they can cause large changes in the Earth's surface.

HEAT AND PRESSURE

Rock can also form when buried sediment is squeezed by the weight of the layers above it. In addition, temperature and pressure can change the minerals in the rocks. In some cases, the rock gets hot enough to melt. This melting produces liquid rock, or *magma*. When the magma cools, it hardens to form new rock. The new rock contains different minerals than the rock that melted.

SECTION 1 The Rock Cycle *continued*

THE ROCK CYCLE

Geologists put rocks into three main groups based on how they form. These groups are igneous rock, sedimentary rock, and metamorphic rock. *Igneous rock* forms when melted rock cools and hardens. *Sedimentary rock* is made of pieces of other rock (sediment). *Metamorphic rock* forms when heat and pressure change the chemical composition of a rock.

Remember that the rock cycle is made of all of the processes that make new rock out of older rock material. Weathering, erosion, deposition, heat, and pressure are some of the processes that are part of the rock cycle. The figure below shows how the processes in the rock cycle can change rocks from one kind to another.

Critical Thinking

4. Compare How are igneous rocks different from metamorphic rocks?

_____

_____

_____

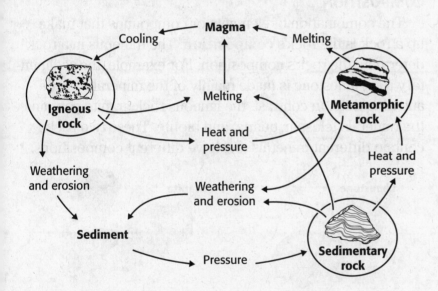

TAKE A LOOK
5. Use a Model Find two paths through the rock cycle that lead from sedimentary rock to igneous rock. Use a colored pen or marker to trace both paths on the figure.

As you can see, rocks do not have to follow a single path through the rock cycle. An igneous rock may be weathered to form sediment, which then forms sedimentary rock. The igneous rock could also melt and cool to form a new igneous rock.

The path that a rock takes through the rock cycle depends on the forces that act on the rock. These forces change depending on where the rock is located. For example, high pressures and temperatures below the Earth's surface can cause metamorphic rock to form.

How Do Geologists Classify Rocks?

Remember that rocks can be divided into three groups based on how they form. Each main group of rock can be divided into smaller groups. These divisions are also based on the ways rocks form. For example, all igneous rock forms when magma cools and hardens. However, different kinds of igneous rock form when magma cools above the ground and when it cools underground.

Each kind of rock has specific features that make it different from other kinds of rock. Geologists can learn how a rock formed by studying its features. Two features that are especially helpful for classifying rocks are composition and texture ☑

COMPOSITION

The combination of elements or compounds that make up a rock is the rock's **composition**. The minerals in a rock determine the rock's composition. For example, the sedimentary rock limestone is made mainly of the minerals calcite and aragonite. In contrast, the igneous rock granite contains the minerals feldspar, quartz, and biotite. These two rocks contain different minerals and have different compositions.

☑ **READING CHECK**

6. Explain How do geologists learn how a rock formed?

_____

_____

Math Focus

7. Calculate Rock A contains 10% quartz and 45% calcite. The rest of the rock is mica. What percentage of the rock is mica?

_____

Limestone, a sedimentary rock, contains the minerals calcite and aragonite.

Granite, an igneous rock, contains the minerals biotite, feldspar, and quartz.

Composition can help geologists classify rocks. This is because different minerals form under different conditions. For example, remember that the mineral garnet forms under high temperatures and pressures. Therefore, a rock with garnet in it probably formed under high temperature and pressure. Such a rock is probably a metamorphic rock.

TEXTURE

The sizes, shapes, and positions of the grains that make up a rock are the rock's **texture**. The texture of a rock can be affected by different things. The texture of a sedimentary rock is mainly affected by the sediment that formed it. For example, a sedimentary rock that forms from small sediment pieces will have a fine-grained texture. The figures below show some examples of sedimentary rock textures.

Fine-grained

Siltstone

Siltstone is made of tiny pieces of sediment, such as silt and clay. Therefore, it has a fine-grained texture. It feels smooth when you touch it.

Medium-grained

Sandstone

Sandstone is made of pieces of sand. It has a medium-grained texture. It feels a bit rough, like sandpaper.

Coarse-grained

Conglomerate

Conglomerate is made of sediment pieces that are large, such as pebbles. Therefore, it has a coarse-grained texture. It feels bumpy.

TAKE A LOOK

8. Explain What determines the texture of a sedimentary rock?

_____

_____

_____

The texture of an igneous rock depends on how fast the melted rock cools. As melted rock cools, mineral crystals form. When melted rock cools quickly, only very small mineral crystals can form. Therefore, igneous rocks that cool quickly tend to have a fine-grained texture. When melted rock cools slowly, large crystals can form, which make a coarse-grained igneous rock.

Fine-grained

Basalt

Basalt forms when melted rock cools quickly on the Earth's surface. It has a fine-grained texture because the mineral crystals in it are very small.

Coarse-grained

Granite

Granite forms when melted rock cools slowly underground. It has a coarse-grained texture because the mineral crystals in it are large.

TAKE A LOOK

9. Describe How does granite form?

_____

_____

Section 1 Review

SECTION VOCABULARY

composition the chemical makeup of a rock; describes either the minerals or other materials in the rock	**rock** a naturally occurring solid mixture of one or more minerals or organic matter
deposition the process in which material is laid down	**rock cycle** the series of processes in which rock forms, changes from one type to another, is destroyed, and forms again by geologic processes
erosion the process by which wind, water, ice, or gravity transports soil and sediment from one location to another	**texture** the quality of a rock that is based on the sizes, shapes, and positions of the rock's grains

1. Compare What is the difference between weathering and erosion?

_____

_____

_____

2. Identify Complete the diagram to show how igneous rock can turn into sedimentary rock.

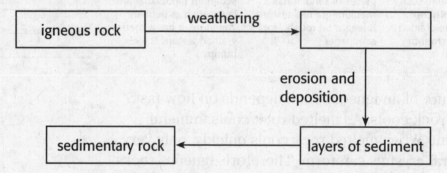

3. List What are two features that geologists use to classify rocks?

_____

_____

4. Describe What determines the texture of an igneous rock?

_____

_____

5. Explain How can a rock's composition help geologists to classify the rock?

_____

_____

_____

CHAPTER 4 Rocks: Mineral Mixtures

SECTION
2 Igneous Rock

National Science Education Standards
ES 1c, 1d

BEFORE YOU READ

After you read this section, you should be able to answer these questions:

• How do igneous rocks form?

• What factors affect the texture of igneous rock?

How Does Igneous Rock Form?

Igneous rocks form when hot, liquid rock, or *magma*, cools and hardens. There are three main ways that magma can form.

• An *increase in temperature:* when temperature increases, the minerals in a rock can melt.

• A *decrease in pressure:* hot rock can remain solid if it is under high pressure deep within the Earth. When the hot rock rises to the surface, the pressure goes down, and the rock can melt.

• An *addition of fluids:* when fluids, such as water, mix with rock, the melting temperature of the rock decreases and the rock can melt. ☑

When magma cools enough, mineral crystals form. This is similar to how water freezes. When you put water into the freezer, the water cools. When its temperature gets low enough, crystals of ice form. In the same way, crystals of different minerals can form as magma cools.

Water is made of a single compound. Therefore, all water freezes at the same temperature (0°C). However, magma is made of many different compounds. These compounds can combine to form different minerals. Each mineral becomes solid at a different temperature. Therefore, as magma cools, different parts of it become solid at different temperatures. Magma can become solid, or freeze, between 700°C and 1,250°C. ☑

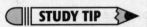

STUDY TIP

Compare After you read this section, make a table comparing the properties of intrusive igneous rock and extrusive igneous rock.

READING CHECK

1. Identify Give three ways that magma can form.

_____

_____

_____

READING CHECK

2. Explain Why do different parts of magma become solid at different times?

_____

_____

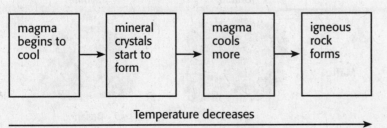

| magma begins to cool | → | mineral crystals start to form | → | magma cools more | → | igneous rock forms |

Temperature decreases

How Do Geologists Classify Igneous Rocks?

Geologists group igneous rocks by how they form. Geologists use clues from the rocks' compositions and textures to guess how they formed.

COMPOSITION

Based on composition, there are two main groups of igneous rocks—felsic rocks and mafic rocks. *Felsic* igneous rocks are rich in elements such as sodium, potassium, and aluminum. These elements combine to form light-colored minerals. Therefore, most felsic igneous rocks are light-colored. Granite and rhyolite are examples of felsic rocks.

Mafic igneous rocks are rich in elements such as iron, magnesium, and calcium. These elements combine to form dark-colored minerals. Therefore, most mafic igneous rocks are dark-colored. Gabbro and basalt are examples of mafic rocks.

TEXTURE

Remember that the texture of a rock is determined by the sizes of the grains in the rock. The texture of an igneous rock depends on how fast the magma cooled.

When magma cools quickly, mineral crystals do not have time to grow very large. Therefore, the rock that forms has a fine-grained texture. When magma cools slowly, large mineral crystals can form. Therefore, the rock that forms has a coarse-grained texture.

Critical Thinking

3. Compare Give two differences between felsic and mafic igneous rocks.

_____

_____

_____

_____

TAKE A LOOK
4. Identify Give an example of a felsic, fine-grained igneous rock.

_____

5. Identify Give an example of a mafic, coarse-grained igneous rock.

_____

	Coarse-grained	Fine-grained
Felsic	Granite	Rhyolite
Mafic	Gabbro	Basalt

SECTION 2 Igneous Rock *continued*

Rock's Texture?

Many people know that volcanoes form from melted rock. Therefore, they may think that igneous rocks only form at volcanoes on the Earth's surface. However, some igneous rocks form deep within the Earth's crust.

INTRUSIVE IGNEOUS ROCKS

Intrusive igneous rock forms when magma cools below the Earth's surface. Because the magma cools slowly, intrusive igneous rock usually has a coarse-grained texture. The minerals can grow into large, visible crystals. Bodies of intrusive igneous rock are grouped by their sizes and shapes. ☑

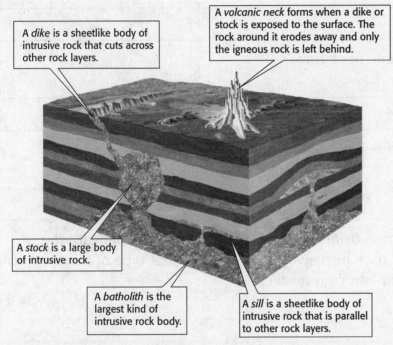

A *dike* is a sheetlike body of intrusive rock that cuts across other rock layers.

A *volcanic neck* forms when a dike or stock is exposed to the surface. The rock around it erodes away and only the igneous rock is left behind.

A *stock* is a large body of intrusive rock.

A *batholith* is the largest kind of intrusive rock body.

A *sill* is a sheetlike body of intrusive rock that is parallel to other rock layers.

✓ **READING CHECK**

6. Define Write your own definition for intrusive igneous rock.

_____

_____

_____

TAKE A LOOK

7. Identify Give four kinds of intrusive rock bodies.

_____

_____

_____

_____

EXTRUSIVE IGNEOUS ROCKS

Magma that reaches the Earth's surface is called *lava*. **Extrusive igneous rock** forms when lava cools. Extrusive igneous rock is common around volcanoes. Because extrusive rock cools quickly, it contains very small crystals or no crystals. ☑

When lava erupts from a volcano, it forms a *lava flow*. Lava flows can cover the land and bury objects on the Earth's surface.

Sometimes, lava erupts and flows along long cracks in Earth's crust called *fissures*. Many fissures are found on the ocean floor. Lava can also flow out of fissures onto land and form a *lava plateau*.

✓ **READING CHECK**

8. Explain Why do extrusive rocks have very small crystals or no crystals?

_____

_____

Section 2 Review

SECTION VOCABULARY

extrusive igneous rock rock that forms from the cooling and solidification of lava at the Earth's surface	intrusive igneous rock rock formed from the cooling and solidification of magma beneath the Earth's surface

1. Compare How are intrusive and extrusive igneous rocks different?

_____

_____

_____

2. Identify Give two examples of fine-grained igneous rocks.

_____

_____

_____

3. Describe How does a volcanic neck form?

_____

_____

_____

4. Compare What is the difference between a dike and a sill?

_____

_____

_____

5. Predict An igneous rock forms from slowly cooled magma deep beneath the surface of the Earth. Is the rock intrusive or extrusive? What type of texture does the rock probably have? Explain your answer.

_____

_____

_____

6. Apply Concepts Complete the table below. (Hint: What is the texture of each rock?)

Rock Name	Composition	Intrusive or Extrusive?
basalt	mafic	
gabbro	mafic	
granite	felsic	
rhyolite	felsic	

CHAPTER 4 Rocks: Mineral Mixtures

SECTION 3 Sedimentary Rock

BEFORE YOU READ

After you read this section, you should be able to answer these questions:

• How do sedimentary rocks form?

• How do geologists classify sedimentary rocks?

• What are some sedimentary structures?

National Science Education Standards

ES 1c, 1d

How Does Sedimentary Rock Form?

Remember that wind, water, ice, and gravity can cause rock to break down into fragments. These fragments are called *sediment*. During erosion, sediment is moved across the Earth's surface. Then the sediment is deposited in layers on the Earth's surface. As new layers are deposited, they cover older layers. The weight of the new layers *compacts*, or squeezes, the sediment in the older layers.

Water within the sediment layers can contain dissolved minerals, such as calcite and quartz. As the sediment is compacted, these minerals can crystallize between the sediment pieces. The minerals act as a natural glue and hold the sediment pieces together. As the loose sediment grains become bound together, a kind of sedimentary rock forms.

Unlike igneous and metamorphic rocks, sedimentary rock does not form at high temperatures and pressures. Sedimentary rock forms at or near the Earth's surface. ☑

Sediment is deposited in layers. Therefore, most sedimentary rocks contain layers called **strata** (singular, *stratum*).

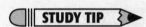

STUDY TIP

Reading Organizer As you read this section, create an outline of this section using the headings from this section.

Say It

Infer and Discuss In what kinds of areas are you likely to find sediment? Write down four places that sediment can be found. Think about the size of the sediment pieces that may be found at each place. Then, discuss your ideas with a small group.

READING CHECK

1. Describe Where does sedimentary rock form?

_____

_____

These "monuments" in Monument Valley, Arizona, formed as sedimentary rock eroded over millions of years.

SECTION 3 Sedimentary Rock *continued*

How Do Geologists Classify Sedimentary Rock?

Like other kinds of rock, sedimentary rock is classified by how it forms. Some sedimentary rock forms when rock or mineral fragments are stuck together. Some forms when minerals crystallize out of water. Other sedimentary rock forms from the remains of plants and animals.

CLASTIC SEDIMENTARY ROCK

Most sedimentary rock is clastic sedimentary rock. *Clastic sedimentary rock* forms when fragments of other rocks are cemented together. In most cases, the cement is a mineral such as calcite or quartz. The sediment pieces in different rocks can be of different sizes. Geologists group clastic sedimentary rocks by the sizes of the sediment pieces in them. ☑

| Conglomerate | Sandstone | Siltstone | Shale |

Coarse-grained ←――――――――――――――→ Fine-grained

Coarse-grained sedimentary rocks, such as conglomerate, contain large sediment pieces. Fine-grained rocks, such as shale, are made of tiny sediment pieces.

CHEMICAL SEDIMENTARY ROCK

Chemical sedimentary rock forms when minerals crystallize out of water. Water moves over rocks on the Earth's surface. As the water moves, it dissolves some of the minerals in the rocks. When the water evaporates, the dissolved minerals can crystallize to form chemical sedimentary rocks. ☑

Many chemical sedimentary rocks contain only one or two kinds of mineral. For example, evaporite is a chemical sedimentary rock. Evaporite is made mainly of the minerals halite and gypsum. These minerals crystallize when water evaporates.

READING CHECK

2. Identify Give two minerals that can act as cement in sedimentary rocks.

_____

_____

TAKE A LOOK

3. Describe What is the texture of conglomerate?

_____

_____

READING CHECK

4. Explain How do chemical sedimentary rocks form?

_____

_____

ORGANIC SEDIMENTARY ROCK

Organic sedimentary rock forms from the remains of plants and animals. Coal is one type of organic sedimentary rock. Coal forms from plant material that has been buried deep underground. Over millions of years, the buried plant material turns into coal.

Some organic sedimentary rock forms from the remains of sea creatures. For example, some limestone is made from the skeletons of creatures called *coral*. Coral are tiny creatures that make hard skeletons out of calcium carbonate. These skeletons and the shells of other sea creatures can be glued together to form *fossiliferous limestone*.

The shells of sea creatures can be cemented together to form fossiliferous limestone.

What Are Some Features of Sedimentary Rock?

The features of sedimentary rocks can give you clues about how the rocks formed. For example, many clastic sedimentary rocks show **stratification**. This means that they contain strata. Clastic sedimentary rocks show stratification because sediment is deposited in layers. ☑

Some sedimentary rock features show the motions of wind and water. For example, some sedimentary rocks show ripple marks or mud cracks. *Ripple marks* are parallel lines that show how wind or water has moved sediment. *Mud cracks* form when fine-grained sediment dries out and cracks.

These ripple marks show how sediment was moved by flowing water.

TAKE A LOOK

5. Define What is fossiliferous limestone?

_____

_____

☑ **READING CHECK**

6. Explain Why do many clastic sedimentary rocks show stratification?

_____

_____

Name ______________________ Class ______________ Date ______________

Section 3 Review

NSES ES 1c, 1d

SECTION VOCABULARY

strata layers of rock (singular, *stratum*)	**stratification** the process in which sedimentary rocks are arranged in layers

1. Define Write your own definition for stratification.

____

2. List Give three examples of clastic sedimentary rocks.

____

____

____

3. Compare How are clastic and organic sedimentary rocks different?

____

____

____

4. Describe How does evaporite form?

____

____

5. Describe How does fossiliferous limestone form?

____

____

____

____

6. Infer Imagine that a geologist finds a sedimentary rock with ripple marks in it. What can the geologist guess about the environment in which the sediment was deposited? Explain your answer.

____

____

____

____

Metamorphic Rock

BEFORE YOU READ

After you read this section, you should be able to answer these questions:

• How do metamorphic rocks form?

• How do geologists classify metamorphic rocks?

National Science Education Standards
ES 1c

How Does Metamorphic Rock Form?

Metamorphic rock forms when the chemical composition of a rock changes because of heat and pressure. This change is called *metamorphism*. Metamorphism can happen to any kind of rock.

Most metamorphism happens at temperatures between 150°C and 1,000°C. Some metamorphism happens at even higher temperatures. Many people think that all rocks must melt at such high temperatures. However, these rocks are also under very high pressure, so they do not melt.

High pressure can keep a hot rock from melting. Even very hot rocks may not melt if the pressure is high. Instead of melting, the minerals in the rock react with each other to form new minerals. In this way, the composition of the rock can change, even though the rock remains solid. ☑

High pressure can also affect the minerals in a rock. It can cause minerals to react quickly. It can also cause minerals to move slowly through the rock. In this way, different minerals can separate into stripes in the rock. The figure below shows an example of these stripes.

STUDY TIP

Ask Questions Read this section quietly to yourself. As you read, write down any questions you have. When you finish reading, try to figure out the answer to your questions in a small group.

READING CHECK

1. Describe How does the composition of a rock change during metamorphism?

_____

_____

The bands in this metamorphic rock formed as molecules of different minerals moved together.

TAKE A LOOK
2. Identify How did the bands in the rock in the figure form?

_____

_____

_____

CONTACT METAMORPHISM

There are two main ways that rock can go through metamorphism—contact metamorphism and regional metamorphism. *Contact metamorphism* happens when rock is heated by nearby magma. As the magma moves through the crust, the rocks in the crust heat up. The minerals in those rocks can react to produce new minerals.

TAKE A LOOK
3. Define What is contact metamorphism?

_____

_____

_____

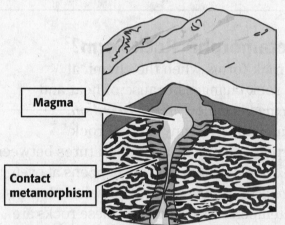

Contact metamorphism happens when magma heats nearby rock.

Rock that is very near the magma changes the most during contact metamorphism. The farther the rock is from the magma, the smaller the changes. This is because the temperature decreases with distance from the magma. Contact metamorphism usually only affects rock in a small area.

REGIONAL METAMORPHISM

During *regional metamorphism*, high pressures and temperatures cause the rock in a large area to change. Regional metamorphism can happen where rock is buried deep below the surface or where pieces of the Earth's crust collide.

TAKE A LOOK
4. Describe Give two places where regional metamorphism can happen.

_____

_____

_____

_____

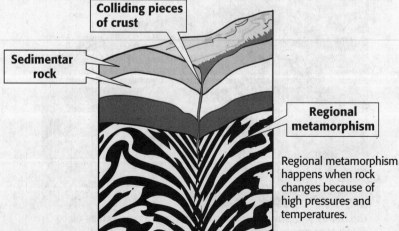

Regional metamorphism happens when rock changes because of high pressures and temperatures.

SECTION 4 Metamorphic Rock *continued*

METAMORPHIC STRUCTURES

Both contact and regional metamorphism can cause deformation. *Deformation* is a change in the shape of a rock. When forces act on a rock, they may cause the rock to be squeezed or stretched.

Folds are features of a rock that show that the rock has been deformed. Some folds are so small that they can only be seen with a microscope. Other folds, like the ones below, are visible to the naked eye.

These folds formed during metamorphism. The rocks in this picture are found in Labrador, Canada.

TAKE A LOOK
5. Infer Were these folds probably caused by squeezing the rock or by stretching it?

_____

_____

What Are Metamorphic Rocks Made Of?

Remember that different minerals form under different conditions. Minerals that form near the Earth's surface, such as calcite, may not be stable under higher temperatures and pressures. During metamorphism, these minerals are likely to react and produce new minerals. The new minerals are stable under high temperatures and pressures. The figure below shows how new minerals can form from unstable minerals.

Critical Thinking

6. Predict The mineral gypsum forms at low temperatures and pressures. The mineral sillimanite forms at high temperatures and pressures. Which mineral would most likely be found in a metamorphic rock? Explain your answer.

_____

_____

_____

_____

_____

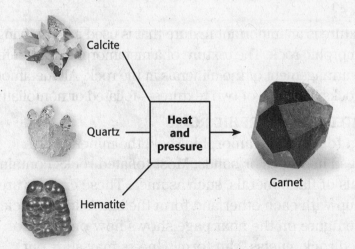

Calcite, quartz, and hematite are not stable under high temperatures and pressures. They react to form garnet in metamorphic rocks.

INDEX MINERALS

Some minerals, such as quartz, can form at many different temperatures and pressures. Other minerals, such as garnet, form only at certain temperatures and pressures. Therefore, rocks that contain minerals like garnet probably also formed at those temperatures and pressures. Geologists can use such minerals as index minerals.

Index minerals can indicate the temperature and pressure or depth at which a rock formed. These minerals help geologists learn the temperature and pressure at which a rock formed. Chlorite, muscovite, and garnet are index minerals for metamorphic rocks.

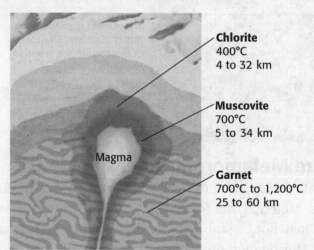

Geologists can use some minerals as index minerals. These minerals help geologists learn the temperature and pressure at which a rock formed. For example, a rock containing garnet most likely formed at a higher temperature and pressure than a rock containing chlorite.

How Do Geologists Classify Metamorphic Rocks?

Texture is an important feature that is used in classifying metamorphic rock. The texture of a metamorphic rock refers to the arrangement of the minerals in the rock. All metamorphic rocks have one of two textures—foliated or nonfoliated.

FOLIATED METAMORPHIC ROCK

In a **foliated** metamorphic rock, the minerals are arranged in stripes or bands. Most foliated rocks contain crystals of flat minerals, such as mica. These crystals are lined up with each other and form the bands in the rock. ☑

The figure on the next page shows how one kind of foliated rock, gneiss, can form. Gneiss may start out as the sedimentary rock shale. Heat and pressure can change shale to slate, phyllite, schist, or gneiss.

Critical Thinking

7. Infer Why can't geologists use minerals like quartz to determine the temperature and pressure that a rock formed at?

_____

_____

_____

TAKE A LOOK

8. Identify Which index mineral in the figure forms at the lowest temperature?

_____

_____

 **READING CHECK**

9. Define What is a foliated metamorphic rock?

_____

_____

_____

SECTION 4 Metamorphic Rock *continued*

Sedimentary shale

The metamorphic rock slate can form when shale is placed under heat and pressure.

Slate

Phyllite can form when slate is put under heat and pressure.

Phyllite

Gneiss can form when schist is put under heat and pressure. The minerals in gneiss line up in bands, so gneiss is a foliated rock.

When phyllite is put under heat and pressure, schist can form.

Schist

Gneiss

UNFOLIATED METAMORPHIC ROCK

In a **nonfoliated** metamorphic rock, the mineral crystals are not arranged in bands or stripes. Most nonfoliated rocks are made of only a few minerals. Metamorphism can cause the mineral crystals in a rock to get bigger. ☑

Quartzite is an example of a nonfoliated metamorphic rock. Quartzite can form from the sedimentary rock quartz sandstone. Quartz sandstone is made of grains of quartz sand that have been cemented together. The quartz crystals in these grains can grow larger during metamorphism. The quartz crystals in quartzite can be much larger than those in quartz sandstone.

✓ **READING CHECK**

11. Describe What can happen to the sizes of mineral crystals during metamorphism?
_____
_____

Type of Metamorphic Rock	Description	Example
Foliated		gneiss
Nonfoliated		quartzite

TAKE A LOOK
12. Define Fill in the blank spaces in the table.

Section 4 Review

SECTION VOCABULARY

foliated describes the texture of metamorphic rock in which the mineral grains are arranged in planes or bands	**nonfoliated** describes the texture of metamorphic rock in which the mineral grains are not arranged in planes or bands

1. **Compare** How are foliated metamorphic rocks different from nonfoliated metamorphic rocks?

_____

_____

_____

2. **Define** What is regional metamorphism?

_____

_____

3. **Describe** What is an index mineral? Give two examples of index minerals for metamorphic rocks.

_____

_____

_____

4. **Explain** How do index minerals help geologists?

_____

_____

_____

5. **Describe** How does quartzite form?

_____

_____

_____

6. **Apply Concepts** A geologist finds two metamorphic rocks. One contains chlorite. The other contains garnet. Which rock probably formed at the greatest depth? Explain your answer.

_____

_____

_____

CHAPTER 5 Energy Resources
SECTION 1 # Natural Resources

National Science
Education Standards
ES 3d

BEFORE YOU READ

After you read this section, you should be able to answer these questions:

- What is the difference between a renewable resource and a nonrenewable resource?
- How can you protect natural resources?

What Are Earth's Resources?

Earth provides what you need to survive. You breathe air from Earth's atmosphere. You drink water from Earth's rivers, lakes, and other water bodies. You eat food from Earth's living things.

A **natural resource** is any material from Earth that is used by people. Air, soil, fresh water, petroleum, rocks, minerals, forests, and wildlife are examples of natural resources. People use some natural resources, such as coal and wind, for energy. The energy in these resources comes from energy from the sun. The figure below shows some examples of natural resources. ☑

STUDY TIP

Summarize After you read this section, make a chart giving the definitions of renewable and nonrenewable resources. In the chart, include two examples of each kind of resource.

✓ **READING CHECK**

1. Define In your own words, write a definition of *natural resource*.

_____

_____

Examples of Natural Resources

Gasoline and plastic are both made from oil that is pumped out of Earth's crust.

This pile of lumber is made of wood, which comes from trees.

These windmills produce electricity from the energy in wind. The energy in wind comes mainly from the sun.

TAKE A LOOK
2. Identify Give two examples of natural resources that are not shown in the figure.

_____

_____

SECTION 1 Natural Resources *continued*

What Types of Resources Exist on Earth?

Natural resources are grouped based on how fast they can be replaced. Some natural resources are nonrenewable. Others are renewable.

NONRENEWABLE RESOURCES

Some resources, such as coal, petroleum, and natural gas, take millions of years to form. A **nonrenewable resource** is a resource that is used much faster than it can be replaced. *Renew* means "to begin again." When nonrenewable resources are used up, people can no longer use them. ☑

☑ **READING CHECK**

3. Define What is a nonrenewable resource?

_____

_____

RENEWABLE RESOURCES

Some natural resources, such as trees and fresh water, can grow or be replaced quickly. A **renewable resource** is a natural resource that can be replaced as quickly as people use it.

Many renewable resources are renewable only if people do not use them too quickly. For example, wood is usually considered a renewable resource. However, if people cut down trees faster than the trees can grow back, wood is no longer a renewable resource. Some renewable resources, such as the sun, will never be used up, no matter how fast people use them.

TAKE A LOOK

4. Explain Describe how some renewable resources can become nonrenewable resources.

_____

_____

_____

_____

_____

Fresh water and trees are both renewable resources. However, they can be used up if people use them too quickly.

| SECTION 1 | Natural Resources *continued* |

How Can We Protect Natural Resources?

Whether the natural resources you use are renewable or nonrenewable, you should be careful how you use them. In order to *conserve* natural resources, you should try to use them only when you have to. For example, leaving the water running while you are brushing your teeth wastes clean water. Turning the water off while you brush your teeth saves water so that it can be used in the future.

The energy we use to heat our homes, drive our cars, and run our computers comes from natural resources. Most of these resources are nonrenewable. If we use too much energy now, we might use up these resources. Therefore, reducing the amount of energy you use can help to conserve natural resources. You can conserve energy by being careful to use it only when you need to. The table shows some ways you can conserve energy.

Instead of...	You can...
...leaving the lights on all the time	...turn them off when you're not in the room
...running the washing machine when it is only half full	...run it only when it is full
...using a car to travel everywhere	...walk, ride a bike, or use public transportation when you can

Recycling is another important way that you can help to conserve natural resources. **Recycling** means using things that have been thrown away to make new objects. Objects made from recycled materials use fewer natural resources than objects made from new materials. Recycling also helps to conserve energy. For example, it takes less energy to recycle an aluminum can than to make a new one. ☑

Newspaper, aluminum cans, some plastic containers, and many types of paper can be recycled. Check with your community's recycling center to see what kinds of materials you can recycle.

Conserving resources also means taking care of them even when you are not using them. For example, it is important to keep our drinking water clean. Polluted water can harm the living things, including humans, that need water in order to live.

Critical Thinking

5. Explain Why is it important to conserve all natural resources, even if they are renewable resources?

_____

_____

_____

_____

_____

_____

TAKE A LOOK

6. Brainstorm Fill in the blank spaces in the table with some other ways you can conserve natural resources.

☑ READING CHECK

7. Identify How does recycling conserve natural resources?

_____

_____

_____

_____

_____

Name ______________________ Class ______________ Date ______________

Section 1 Review

NSES ES 3d

SECTION VOCABULARY

natural resource any natural material that is used by humans, such as water, petroleum, minerals, forests, and animals **nonrenewable resource** a resource that forms at a rate that is much slower than the rate at which the resource is consumed	**recycling** the process of recovering valuable or useful materials from waste or scrap **renewable resource** a natural resource that can be replaced at the same rate at which the resource is consumed

1. Identify What is the difference between a renewable resource and a nonrenewable resource?

____

____

____

2. List Give four ways to conserve natural resources.

____

____

____

____

3. Explain Why is wood usually considered a renewable resource? When would it be considered a nonrenewable resource?

____

____

____

____

4. Describe What does it mean to conserve natural resources?

____

____

____

5. Explain Why are coal, oil, and natural gas considered nonrenewable resources, even though they come from living things that can reproduce?

____

____

CHAPTER 5 | Energy Resources

SECTION 2 | # Fossil Fuels

National Science Education Standards
ES 1d, 1e, 1k, 3d

> **BEFORE YOU READ**
>
> **After you read this section, you should be able to answer these questions:**
> • What are the different kinds of fossil fuels?
> • How do fossil fuels form?
> • What are the problems with using fossil fuels?

What Are Fossil Fuels?

How do plants and animals that lived hundreds of millions of years ago affect your life today? Plants and animals that lived long ago provide much of the energy we use. If you turned on the lights or traveled to school in a car or bus, you probably used some of this energy.

Energy resources are natural resources that people use to produce energy, such as heat and electricity. Most of the energy we use comes from fossil fuels. A **fossil fuel** is an energy resource made from the remains of plants and tiny animals that lived long ago. The different kinds of fossil fuels are petroleum, coal, and natural gas. ☑

Fossil fuels are an important part of our everyday life. When fossil fuels burn, they release a lot of energy. Power plants use the energy to produce electricity. Cars use the energy to move.

However, there are also some problems with using fossil fuels. Fossil fuels are nonrenewable, which means that they cannot be replaced once they have been used. Also, when they burn, they release pollution.

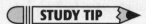

STUDY TIP

Compare In your notebook, make a table to show the similarities and differences between different kinds of fossil fuels.

READING CHECK

1. Identify Where do we get most of the energy we use?

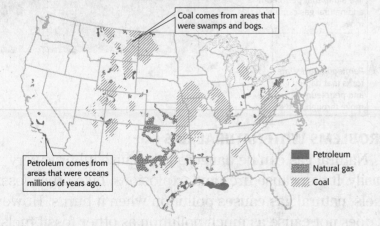

Coal comes from areas that were swamps and bogs.

Petroleum comes from areas that were oceans millions of years ago.

Petroleum
Natural gas
Coal

TAKE A LOOK

2. Describe In general, where are the natural gas deposits in the United States?

_____

_____

 **Say It**

Discuss Have you ever seen or used methane, propane, or butane? In a small group, talk about what these different gases are used for.

 **READING CHECK**

3. Identify What is most natural gas used for?

_____

_____

STANDARDS CHECK

ES 1k Living organisms have played many <u>roles</u> in the earth system, including <u>affecting</u> the composition of the atmosphere, producing some types of rocks, and contributing to the weathering of rocks.

Word Help: <u>role</u>
a part or function; purpose

Word Help: <u>affect</u>
to change; to act upon

4. Explain How does natural gas form?

_____

_____

_____

_____

_____

What Is Natural Gas?

A *hydrocarbon* is a compound that contains the elements carbon, hydrogen, and oxygen. **Natural gas** is a mixture of hydrocarbons that are in the form of gases. Natural gas includes methane, propane, and butane, which can be separated from one another.

Most natural gas is used for heating. Your home may be heated by natural gas. Your kitchen stove may run on natural gas. Some natural gas is used for creating electrical energy, and some cars are able to run on natural gas, too. ☑

HOW NATURAL GAS FORMS

When tiny sea creatures die, their remains settle to the ocean floor and are buried in sediment. The sediment slowly becomes rock. Over millions of years, heat and pressure under the ground chemically change the remains. The carbon, hydrogen, and oxygen in them can become natural gas.

Natural gas is always forming. Some of the sea life that dies today will become natural gas millions of years from now.

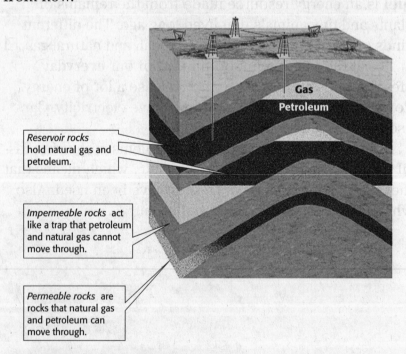

Reservoir rocks hold natural gas and petroleum.

Impermeable rocks act like a trap that petroleum and natural gas cannot move through.

Permeable rocks are rocks that natural gas and petroleum can move through.

PROBLEMS WITH NATURAL GAS

Natural gas can be dangerous because it burns very easily. It can cause fires and explosions. Like other fossil fuels, natural gas causes pollution when it burns. However, it does not cause as much pollution as other fossil fuels.

SECTION 2 | Fossil Fuels *continued*

What Is Petroleum?

Petroleum is a mixture of hydrocarbons that are in the form of liquids. It is also known as crude oil. At a *refinery*, petroleum is separated into many different products, including gasoline, jet fuel, kerosene, diesel fuel, and fuel oil. ☑

Petroleum products provide more than 40% of the world's energy, including fuel for airplanes, trains, boats, ships, and cars. Petroleum is so valuable that it is often called "black gold."

HOW PETROLEUM FORMS

Petroleum forms the same way natural gas does. Tiny sea creatures die and then get buried in sediment, which turns into rock. Some of their remains become petroleum, which is stored in permeable rock within Earth's crust.

PROBLEMS WITH PETROLEUM

Petroleum can be harmful to animals and their environment. For example, in June 2000, the carrier ship *Treasure* sank off the coast of South Africa. The ship spilled more than 400 tons of oil into the ocean. The oil covered penguins and other sea creatures, making it hard for them to swim, breathe, and eat.

Burning petroleum causes smog. **Smog** is a brownish haze that forms when sunlight reacts with pollution in the air. Smog can make it hard for people to breathe. Many cities in the world have problems with smog.

READING CHECK

5. Describe What form is petroleum found in naturally?

_____

_____

Math Focus

6. Make a Graph Use the information below to show where the world's crude oil comes from.

Middle East: 66%

North America and South America: 15%

Europe and Asia: 12%

Africa: 7%

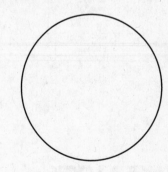

What Is Coal?

Coal is a solid fossil fuel that is made of partly decayed plant material. Coal was once the main source of energy in the United States. Like other fossil fuels, coal releases heat when it is burned. Many people used to burn coal in stoves to heat their homes. Trains in the 1800s and 1900s were powered by coal-burning steam locomotives. Coal is now used in power plants to make electricity.

HOW COAL FORMS

When swamp plants die, they sink to the bottom of the swamp. If they do not decay completely, coal can start to form. Coal forms in several different stages.

TAKE A LOOK
7. Identify Name the three types of coal.

_____

_____

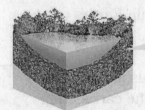

Stage 1: Formation of Peat
Dead swamp plants that have not decayed can turn into *peat*, a crumbly brown material made mostly of plant material and water. Dried peat is about 60% carbon. In some parts of the world, peat is dried and burned for fuel. Peat is not coal, but it can turn into coal.

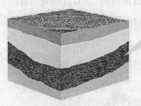

Stage 2: Formation of Lignite
If sediment buries the peat, pressure and temperature increase. The peat slowly changes into a type of coal called *lignite*. Lignite coal is harder than peat. Lignite is about 70% carbon.

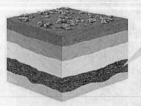

Stage 3: Formation of Bituminous Coal
If more sediment is added, pressure and temperature force more water and gases out of the lignite. Lignite slowly changes into *bituminous* coal. Bituminous coal is about 80% carbon.

TAKE A LOOK
8. Identify Which kind of coal contains the most carbon?

_____

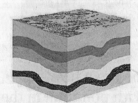

Stage 4: Formation of Anthracite
If more sediment accumulates, temperature and pressure continue to increase. Bituminous coal slowly changes into *anthracite*. Anthracite coal is the hardest type of coal. Anthracite coal is about 90% carbon.

SECTION 2 Fossil Fuels *continued*

PROBLEMS WITH COAL

Mining coal can create environmental problems. When coal is mined from Earth's surface, people remove the layers of soil above the coal. This can harm the plants that need soil to grow and the animals that need soil for shelter. If the land is not restored after mining, wildlife habitats can be destroyed for years.

Coal that is on Earth's surface can cause pollution. Water that flows through coal can pick up poisonous metals. That water can then flow into streams and lakes and pollute water supplies.

When coal is burned without pollution controls, a gas called sulfur dioxide is released. Sulfur dioxide can combine with the water in the air to produce sulfuric acid. Sulfuric acid is one of the acids in acid precipitation. **Acid precipitation** is rain, sleet, or snow that contains a lot of acids, often because of air pollutants. Acid precipitation is also called "acid rain." Acid precipitation can harm wildlife, plants, and buildings.

In 1935, this statue had not been damaged by acid precipitation.

By 1994, acid precipitation had caused serious damage to the statue.

Critical Thinking

9. Infer What do you think is the reason that fossil fuels are still used today, even though they create many environmental problems?

_____

_____

_____

_____

_____

TAKE A LOOK
10. Define What is acid precipitation?

_____

_____

How Do We Obtain Fossil Fuels?

People remove fossil fuels from the Earth in different ways. The way that a fossil fuel is removed depends on the kind of fuel and where it is located. Remember that people remove coal from the Earth by mining it. People remove petroleum and natural gas by drilling into the rocks that contain the fuels. Then, the petroleum or natural gas is removed through a well. These wells may be on land or in the ocean.

Section 2 Review

NSES   ES 1d, 1e, 1k, 3d

SECTION VOCABULARY

acid precipitation rain, sleet, or snow that contains a high concentration of acids	**natural gas** a mixture of gaseous hydrocarbons located under the surface of Earth, often near petroleum deposits; used as a fuel
coal a fossil fuel that forms underground from partially decomposed plant material	**petroleum** a liquid mixture of complex hydrocarbon compounds; used widely as a fuel source
fossil fuel a nonrenewable energy resource formed from the remains of organisms that lived long ago	**smog** photochemical haze that forms when sunlight acts on industrial pollutants and burning fuels

1. Compare How are petroleum and natural gas different?

_____

_____

_____

2. Compare Fill in the table to compare the different kinds of fossil fuels.

Kind of fossil fuel	What it is	How it forms
Coal		
	a mixture of gases containing carbon, hydrogen, and oxygen	

3. Summarize What are some of the problems with using fossil fuels for energy?

_____

_____

_____

_____

_____

CHAPTER 5 Energy Resources

SECTION
3 **Alternative Resources**

BEFORE YOU READ

After you read this section, you should be able to answer these questions:

• What are some kinds of alternative energy?

• What are some benefits of alternative energy?

• What are some problems with alternative energy?

National Science Education Standards
ES 3d

What Is Alternative Energy?

What would your life be like if you couldn't turn on the lights, microwave your dinner, take a hot shower, or ride the bus to school? We get most of the energy we use for heating and electricity from fossil fuels. However, fossil fuels can be harmful to the environment and to living things. In addition, they are nonrenewable resources, so we cannot replace them when they are used up.

Many scientists are trying to find alternative energy sources. *Alternative energy sources* are sources of energy that are not fossil fuels. Some sources can be converted easily into usable energy. Others are not as easy to use.

What Is Nuclear Energy?

One kind of alternative energy source is nuclear energy. **Nuclear energy** is the energy that is released when atoms come together or break apart. Nuclear energy can be obtained in two main ways: fission and fusion. ☑

FISSION

Fission happens when an atom splits into two or more lighter atoms. Fission releases a large amount of energy. This energy can be used to generate electricity. All nuclear power that people use today is generated by fission.

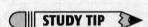

STUDY TIP
Compare and Contrast In your notebook, make a chart to show each kind of alternative energy source and its benefits and problems.

READING CHECK
1. List What are the two ways in which nuclear energy is produced?

_____

_____

During nuclear fission, a neutron collides with a uranium-235 atom. The uranium is the fuel for the reaction.

Neutron

Barium-142 atom

After colliding with a neutron, the uranium atom splits into two smaller atoms, called *fission products*, and two or more neutrons.

Neutron

ENERGY

Uranium-235 atom

Krypton-91 atom

TAKE A LOOK
2. Identify What are the fission products in the figure?

_____

_____

FISSION'S BENEFITS AND PROBLEMS

One benefit of fission is that it does not cause air pollution. Mining uranium, the fuel for nuclear power, is less harmful to the environment than mining other energy sources, such as coal. ☑

However, nuclear fission power has several problems. The fission products created in nuclear power plants are poisonous. They must be stored for thousands of years. Nuclear fission plants can release harmful radiation into the environment. Also, nuclear power plants must release extra heat from the fission reaction. This extra heat cannot be used to make electricity. The extra heat can harm the environment.

FUSION

Fusion happens when two or more atoms join to form a heavier atom. This process occurs naturally in the sun. Fusion releases a lot of energy.

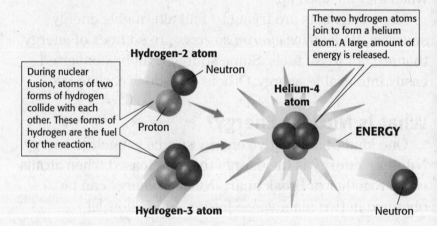

During nuclear fusion, atoms of two forms of hydrogen collide with each other. These forms of hydrogen are the fuel for the reaction.

Hydrogen-2 atom
Neutron
Proton
Hydrogen-3 atom

The two hydrogen atoms join to form a helium atom. A large amount of energy is released.

Helium-4 atom
ENERGY
Neutron

FUSION'S BENEFITS AND PROBLEMS

Fusion has two main benefits. First, fusion does not create a lot of dangerous wastes. Second, the fuels used in fusion are renewable.

The main problem with fusion is that it can take place only at high temperatures. The reaction is difficult to control and keep going. Right now, people cannot control fusion reactions or use them to create usable energy. ☑

✓ READING CHECK

3. Explain Why is nuclear energy called a "clean" energy source?

_____

_____

_____

TAKE A LOOK

4. Identify How many protons and how many neutrons are there in the helium-4 nucleus?

_____

_____

✓ READING CHECK

5. Describe What is the main problem with fusion?

_____

_____

_____

What Is Wind Power?

Wind is air that is moving. Moving air contains energy. People can use windmills to turn the energy in wind into electricity. The electricity that is produced by windmills is called **wind power**. Large groups of windmills can make a lot of electricity.

Like all energy sources, wind power has benefits and problems. Since the wind can't be used up, wind energy is renewable. Wind power does not cause air pollution. However, in many areas, the wind isn't strong or regular enough to generate enough electricity for people to use.

This pickup truck shows how large the windmills are.

These windmills near Livermore, California, produce electricity.

What Are Fuel Cells?

What powers a car? You probably thought of gasoline. However, not all cars are powered by gasoline. Some cars are powered by fuel cells. *Fuel cells* are devices that change chemical energy into electrical power. The **chemical energy** is released when hydrogen and oxygen react to form water.

Fuel cells have been used in space travel since the 1960s. They have provided space crews with electrical energy and drinking water. Today, fuel cells are used to create electrical energy in some buildings and ships. ☑

The only waste product of fuel cells is water, so they do not create pollution. However, not very many cars today use fuel cells. The hydrogen and oxygen used in fuel cells can be expensive to make and to store. Many people hope that we will be able to use fuel cells to power cars in the future.

Critical Thinking

6. Infer In most cases, people use a large number of windmills to create electricity. What do you think is the reason a lot of windmills are used, instead of just one or two?

_____

_____

_____

TAKE A LOOK

7. Explain Based on what you see in the figure, what do you think is the reason windmills are not used in cities or other crowded areas?

_____

_____

READING CHECK

8. Explain How could fuel cells give space crews electricity and water?

_____

_____

_____

What Is Solar Energy?

Most forms of energy originally come from the sun. For example, the fossil fuels we use today were made from plants. The plants got their energy from the sun. The heat and light that Earth gets from the sun is **solar energy**. This type of energy is a renewable resource.

People can use solar energy to create electricity. *Photovoltaic cells*, or solar cells, can change sunlight into electrical energy. Solar energy can also be used to heat buildings.

Solar energy does not produce pollution and is renewable. The energy from the sun is free. However, some climates don't have enough sunny days to be able to use solar energy all the time. Also, even though sunlight is free, solar cells are expensive to make.

What Is Hydroelectric Energy?

Water wheels have been used since ancient times to help people do work. Today, the energy of falling water is used to generate electrical energy. **Hydroelectric energy** is electrical energy produced from moving water.

Hydroelectric energy causes no air pollution and is considered renewable. Hydroelectric energy is generally not very expensive to produce. ☑

However, hydroelectric energy can be produced only in places that have a lot of fast-moving water. In addition, building a dam and a power plant to generate hydroelectric energy can be expensive. Dams can harm wildlife living in and around the river. Damming a river can cause flooding and erosion.

Say It

Share Experiences Have you ever used an object that was powered by sunlight? In a small group, talk about the different ways that sunlight can be used for energy.

☑ **READING CHECK**

9. Identify What are two benefits of hydroelectric energy?

_____

_____

This dam in California can create electricity because a lot of water moves through it every day.

SECTION 3 Alternative Resources *continued*

How Can Plants Be Used for Energy?

Plants store energy from the sun. Leaves, wood, and stems contain stored energy. Even the dung of plant-eating animals has a lot of stored energy. These sources of energy are called biomass. **Biomass** is organic matter that can be a source of energy.

Biomass is commonly burned in its solid form to release heat. However, biomass can also be changed into a liquid form. The sugar and starch in plants can be made into alcohol and used as fuel. Alcohol can be mixed with gasoline to make a fuel called **gasohol**.

Biomass is not very expensive. It is available almost everywhere. Since biomass grows quickly, it is considered a renewable resource. However, people must be careful not to use up biomass faster than it can grow back.

What Is Geothermal Energy?

Geothermal energy is energy produced by the heat within Earth. This heat makes solid rocks get very hot. If there is any water contained within the solid rock, the water gets hot, too. The hot water can be used to generate electricity and to heat buildings. ☑

Geothermal energy is considered renewable because the heat inside Earth will last for millions of years. Geothermal energy does not create air pollution or harm the environment. However, this kind of energy can be used only where hot rock is near the surface.

Critical Thinking

10. Infer What would happen if biomass were used at a faster rate than it was produced?

_____

_____

_____

☑ READING CHECK

11. List What are two uses for water that has been heated by hot rock?

_____

_____

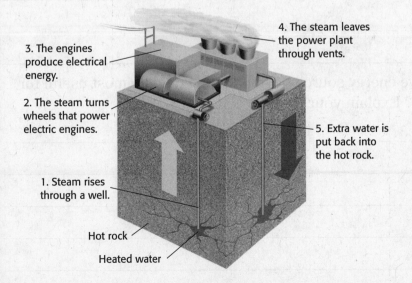

3. The engines produce electrical energy.

4. The steam leaves the power plant through vents.

2. The steam turns wheels that power electric engines.

5. Extra water is put back into the hot rock.

1. Steam rises through a well.

Hot rock

Heated water

TAKE A LOOK

12. Describe On the figure, draw arrows showing the path that the steam takes as it moves through the power plant.

Section 3 Review

SECTION VOCABULARY

biomass plant material, manure, or any other organic matter that is used as an energy source	**hydroelectric energy** electrical energy produced by the flow of water
chemical energy the energy released when a chemical compound reacts to produce new compounds	**nuclear energy** the energy released by a fission or fusion reaction; the binding energy of the atomic nucleus
gasohol a mixture of gasoline and alcohol that is used as a fuel	**solar energy** the energy received by Earth from the sun in the form of radiation
geothermal energy the energy produced by heat within Earth	**wind power** the use of a windmill to drive an electric generator

1. Explain Why is solar energy considered a renewable resource?

_____

_____

2. Identify When would biomass not be considered a renewable resource?

_____

_____

3. Apply Concepts Which place is *more likely* to be able to use geothermal energy: a city near a volcano or a city near a waterfall? Explain your answer.

_____

_____

_____

4. Identify Why is wind a useful energy source in some places, but not in others?

_____

_____

_____

5. Analyze Which alternative energy source do you think would be most useful for the place where you live? Explain your answer.

_____

_____

_____

_____

_____

SECTION 1 Earth's Story and Those Who First Listened

BEFORE YOU READ

After you read this section, you should be able to answer these questions:

- How fast do changes on Earth happen?
- What is paleontology?

National Science Education Standards
ES 2a, 2b

How Fast Do Changes on Earth Happen?

Earth has not always looked the way it does today. Our planet is slowly changing all the time. Through history, many people have studied these changes. Many different ideas have been put forward to explain how Earth changes with time.

Until about 200 years ago, most people believed that Earth changes because of sudden events, such as floods. The belief that Earth changes only because of sudden events is called **catastrophism**. However, scientists soon realized that catastrophism could not explain all of their observations about the things that happen on Earth.

James Hutton was one of the scientists who first realized that geologic changes can happen very slowly. Hutton observed the processes that were happening around him. He hypothesized that the same processes have been happening for all of the Earth's history. The figure below shows some examples of these processes.

STUDY TIP

Graphic Organizer As you read this section, make a table comparing catastrophism, uniformitarianism, and modern ideas about how the Earth changes.

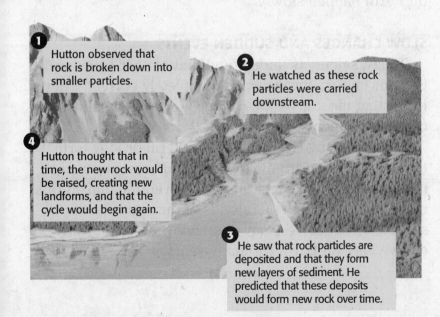

1 Hutton observed that rock is broken down into smaller particles.

2 He watched as these rock particles were carried downstream.

4 Hutton thought that in time, the new rock would be raised, creating new landforms, and that the cycle would begin again.

3 He saw that rock particles are deposited and that they form new layers of sediment. He predicted that these deposits would form new rock over time.

TAKE A LOOK

1. Describe What did Hutton predict would form new rock over time?

_____

_____

UNIFORMITARIANISM

James Hutton reasoned that the slow processes that shape Earth now have been the same through all of Earth's history. Over a very long time, these processes have added up to form the features we see on Earth today. The theory that the Earth's features form only because of small changes over long periods of time is called **uniformitarianism**.

TAKE A LOOK

2. Identify Fill in the blank spaces in the table to show how catastrophism is different from uniformitarianism.

Catastrophism	
Uniformitarianism	

Hutton published his ideas in the late 1700s, but they were not accepted by most scientists at that time. However, in the early 1800s, a scientist named Charles Lyell reintroduced the idea of uniformitarianism. Lyell provided more evidence to support uniformitarianism. His work helped to convince many scientists that uniformitarianism was valid.

By the mid-1800s, many scientists had accepted that uniformitarianism can explain many of the Earth's features. However, they also saw that sudden events can change Earth's surface. For example, large storms can cause the Earth's surface to change very quickly. These observations helped scientists realize that not all changes on Earth happen slowly.

SLOW CHANGES AND SUDDEN EVENTS

Today, scientists understand that neither catastrophism nor uniformitarianism is completely correct. They know that most geologic change is slow, but sudden changes happen sometimes. ☑

Sudden changes can have short-term or long-term effects. The wind from a hurricane affects only a small part of Earth for a short time. However, the impact of a comet on Earth may put clouds of dust into the atmosphere. These clouds may decrease the temperature everywhere on Earth for many years.

✓ READING CHECK

3. Explain According to scientists today, how fast do changes on Earth happen?

_____

_____

_____

_____

SECTION 1 Earth's Story and Those Who First Listened *continued*

How Do Scientists Study Earth's Past?

Scientists can use fossils to learn about what Earth was like in the past. A *fossil* is any evidence of past life. Some fossils are made from the remains, such as shells, of dead organisms. Other fossils are simply signs, such as footprints, that an organism once existed. The study of fossils and ancient life is called **paleontology**. The root *paleo* means "old." The root *onto* means "life."

Scientists who study paleontology are called *paleontologists*. Different paleontologists study different certain kinds of organisms. For example, *vertebrate paleontologists* study the remains of *vertebrates*, or animals with backbones. *Paleobotanists* study fossils of ancient plants.

Fossils provide evidence that life on Earth has changed with time. Different organisms have appeared and disappeared throughout Earth's history. For example, fossils show that dinosaurs once lived on Earth, even though none are alive today.

Fossils of dinosaurs have been found in many places on Earth. However, no dinosaurs are alive today. The fossils show that the kinds of life on Earth have changed over time.

Fossils also provide evidence of how Earth has changed over time. For example, there are fossils of sea life from millions of years ago in deserts and on the tops of mountains. The fossils show that some areas that are now deserts or mountains were once parts of an ocean.

Critical Thinking

4. Infer Paleobotanists study the remains of ancient plants. What do botanists most likely study?

_____

_____

STANDARDS CHECK

ES 2b Fossils provide important evidence of how life and environmental conditions have changed.

Word Help: evidence information showing whether an idea or belief is true or valid

Word Help: environment the surrounding natural conditions that affect an organism

5. Explain What are two things that paleontologists can learn from fossils?

_____

_____

_____

_____

Name ____________________ Class ____________ Date ____________

Section 1 Review

NSES ES 2a, 2b

SECTION VOCABULARY

catastrophism a principle that states that geologic change occurs suddenly **paleontology** the scientific study of fossils	**uniformitarianism** a principle that geologic processes that occurred in the past can be explained by current geologic processes

1. Identify How can sudden events affect Earth?

2. Describe What were two processes that James Hutton observed that helped him develop the idea of uniformitarianism?

3. Define What is a fossil?

4. Describe One kind of fossil forms from the body parts of organisms. What is another kind of fossil?

5. Apply Concepts Imagine that you find a layer of rock containing many fossil clams. The layer of rock is 50 km from the ocean. The fossils are about 5 million years old. Clams usually live in shallow ocean water. Based on the fossils, what can you guess about the environment in this area 5 million years ago? Explain your answer.

CHAPTER 6 | The Rock and Fossil Record

SECTION 2 Relative Dating: Which Came First?

National Science
Education Standards
ES 2b

BEFORE YOU READ

After you read this section, you should be able to answer these questions:

- What is relative dating?
- How can rock layers be disturbed?

What Is Relative Dating?

Imagine that you get a newspaper every day. At the end of the day, you stack the day's paper on top of the paper from yesterday. In time, you build up a large stack of newspapers. Where are the oldest newspapers in the pile? Where are the newest ones? The oldest papers are at the bottom of the pile, and the newest ones are at the top.

Layers of rock are similar to your stack of newspapers. In most cases, the oldest layers of rock are found below the youngest layers. The idea that younger rocks lie above older rocks is called **superposition**.

The idea of superposition can help geologists learn the order in which different rock layers formed. In general, rock layers near the top of a rock sequence formed after layers of rock lower in the sequence. Therefore, the layers at the top of the sequence are younger than the layers lower down. Figuring out whether a rock layer is older or younger than the layers around it is called **relative dating**. ☑

STUDY TIP

Compare In your notebook, make a chart explaining different ways that rock layers can be changed after they form.

READING CHECK

1. Define What is relative dating?

_____

_____

_____

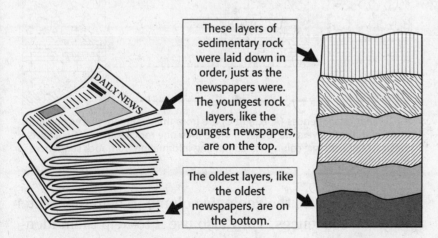

These layers of sedimentary rock were laid down in order, just as the newspapers were. The youngest rock layers, like the youngest newspapers, are on the top.

The oldest layers, like the oldest newspapers, are on the bottom.

The idea of _____ says that rock layers at the bottom of a body of rock are older than layers at the top. Geologists can use this idea to determine the relative age of different rock layers.

TAKE A LOOK

2. Identify Fill in the blank line in the figure.

THE GEOLOGIC COLUMN

The idea of superposition only applies to rock layers that have not been changed after they formed. However, not all rock layers are undisturbed. Forces from inside the Earth and processes on the Earth's surface can affect rock layers. These forces and processes can break rock layers apart or cause them to bend or tilt. Sometimes, the forces can even turn the rock layers upside down!

These disruptions can make it difficult for a geologist to determine the relative ages of different rocks. However, geologists have an important tool that can help them in relative dating: the geologic column.

The **geologic column** is a detailed series of rock layers. It contains all the known fossils and rock formations on Earth, ordered from oldest to youngest. Geologists have created the geologic column by combining information from all over the world.

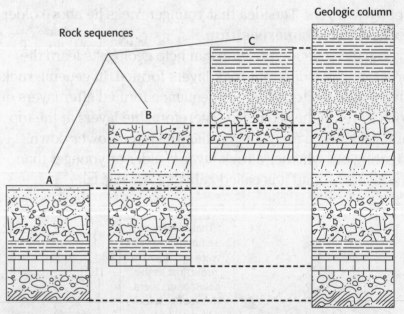

The three rock sequences (A, B, and C) are from three different places on Earth. Some of the rock layers are found in more than one rock sequence. Geologists construct the geologic column by combining information from rock sequences around the world.

Geologists use the geologic column to help them interpret rock sequences. They also use it to help them identify rock layers in complicated rock sequences.

Say It

Share Experiences Have you ever been to a place where you could see many rock layers stacked up? In a group, talk about what you observed.

TAKE A LOOK

3. Explain How do geologists construct the geologic column?

_____

_____

_____

_____

SECTION 2 Relative Dating: Which Came First? *continued*

How Can Rock Layers Be Disturbed?

Gravity causes sediment to be deposited in flat, horizontal layers. Flat, horizontal layers of sediment should form flat, horizontal layers of rock. If rock layers are not horizontal, then some force must have disturbed them after they were formed. ☑

CHANGED ROCK LAYERS

Folding and tilting are two ways that rock layers can be disturbed. *Folding* happens when rock layers are bent because of pressure. *Tilting* happens when forces from inside Earth move rock layers so that they are slanted.

Folding happens when rock layers bend and buckle under pressure.

Tilting happens when forces from inside Earth cause rock layers to become slanted.

Faults and intrusions can cut across many rock layers. A *fault* is a break or crack in Earth's crust. Large pieces of rock can move or slide along a fault. An *intrusion* forms when melted rock moves into cracks in rock layers and then cools. ☑

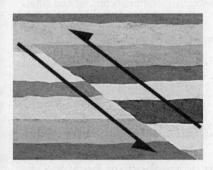

A fault is a break in Earth's crust. Rock can slide along a fault and disturb rock layers.

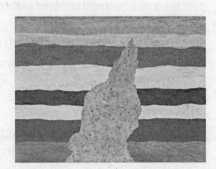

Intrusions form when melted rock moves through cracks in rock layers. The melted rock cools and hardens to form igneous rock.

READING CHECK

4. Describe What is one thing that tells a geologist that rock layers have been disturbed?

_____

_____

TAKE A LOOK

5. Compare How is folding different from tilting?

_____

_____

_____

_____

_____

READING CHECK

6. Identify What kind of rock is an intrusion made of?

_____

_____

TAKE A LOOK

7. Define What is a fault?

_____

_____

SECTION 2 Relative Dating: Which Came First? *continued*

MISSING ROCK LAYERS

Think back to your stack of newspapers. Imagine that you want to read something in the paper from Valentine's Day, February 14. However, when you look, the paper from February 14 is not there. The papers go from February 13 to February 15. What happened? Maybe you didn't put that day's newspaper on the stack. Maybe someone took that paper out of the stack.

The same ideas that apply to a missing newspaper apply to a missing rock layer. An **unconformity** is a break in, or a missing part of, the geologic record. Unconformities can form when sediment is not deposited in an area for a long time. If sediment is not deposited, no new layer of rock can form. This is like your forgetting to put a newspaper onto the stack.

Unconformities can also form when erosion removes a layer of rock after it formed. This is like someone taking a paper out of the stack.

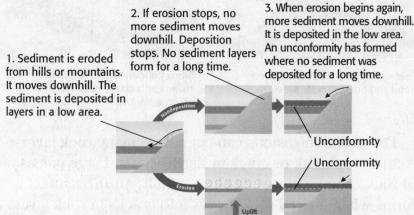

1. Sediment is eroded from hills or mountains. It moves downhill. The sediment is deposited in layers in a low area.

2. If erosion stops, no more sediment moves downhill. Deposition stops. No sediment layers form for a long time.

3. When erosion begins again, more sediment moves downhill. It is deposited in the low area. An unconformity has formed where no sediment was deposited for a long time.

Unconformity

Unconformity

Nondeposition

Erosion

Uplift

2. Erosion removes some of the mountain and sediment layers. The material is moved farther away.

3. More sediment begins to be deposited. An unconformity has formed where material was eroded.

Geologists put all unconformities into three main groups: disconformities, nonconformities, and angular unconformities.

DISCONFORMITIES

A *disconformity* is a place where part of a sequence of parallel rock layers is missing. Disconformities can form when a sequence of rock layers is pushed up because of forces inside the Earth. Erosion removes the younger layers of rock at the top of the sequence, forming an *erosion surface*. Later, deposition starts happening again, and sediment buries the erosion surface.

Critical Thinking

8. Infer Imagine that you are a geologist and you find an unconformity between two rock layers. What can you guess about the environment at the time the unconformity was forming?

_____

_____

_____

_____

TAKE A LOOK

9. Identify Give two ways that an unconformity can form.

_____

_____

Interactive Textbook

The Rock and Fossil Record

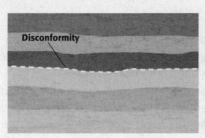

Disconformities are the most common kind of unconformity. The rock layers above the disconformity may be thousands to many millions of years younger than those below it.

TAKE A LOOK
10. **Define** What is a disconformity?

_____

_____

_____

NONCONFORMITIES

A *nonconformity* is a place where sedimentary rocks are found on top of eroded igneous or metamorphic rocks. The igneous or metamorphic rocks can be pushed up by forces inside the Earth. Then, erosion can remove some of the rock. Later, sediment may be deposited on top of the eroded rock. ☑

✓ READING CHECK
11. **Explain** How do nonconformities form?

_____

_____

_____

_____

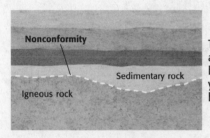

The rock layers above a nonconformity may be millions of years younger than those below it.

ANGULAR UNCONFORMITIES

An *angular unconformity* is a place where horizontal rock layers lie on top of tilted or folded rock layers. The tilted or folded layers were eroded before the horizontal layers formed above them.

TAKE A LOOK
12. **Compare** How is a nonconformity different from an angular unconformity?

_____

_____

_____

_____

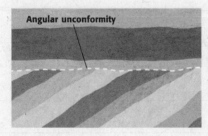

The rock layers above an angular unconformity may be millions of years younger than those below it.

Folding, tilting, faults, intrusions, and unconformities all disturb rock layers. Sometimes, a single rock body may have been disturbed many times. Geologists must use their knowledge of the things that disturb rock layers to piece together the Earth's history.

Section 2 Review

SECTION VOCABULARY

geologic column an ordered arrangement of rock layers that is based on the relative ages of the rocks and in which the oldest rocks are at the bottom	**superposition** a principle that states that younger rocks lie above older rocks if the layers have not been disturbed
relative dating any method of determining whether an event or object is older or younger than other events or objects	**unconformity** a break in the geologic record created when rock layers are eroded or when sediment is not deposited for a long period of time

1. Identify Give two ways that an unconformity can form.

_____

_____

2. Apply Concepts If the tops of the rock layers in the figure below were eroded and then deposition started happening again, what kind of unconformity would have formed?

_____

3. Describe What does the idea of superposition say about rock layers that have not been disturbed?

_____

_____

4. Identify Give two ways in which geologists use the geologic column.

_____

_____

5. Explain How does a disconformity form?

_____

_____

_____

_____

CHAPTER 6 The Rock and Fossil Record

SECTION 3 Absolute Dating: A Measure of Time

National Science
Education Standards
ES 2b

BEFORE YOU READ

After you read this section, you should be able to answer these questions:

- How can geologists learn the exact age of a rock?
- What is radiometric dating?

What Is Radioactive Decay?

Geologists can use the methods of relative dating to learn whether a rock is older or younger than another rock. However, they often also need to know exactly how old a rock is. Finding the exact age of an object is called **absolute dating**. One way to learn the age of a rock is to use unstable atoms.

All matter, including rock, is made of atoms. All atoms are made of three kinds of particles: protons, neutrons, and electrons. All of the atoms of an element, such as uranium, have the same number of protons. However, some atoms of an element have different numbers of neutrons. Atoms of an element that have different numbers of neutrons are called **isotopes**. ☑

Many isotopes are stable and are always in the same form. However, other isotopes are unstable and can break down into new isotopes of different elements. An unstable isotope is also called a *radioactive isotope*. **Radioactive decay** happens when a radioactive isotope breaks down into a new isotope.

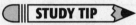

STUDY TIP

Learn New Words As you read, underline words that you don't understand. When you learn what they mean, write the words and their definitions in your notebook.

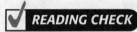

READING CHECK

1. Define What are isotopes?

_____

_____

_____

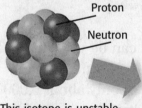

Proton

Neutron

This isotope is unstable, or radioactive.

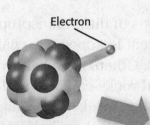

Electron

Radioactive isotopes can decay in different ways. During one kind of radioactive decay, a neutron becomes a proton and an electron. The electron moves to a different part of the atom.

After radioactive decay, an isotope of a new element is left. The new isotope is stable.

TAKE A LOOK

2. Compare How is a radioactive isotope different from a stable isotope?

_____

_____

_____

_____

SECTION 3 Absolute Dating: A Measure of Time *continued*

RADIOMETRIC DATING

A radioactive isotope is also called a *parent isotope*. Parent isotopes break down into *daughter isotopes*. Because of radioactive decay, the amounts of parent and daughter isotopes in a rock are always changing. However, they change at a constant, known rate. Therefore, scientists can learn the age of a rock by studying the amounts of parent and daughter isotopes in it.

Radiometric dating is the process of determining the absolute age of a sample based on the ratio of parent isotope to daughter isotope. In order to use radiometric dating, you need to know the half-life of the parent isotope.

The **half-life** of a radioactive isotope is how long it takes for half of a sample of the isotope to decay. For example, imagine that a parent isotope has a half-life of 10,000 years. A sample of this isotope has a mass of 12 mg. After 10,000 years, only one-half, or 6 mg, of the sample will be left.

0 years
Parent isotope = 16 mg
This sample contains 16 mg of a parent isotope. The isotope has a half-life of 10,000 years.

10,000 years
Parent isotope =
(1/2) × (16 mg) = _____
After one half-life, 1/2 of the original mass of parent isotope is left.

20,000 years
Parent isotope =
(1/2) × (1/2) × (16 mg) = _____
After two half-lives, (1/2) × (1/2), or 1/4 of the original mass of parent isotope is left.

30,000 years
Parent isotope =
(1/2) × (1/2) × (1/2) × (16 mg) = _____
After three half-lives, (1/2) × (1/2) × (1/2), or 1/8 of the original mass of parent isotope is left.

The half-lives of different isotopes can be very different. Some parent isotopes have half-lives of more than 4 billion years. Others have half-lives of only about 6,000 years. Very old rocks can be dated only if isotopes with long half-lives are used. Very young rocks can be dated only if isotopes with short half-lives are used.

How do scientists know which isotope to use to find the age of a rock? They use information about the relative age of the rock to guess about how old the rock is. Then, they find its age, using an isotope that is useful for dating rocks of that age.

Critical Thinking

3. Infer What happens to the amount of parent isotope in a rock with time? What happens to the amount of daughter isotope?

_____

_____

_____

_____

Math Focus

4. Calculate Fill in the blank lines in the figure with the mass of parent isotope that is left at each step.

What Isotopes Can Be Used for Radiometric Dating?

Remember that different parent isotopes have different half-lives. Each parent isotope can be used to date rocks of different ages.

POTASSIUM-ARGON METHOD

Potassium-40 is one isotope that is often used in radiometric dating. It has a half-life of 1.3 billion years. It decays to produce the daughter isotope argon-40. Scientists usually use the potassium-argon method to date rocks that are older than about 1 million years. ☑

URANIUM-LEAD METHOD

Uranium-238 is also used for radiometric dating. It has a half-life of 4.5 billion years. It decays to produce lead-206. Scientists use the uranium-lead method to date rocks that are older than about 10 million years.

RUBIDIUM-STRONTIUM METHOD

Rubidium-87 is also used for radiometric dating. It has a half-life of about 48 billion years. It decays to produce the daughter isotope strontium-87. The half-life of rubidium-87 is very long. Therefore, this method is only useful for dating rocks older than about 10 million years.

CARBON-14 METHOD

Carbon-14 is a radioactive isotope of the element carbon. Carbon-14, along with the other isotopes of carbon, combines with oxygen to form the gas carbon dioxide. Plants use carbon dioxide to make food. Therefore, living plants are always taking in small amounts of carbon-14. Animals that eat plants also take in carbon-14 from the plants. ☑

When a plant or animal dies, it stops taking in carbon-14. The carbon-14 already in its body starts to decay to produce nitrogen-14. Carbon-14 has a short half-life: only 5,730 years. Therefore, this method can be used to date the remains of organisms that died in the last 50,000 years.

Parent isotope	Daughter isotope	Half-life
Potassium-40		
Uranium-238		
Rubidium-87		
Carbon-14		

✔ **READING CHECK**

5. Explain Using relative dating, a scientist learns that a rock is about 50,000 years old. Can the scientist use the potassium-argon method to find the exact age of this rock? Explain your answer.

_____

_____

_____

_____

✔ **READING CHECK**

6. Describe How do animals take in carbon-14?

_____

_____

TAKE A LOOK

7. Identify Fill in the spaces in the chart to show the features of different parent isotopes.

Section 3 Review

SECTION VOCABULARY

absolute dating any method of measuring the age of an event or object in years

half-life the time required for half of a sample of a radioactive isotope to break down by radioactive decay to form a daughter isotope

isotope an atom that has the same number of protons (or the same atomic number) as other atoms of the same element do but that has a different number of neutrons (and thus a different atomic mass)

radioactive decay the process in which a radioactive isotope tends to break down into a stable isotope of the same element or another element

radiometric dating a method of determining the absolute age of an object by comparing the relative percentages of a radioactive (parent) isotope and a stable (daughter) isotope

1. Describe How is radioactive decay related to radiometric dating?

_____

_____

_____

2. Calculate A parent isotope has a half-life of 1 million years. If a rock contained 20 mg of the parent isotope when it formed, how much parent isotope would be left after 2 million years? Show your work.

_____

_____

3. List What are two radioactive isotopes that are useful for dating rocks that are older than 10 million years?

_____

_____

4. Apply Concepts A geologist uses relative dating methods to guess that a rock is between 1 million and 5 million years old. What is one radioactive isotope the geologist can use to learn the exact age of the rock? Explain your answer.

_____

_____

5. Infer Why can't geologists use the carbon-14 method to date igneous rocks? Why can't they use the carbon-14 method to date dinosaur bones?

_____

_____

_____

CHAPTER 6 | The Rock and Fossil Record

SECTION 4 | **Looking at Fossils**

National Science Education Standards
ES 1k, 2b

BEFORE YOU READ

After you read this section, you should be able to answer these questions:

• What are fossils?

• How do fossils form?

• What can fossils tell us about the history of life on Earth?

What Are Fossils?

Scientists can tell us many things about organisms, such as dinosaurs, that lived millions of years ago. How do scientists learn about these organisms if they have never seen them? They study fossils. A **fossil** is the trace or remains of an organism that lived long ago.

Some fossils are made from parts of an organism's body. These fossils are called *body fossils*. Other fossils are simply signs, such as footprints, that an organism was alive. These fossils are called **trace fossils**.

FOSSILS IN ROCKS

Usually, when an organism dies, it begins to decay or it is eaten by other organisms. Sometimes, organisms are quickly buried by sediment when they die. The sediment can help preserve the organism. Hard parts, such as shells, teeth, and bones, are preserved more often than soft parts, such as organs and skin. When the sediment hardens to form sedimentary rock, the parts of the organism that remain can become body fossils. ☑

FOSSILS IN AMBER

Sometimes, organisms such as insects are caught in sticky tree sap. If the sap hardens around the insect, a fossil is created. Hardened tree sap is called *amber*. Some of the best insect fossils are found in amber. Frogs and lizards have also been found in amber.

STUDY TIP

Organize As you read, make a chart comparing the five ways that body fossils can form.

Critical Thinking

1. Compare How is a trace fossil different from a body fossil?

_____

_____

_____

_____

_____

READING CHECK

2. List Give three examples of hard parts of an organism that could become fossils.

_____

_____

_____

This insect is preserved in amber. It is more than 38 million years old.

SECTION 4 Looking at Fossils *continued*

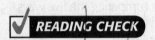**Say It**

Discuss Have you ever seen any fossils? What kind of organisms were they? Where did you see them? How did they probably form? In a small group, talk about the fossils you have seen.

☑ **READING CHECK**

3. Define What is petrifaction?

_____
_____
_____
_____
_____

TAKE A LOOK

4. Explain Why are these tracks considered trace fossils?

_____
_____
_____
_____
_____

FROZEN FOSSILS

Ice and cold temperatures slow down decay and can allow body fossils to form. Woolly mammoths, which are relatives of modern elephants, became extinct about 10,000 years ago. However, scientists have found frozen remains of mammoths preserved in blocks of ice.

PETRIFACTION

Organisms can also become fossils by petrifaction. During *petrifaction*, minerals replace all or part of an organism's tissues. For example, minerals may fill the tiny spaces in an animal's bones. Sometimes, the organism's tissues are completely replaced by minerals. For example, petrified wood forms when minerals replace all of the tissue in a piece of wood. ☑

FOSSILS IN ASPHALT

In some places, asphalt or tar bubbles up to the Earth's surface and forms sticky pools. The La Brea asphalt deposits in Los Angeles, California, are at least 38,000 years old. These pools have trapped and preserved many different organisms. From these fossils, scientists have learned about the ancient environment of southern California.

FOOTPRINTS

Remember that trace fossils are evidence that an organism was once alive. A footprint is an example of a trace fossil. Footprints may be preserved as trace fossils when they are filled with sediment and harden into rock. Footprints can show how big an animal was and how fast it was moving. For example, parallel paths of dinosaur tracks have led scientists to hypothesize that some dinosaurs moved in herds.

These dinosaur tracks are found in Arizona. They show that the dinosaur was running when it made the tracks.

BURROWS AND COPROLITES

Burrows are another kind of trace fossil. Burrows are shelters made by animals, such as clams, that dig into sediment. A burrow can be preserved when it is filled with a different kind of sediment and buried quickly.

Coprolites, or preserved animal dung, are another example of trace fossils.

MOLDS AND CASTS

Molds and casts are two more kinds of fossils. A **mold** is an *impression,* or print, left in sediment where a plant or animal was buried. The figure below shows two types of molds from the same organism. One is an internal mold of the inside of the shell. The other is an external mold of the outside of the shell. ☑

A **cast** is an object that forms when sediment fills a mold and becomes rock. Like a mold, a cast can show what the inside or outside of an organism looked like.

The fossil on the left is the internal mold of an ammonite. It formed when sediment filled the ammonite's shell. On the right is the external mold of the ammonite. The shell later dissolved.

READING CHECK

5. Define What is a mold?

_____

_____

STANDARDS CHECK

ES 2b Fossils provide important <u>evidence</u> of how life and <u>environmental</u> conditions have changed.

Word Help: <u>evidence</u>
information showing whether an idea or belief is true or valid

Word Help: <u>environment</u>
the surrounding natural conditions that affect an organism

6. List Name three things fossils can show scientists.

_____

_____

_____

_____

_____

_____

What Can We Learn from Fossils?

Think about your favorite outdoor place. Imagine the plants and animals around you. Now, imagine that you are a scientist at the same site 65 million years from now. What types of fossils would you dig up? Would you find fossils for every organism that existed? Based on the fossils you found, what would you guess about how this place used to look?

Fossils can show scientists three main things:

• What kind of organisms lived in the past
• How the environment has changed with time
• How organisms have changed with time

SECTION 4 Looking at Fossils *continued*

THE INFORMATION IN THE FOSSIL RECORD

Scientists have used fossils to learn some of the history of life on Earth. However, scientists cannot learn everything about life from fossils. This is because most organisms never became fossils, and many fossils have not been discovered yet.

Scientists know more about some kinds of ancient organisms than others. Remember that hard body parts are more likely to form fossils than soft body parts. Therefore, scientists know more about organisms with hard body parts than about organisms with only soft body parts. Some organisms lived in environments where fossils can form more easily. Scientists know more about these organisms than those that lived in other environments.

TAKE A LOOK
7. Explain Why do scientists know more about some kinds of ancient organisms than others?

_____

_____

_____

_____

_____

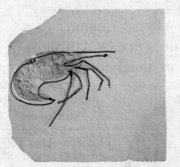

Fossils of organisms with hard parts, such as shells, are more common than fossils of organisms without hard parts.

This organism lived in an environment with a lot of sediment. Fossils form more easily in environments with a lot of sediment. Therefore, organisms that lived in these environments are more likely to be found as fossils.

A HISTORY OF ENVIRONMENTAL CHANGES

Fossils can show evidence of climate change. For example, Antarctica is covered with ice and snow in the present. However, scientists have found fossils of forest and freshwater organisms in Antarctica. They have even found fossils of dinosaurs in Antarctica! These fossils show that Antarctica's climate must have been warmer in the past.

A HISTORY OF CHANGING ORGANISMS

READING CHECK
8. Explain How can scientists find out how life has changed?

_____

_____

_____

To understand how life on Earth has changed, scientists compare fossils. Scientists also look for similarities between fossils and living organisms. However, only a small fraction of the organisms that have existed in Earth's history have been fossilized. As a result, the fossil record is incomplete. This means that scientists do not have a complete record of changes in life on Earth. ☑

How Do Scientists Know How Old Fossils Are?

To understand the history of life on Earth, scientists have put fossils in order based on their ages. Scientists learn the ages of fossils in different ways. In some cases, they can use *absolute dating methods*, such as radiometric dating, to determine the age of fossils. More commonly, scientists use relative dating methods. ☑

Relative dating methods can't tell scientists the exact age of a fossil. However, relative dating can show which fossils are older than others. Fossils found in older layers of rock come from more ancient life forms. Fossils found in younger layers of rock are from more recent organisms.

USING FOSSILS TO DATE ROCKS

Scientists can use fossils of certain types of organisms to learn how old rock layers are. These fossils are called index fossils. **Index fossils** are fossils of organisms that lived during a relatively short period of time. Because they lived for only a short time, their fossils are only found in rocks of a certain age. To be an index fossil, a fossil must have three features:

• The organism must be common in rocks from most of the world.

• The organism must have lived for only a geologically short period of time (a few million years to a few hundred million years).

• The organism must be easy to identify.

Trilobites and ammonites are two kinds of organisms that are used as index fossils. The figures below show examples of these fossils.

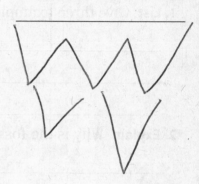

✔ **READING CHECK**

9. Describe How do scientists put fossils in order?

_____

The trilobite *Phacops* is an example of an index fossil. *Phacops* lived about 400 million years ago. Therefore, rocks that contain *Phacops* fossils are probably about 400 million years old.

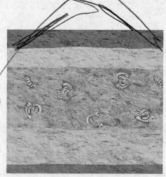

The ammonite *Tropites* is an index fossil. *Tropites* lived from between 230 million and 208 million years ago. Because it lived for such a short time, it is a good index fossil.

TAKE A LOOK

10. Identify What feature of these organisms made them more likely to be preserved as fossils?

_____

_____

Name ______________________ Class ______________ Date ______________

Section 4 Review

NSES ES 1k, 2b

SECTION VOCABULARY

cast a type of fossil that forms when sediments fill in the cavity left by a decomposed organism **fossil** the trace or remains of an organism that lived long ago, most commonly preserved in sedimentary rock **index fossil** a fossil that is used to establish the age of a rock layer because the fossil is distinct, abundant, and widespread and the species that formed that fossil existed for only a short span of geologic time	**mold** a mark or cavity made in a sedimentary surface by a shell or other body **trace fossil** a fossilized mark that formed in sedimentary rock by the movement of an animal on or within soft sediment

1. List Give three examples of trace fossils.

____

____

____

2. Explain Why is the fossil record incomplete?

____

____

____

3. Infer Which organism is more likely to be found as a fossil in amber, a beetle or a rabbit? Explain your answer.

____

____

____

4. Apply Concepts What could you conclude if you found a fossil of a tropical plant in a cold climate?

____

____

5. List What three features must a fossil have in order to be an index fossil?

____

____

____

CHAPTER 6 (The Rock and Fossil Record)

SECTION 5 **Time Marches On**

National Science Education Standards
ES 2b

BEFORE YOU READ

After you read this section, you should be able to answer these questions:

• How do geologists measure time?

• How has life changed during Earth's history?

• What can cause a mass extinction?

How Old Is the Earth?

The fossils in the figure below are dinosaur bones. The dinosaurs that became these fossils lived 150 million years ago. To most people, 150 million years seems like a very long time. However, to geologists, 150 million years is not very long at all. Geologists study the history of the Earth. The Earth is about 4.6 billion years old. Therefore, 150 million years is less than 3% of the age of the Earth!

STUDY TIP

Organize As you read, make a chart showing the eras of geologic time. Include major changes in life that happened during each era.

How do geologists study such long periods of time? They use rocks and fossils to learn about how the Earth has changed with time. Remember that rocks form in layers, and that different rocks form in different environments. By studying very old rocks, geologists can guess what environments were like long ago. ☑

Fossils are also very important in helping geologists learn about the Earth's history. Remember that geologists can learn about an area's environment by studying the fossils that formed there. The fossils can give clues about how the Earth has changed with time.

Geologists have combined information from rocks and fossils to produce a timeline of the Earth's history. This timeline is called the **geologic time scale**.

TAKE A LOOK
1. Identify What kinds of organisms formed the fossils in the picture?

_____

READING CHECK

2. Explain How can geologists figure out what the Earth's environment was like very long ago?

_____

_____

SECTION 5 Time Marches On *continued*

What Is the Geologic Time Scale?

The geologic time scale includes all of the Earth's 4.6 billion years of history. Geologists have divided Earth's history into many shorter sections of time. These sections are shown in the figure below.

Geologic Time Scale

	Era	Period	Epoch	Millions of years ago
PHANEROZOIC EON	**Cenozoic**	Quaternary	Holocene	0.01
			Pleistocene	1.8
		Tertiary	Pliocene	5.3
			Miocene	23.8
			Oligocene	33.7
			Eocene	54.8
			Paleocene	65
	Mesozoic	Cretaceous		144
		Jurassic		206
		Triassic		248
	Paleozoic	Permian		290
		Pennsylvanian		323
		Mississippian		354
		Devonian		417
		Silurian		443
		Ordovician		490
		Cambrian		543

PROTEROZOIC EON	**Proterozoic Eon** The first cells appeared during this eon.
	2,500
ARCHEAN EON	**Archean Eon** The oldest rocks on Earth formed during this eon.
	3,800
HADEAN EON	**Hadean Eon** The only known rocks from this eon are meteorites and rocks from the moon.
	4,600

Math Focus

3. Calculate About how long did the Triassic period last?

_____

_____

TAKE A LOOK

4. List Give the five epochs in the Tertiary period in order from oldest to most recent.

_____

_____

_____

_____

As you can see, the largest divisions of geologic time are the **eons**. Earth's history is divided into four eons: the Hadean eon, the Archean eon, the Proterozoic eon, and the Phanerozoic eon. Most rocks and fossils on the Earth formed during the Phanerozoic eon. Scientists divide the Phanerozoic eon into three **eras**, which are the second-largest divisions of the time scale.

Scientists divide the eras into **periods**, which are the third-largest divisions of the time scale. For example, the Mesozoic era is divided into the Triassic, Jurassic, and Cretaceous periods. Periods are divided into **epochs**, which are the fourth-largest divisions of the time scale.

SECTION 5 Time Marches On *continued*

THE APPEARANCE AND DISAPPEARANCE OF SPECIES

Scientists use changes in life to define many of the boundaries between sections of geologic time. For example, some boundaries are defined by mass extinctions. **Extinction** is the death of every member of a species. A *mass extinction* happens when many species go extinct at one time. ☑

Mass extinctions happen for different reasons. Gradual events, such as global climate change or changes in ocean currents, can cause mass extinctions. Sudden events, such as a large volcanic eruption or a meteorite impact, can also cause many species to go extinct. In many cases, mass extinctions happen because of a combination of sudden and gradual events.

Many geologic time boundaries are defined by the disappearance of species. Others are defined by the appearance of species. For example, the beginning of the Phanerozoic eon was marked by the appearance of many new species of ocean life. Some of these organisms looked similar to organisms that are alive today. However, others, such as the organism in the figure below, looked very different.

This organism, called *Hallucigenia*, lived in the early Cambrian period. Many ocean life forms, including *Hallucigenia*, first appeared at the beginning of the Phanerozoic eon.

READING CHECK

5. Describe How do geologists define many of the boundaries between sections of geologic time?

_____

_____

TAKE A LOOK

6. Apply Concepts During which era did *Hallucigenia* live?

_____

How Has Life Changed During the Phanerozoic Eon?

The Phanerozoic eon is the most recent eon in the Earth's history. It is the eon in which we live. Almost all of the fossils and rocks that are found on Earth today formed during the Phanerozoic eon.

SECTION 5 Time Marches On *continued*

THE PALEOZOIC ERA: BEGINNINGS OF MODERN LIFE

The Paleozoic era lasted from about 542 million to 251 million years ago. *Paleo* means "old," and *zoic* means "life." Therefore, the Paleozoic was the era of "old life."

During the Paleozoic era, many species of organisms lived in the Earth's oceans. However, there were not many species of organisms living on land until the middle of the Paleozoic era. By the end of the era, amphibians and reptiles lived on the land, and many species of insects existed. The figure below shows some of the types of organisms that evolved during the Paleozoic era. ☑

Plants, fish, amphibians, and reptiles evolved during the Paleozoic era.

The end of the Paleozoic era is marked by a huge mass extinction. Ninety percent of all ocean species died out during this extinction. Scientists are not sure what caused this extinction, but it may have been caused by changing ocean currents.

THE MESOZOIC ERA: THE AGE OF REPTILES

The Mesozoic era began about 251 million years ago and ended about 65 million years ago. *Meso* means "middle," so the Mesozoic was the era of "middle life."

Reptiles were the dominant organisms that lived during the Mesozoic. Probably the most famous of these reptiles are the dinosaurs. However, small mammals and birds also evolved during the later parts of the Mesozoic. Many scientists think that birds evolved from a type of dinosaur.

The end of the Mesozoic era is also marked by a mass extinction. About 15% to 20% of all species on Earth, including all of the dinosaurs, went extinct at the end of the Mesozoic era. Most scientists think that global cooling because of a meteorite impact caused this extinction.

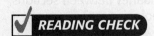

READING CHECK

7. Identify During which era did amphibians and reptiles evolve?

_____

_____

TAKE A LOOK

8. Apply Concepts Why are there no birds or mammals in this image?

_____

_____

_____

Say It

Infer and Discuss How could changes in ocean currents cause mass extinction? Write down some ideas of your own about how this could happen. Then, talk about your ideas with other students.

SECTION 5 Time Marches On *continued*

Dinosaurs were the dominant land organisms during the Mesozoic era. Small mammals and birds also evolved during the Mesozoic.

THE CENOZOIC ERA: THE AGE OF MAMMALS

The Cenozoic era began about 65 million years ago and continues to the present. *Ceno* means "recent," so the Cenozoic is the era of "recent life."

After the dinosaurs went extinct, mammals no longer had to compete with them for resources. As a result, mammals have become more dominant during the Cenozoic. Many features of mammals may have helped them survive the climate changes that caused the extinction of the dinosaurs. These features include being able to control body temperature and bearing young that grow inside the mother.

The Cenozoic era continues today, but many organisms that lived at the beginning of the era are now extinct. The figure below shows some of these organisms.

Thousands of species of mammals evolved during the Cenozoic. Many of the mammals in this figure are now extinct.

TAKE A LOOK

9. Explain Why is the Mesozoic era sometimes called the "Age of Reptiles?"

_____

_____

_____

Critical Thinking

10. Infer How could controlling body temperature and bearing live young have helped mammals survive the events that caused the extinction of the dinosaurs?

_____

_____

_____

_____

_____

_____

_____

Section 5 Review

SECTION VOCABULARY

eon the largest division of geologic time	**extinction** the death of every member of a species
epoch a subdivision of geologic time that is longer than an age but shorter than a period	**geologic time scale** the standard method used to divide the Earth's long natural history into manageable parts
era a unit of geologic time that includes two or more periods	**period** a unit of geologic time that is longer than an epoch but shorter than an era

1. List What are four divisions of geological time?

_____

_____

_____

2. Identify How old is the Earth?

_____

3. Explain How can geologists use rocks and fossils to learn how the Earth's environments have changed?

_____

_____

_____

_____

4. List Write the seven periods of the Paleozoic era in order from oldest to most recent.

_____

_____

_____

_____

5. Describe How do geologists define the ends of the Paleozoic and Mesozoic eras?

_____

6. Identify Give two things that can cause mass extinctions.

_____

_____

CHAPTER 7 | Plate Tectonics

SECTION 1 | # Inside the Earth

BEFORE YOU READ

After you read this section, you should be able to answer these questions:

- What are the layers inside Earth?
- How do scientists study Earth's interior?

What Is Earth Made Of?

Scientists divide the Earth into three layers based on composition: the crust, the mantle, and the core. These divisions are based on the compounds that make up each layer. A *compound* is a substance composed of two or more elements. The densest elements make up the core. Less-dense compounds make up the crust and mantle.

THE CRUST

The thinnest, outermost layer of the Earth is the **crust**. There are two main kinds of crust: continental crust and oceanic crust. *Continental crust* forms the continents. It is thicker and less dense than oceanic crust. Continental crust can be up to 100 km thick. *Oceanic crust* is found beneath the oceans. It contains more iron than continental crust. Most oceanic crust is 5 km to 7 km thick. ☑

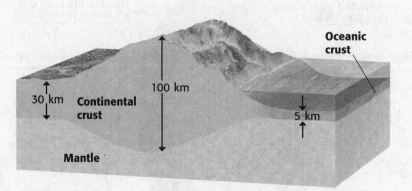

Oceanic crust is thinner and denser than continental crust.

STUDY TIP

Summarize As you read, make a chart showing the features of Earth's layers. Include both the compositional layers and the physical layers.

READING CHECK

1. Compare How is oceanic crust different from continental crust?

_____

_____

_____

Math Focus

2. Identify What fraction of the thickness of the thickest continental crust is the thickness of the oceanic crust? Give your answer as a reduced fraction.

_____

SECTION 1 Inside the Earth *continued*

THE MANTLE

The layer of the Earth between the crust and the core is the **mantle**. The mantle is much thicker than the crust. It contains most of the Earth's mass. The mantle contains more magnesium and less aluminum than the crust. This makes the mantle denser than the crust. ☑

No one has ever visited the mantle. The crust is too thick to drill through to reach the mantle. Therefore, scientists must use observations of Earth's surface to draw conclusions about the mantle. In some places, mantle rock pushes to the surface. This allows scientists to study the rock directly.

Another place scientists look for clues about the mantle is the ocean floor. Melted rock from the mantle flows out from active volcanoes on the ocean floor. These underwater volcanoes have given scientists many clues about the composition of the mantle. ☑

THE CORE

The layer beneath the mantle that extends to the center of the Earth is the **core**. Scientists think the core is made mostly of iron and smaller amounts of nickel. Scientists do not think that the core contains large amounts of oxygen, silicon, aluminum, or magnesium.

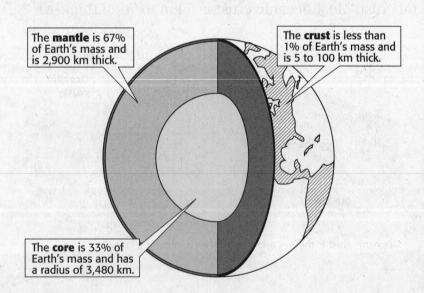

The **mantle** is 67% of Earth's mass and is 2,900 km thick.

The **crust** is less than 1% of Earth's mass and is 5 to 100 km thick.

The **core** is 33% of Earth's mass and has a radius of 3,480 km.

READING CHECK

3. Explain Why is the mantle denser than the crust?

_____

_____

_____

_____

READING CHECK

4. Identify How can scientists learn about the mantle if they cannot study it directly?

_____

_____

_____

TAKE A LOOK

5. List What are the three compositional layers of the Earth?

_____

_____

_____

_____

SECTION 1 Inside the Earth *continued*

EARTH'S PHYSICAL STRUCTURE

Scientists also divide Earth into five layers based on physical properties. The outer layer is the **lithosphere**. It is a cool, stiff layer that includes all of the crust and a small part of the upper mantle. The lithosphere is divided into pieces. These pieces move slowly over Earth's surface. ☑

The **asthenosphere** is the layer beneath the lithosphere. It is a layer of hot, solid rock that flows very slowly. Beneath the asthenosphere is the **mesosphere**, which is the lower part of the mantle. The mesosphere flows more slowly than the asthensosphere.

There are two physical layers in Earth's core. The outer layer is the *outer core*. It is made of liquid iron and nickel. At the center of Earth is the *inner core*, which is a ball of solid iron and nickel. The inner core is solid because it is under very high pressure.

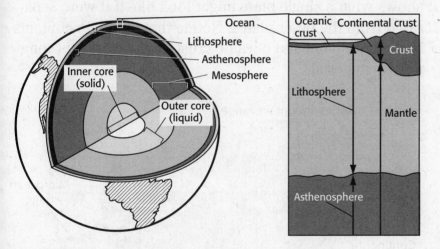

Ocean — Lithosphere — Asthenosphere — Mesosphere — Inner core (solid) — Outer core (liquid) — Oceanic crust — Continental crust — Crust — Lithosphere — Mantle — Asthenosphere

What Are Tectonic Plates?

Pieces of the lithosphere that move around on top of the asthenosphere are called **tectonic plates**. Tectonic plates can contain different kinds of lithosphere. Some plates contain mostly oceanic lithosphere. Others contain mostly continental lithosphere. Some contain both continental and oceanic lithosphere. The figure on the top of the next page shows Earth's tectonic plates.

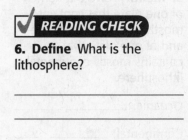

✓ **READING CHECK**

6. Define What is the lithosphere?

_____

_____

_____

Critical Thinking

7. Infer What do you think is the reason that scientists divide the Earth into two different sets of layers?

_____

_____

_____

TAKE A LOOK

8. Describe What are the five layers of Earth, based on physical properties?

_____

_____

_____

_____

_____

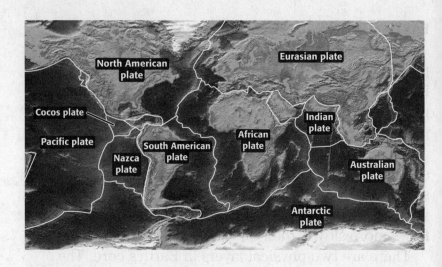

TAKE A LOOK

9. Identify Give the name of one plate that contains mostly oceanic lithosphere and of one plate that contains mostly continental lithosphere.

Oceanic: _____

Continental: _____

STRUCTURE OF A TECTONIC PLATE

The tectonic plates that make up the lithosphere are like pieces of a giant jigsaw puzzle. The figure below shows what a single plate might look like it if were separated from the other plates. Notice that the plate contains both continental and oceanic crust. It also contains some mantle material.

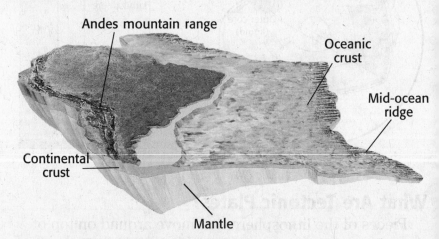

TAKE A LOOK

10. Compare Which type of crust is thicker, oceanic crust or continental crust?

_____

This figure shows what the South American plate might look like if it were lifted off the asthenosphere. Notice that the plate is thickest where it contains continental crust and thinnest where it contains oceanic crust.

How Do Scientists Study Earth's Interior?

How do scientists know things about the deepest parts of the Earth? No one has ever been to these places. Scientists have never even drilled through the crust, which is only a thin layer on the surface of the Earth. So how do we know so much about the mantle and the core?

Much of what scientists know about Earth's layers comes from studying earthquakes. Earthquakes create vibrations called *seismic waves*. Seismic waves travel at different speeds through the different layers of Earth. Their speed depends on the density and composition of the material that they pass through. Therefore, scientists can learn about the layers inside the Earth by studying seismic waves. ☑

Scientists detect seismic waves using instruments called *seismometers*. Seismometers measure the times at which seismic waves arrive at different distances from an earthquake. Seismologists can use these distances and travel times to calculate the density and thickness of each physical layer of the Earth. The figure below shows how seismic waves travel through the Earth.

READING CHECK

11. Define What are seismic waves?

_____

_____

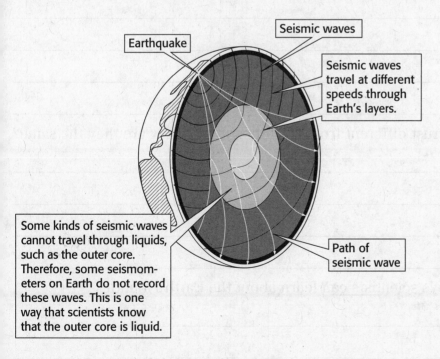

Earthquake

Seismic waves

Seismic waves travel at different speeds through Earth's layers.

Some kinds of seismic waves cannot travel through liquids, such as the outer core. Therefore, some seismometers on Earth do not record these waves. This is one way that scientists know that the outer core is liquid.

Path of seismic wave

TAKE A LOOK

12. Explain What is one way that scientists know the outer core is liquid?

_____

_____

_____

Section 1 Review

SECTION VOCABULARY

asthenosphere the soft layer of the mantle on which the tectonic plates move	**mantle** the layer of rock between the Earth's crust and core
core the central part of the Earth below the mantle	**mesosphere** the strong, lower part of the mantle between the asthenosphere and the outer core
crust the thin and solid outermost layer of the Earth above the mantle	**tectonic plates** a block of lithosphere that consists of the crust and the rigid, outermost part of the mantle
lithosphere the solid, outer layer of Earth that consists of the crust and the rigid upper part of the mantle	

1. Describe Complete the table below.

	Crust	Mantle	Core
Thickness or radius			3,430 km
Location	outer layer of the Earth		
Percent of Earth's mass			

2. Compare How is the inner core similar to the outer core? How are they different?

_____

_____

_____

_____

3. Compare How is the crust different from the lithosphere? How are they the same?

_____

_____

_____

_____

4. Identify Give three ways scientists can learn about the Earth's mantle.

_____

_____

_____

_____

CHAPTER 7 | Plate Tectonics
SECTION
2 **Restless Continents**

National Science
Education Standards
ES 1b, 2a

BEFORE YOU READ

After you read this section, you should be able to answer these questions:

• What is continental drift?

• How are magnetic reversals related to sea-floor spreading?

What Is Continental Drift?

Look at the map below. Can you see that South America and Africa seem to fit together, like the pieces of a jigsaw puzzle? In the early 1900s, a German scientist named Alfred Wegener made this same observation. Based on his observations, Wegener proposed the hypothesis of **continental drift**. According to this hypothesis, the continents once formed a single landmass. Then, they broke up and drifted to their current locations.

Continental drift can explain why the continents seem to fit together. For example, South America and Africa were once part of a single continent. They have since broken apart and moved to their current locations. ☑

Evidence for continental drift can also be found in fossils and rocks. For example, similar fossils have been found along the matching coastlines of South America and Africa. The organisms that formed these fossils could not have traveled across the Atlantic Ocean. Therefore, the two continents must once have been joined together.

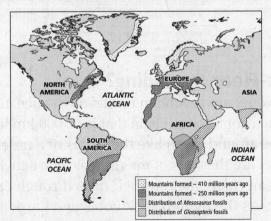

Similar fossils and rocks are found on widely separated continents. For example, *Glossopteris* and *Mesosaurus* fossils are found in Africa and in South America. These fossils and rocks indicate that, at one time, all of the continents were joined together.

 **STUDY TIP**
Paired Summarizing Read this section silently. In pairs, take turns summarizing the material. Stop to discuss ideas that seem confusing.

READING CHECK
1. Explain Why do South America and Africa seem to fit together?

_____

_____

Critical Thinking
2. Infer Which continent was once joined with Greenland? How do you know?

_____

_____

_____

TAKE A LOOK
3. Explain How do fossils indicate that the continents have moved with time?

_____

_____

_____

BREAKUP OF PANGAEA

About 245 million years ago, all of the continents were joined into a single *supercontinent*. This supercontinent was called *Pangaea*. The word *Pangaea* means "all Earth" in Greek. About 200 million years ago, Pangaea began breaking apart. It first separated into two large landmasses called Laurasia and Gondwana. The continents continued to break apart and slowly move to where they are today. ☑

As the continents moved, some of them collided. These collisions produced many of the landforms that we see today, such as mountain ranges and volcanoes.

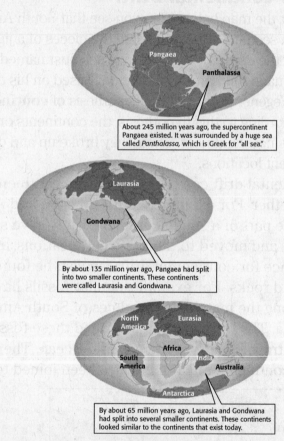

About 245 million years ago, the supercontinent Pangaea existed. It was surrounded by a huge sea called *Panthalassa*, which is Greek for "all sea."

By about 135 million year ago, Pangaea had split into two smaller continents. These continents were called Laurasia and Gondwana.

By about 65 million years ago, Laurasia and Gondwana had split into several smaller continents. These continents looked similar to the continents that exist today.

READING CHECK

4. Identify When did Pangaea start to break apart?

_____

_____

TAKE A LOOK

5. Describe How were the locations of the continents 65 million years ago different from the locations of the continents today? Give two ways.

_____

_____

_____

_____

What Is Sea-Floor Spreading?

Mid-ocean ridges are mountain chains on the ocean floor. They form a continuous chain that is 50,000 km long. The chain wraps around Earth like the seams of a baseball. Mid-ocean ridges are the sites of intense volcanic activity. ☑

At a mid-ocean ridge, melted rock rises through cracks in the sea floor. As the melted rock cools and hardens, it forms new crust. The newly formed crust pushes the older crust away from the mid-ocean ridge. This process is called **sea-floor spreading**.

READING CHECK

6. Define What is a mid-ocean ridge?

_____

_____

SECTION 2 Restless Continents *continued*

SEA-FLOOR SPREADING AND MAGNETISM

In the 1960s, scientists studying the ocean floor discovered an interesting property of mid-ocean ridges. Using a tool that can record magnetism, they found magnetic patterns on the sea floor! The pattern on one side of a mid-ocean ridge was a mirror image of the pattern on the other side of the ridge. What caused the rocks to have these magnetic patterns?

Throughout Earth's history, the north and south magnetic poles have switched places many times. This process is called magnetic reversal. This process, together with sea-floor spreading, can explain the patterns of magnetism on the sea floor. ☑

☑ **READING CHECK**

7. Define What is a magnetic reversal?

_____

_____

_____

Normal   **Reverse**

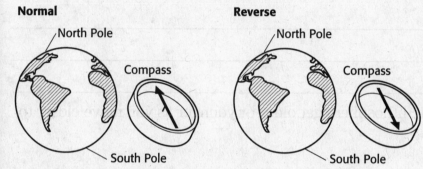

During times of normal polarity, such as today, a compass needle points toward the North Pole. During times of reverse polarity, a compass needle points toward the South Pole.

As ocean crust forms from melted rock, magnetic minerals form. These minerals act as compasses. As they form, they line up with Earth's magnetic north pole. When the melted rock cools, the minerals are stuck in place.

After Earth's magnetic field reverses, these minerals point to Earth's magnetic south pole. However, new rock that forms will have minerals that point to the magnetic north pole. Therefore, the ocean floor contains "stripes" of rock whose magnetic minerals point to the north or south magnetic poles.

This part of the sea floor formed when Earth's magnetic field was reversed.

This part of the sea floor formed when Earth's magnetic field was the same as it is today. As new sea floor formed, the rock was pushed away from the ridge.

Magma

▬ Normal polarity lithosphere
▬ Reversed polarity lithosphere

Sea-floor spreading produces new oceanic lithosphere at mid-ocean ridges. The oldest oceanic crust is found far from the ridges, and the youngest crust is found very close to the ridges.

TAKE A LOOK
8. Describe How are the "stripes" of magnetism on each side of the ridge related?

_____

_____

Section 2 Review

NSES   ES 1b, 2a

SECTION VOCABULARY

continental drift the hypothesis that states that the continents once formed a single landmass, broke up, and drifted to their present locations	**sea-floor spreading** the process by which new oceanic lithosphere (sea floor) forms as magma rises to Earth's surface and solidifies at a mid-ocean ridge

1. Identify Give three pieces of evidence that support the idea of continental drift.

_____

_____

_____

2. Describe How does oceanic lithosphere form?

_____

_____

_____

3. Identify Does the oceanic lithosphere get older or younger as you move closer to the mid-ocean ridge?

_____

4. Explain How do the parallel magnetic "stripes" near mid-ocean ridges form?

_____

_____

_____

_____

_____

5. Apply Concepts The Earth is about 4.6 billion years old. However, the oldest sea floor is only about 180 million years old. What do you think is the reason for this? (Hint: Remember that new seafloor is constantly being created, but the Earth is not getting bigger with time.)

_____

_____

_____

CHAPTER 7 | Plate Tectonics

SECTION 3

The Theory of Plate Tectonics

BEFORE YOU READ

After you read this section, you should be able to answer these questions:

- What is the theory of plate tectonics?
- What are the three types of tectonic plate boundaries?

National Science Education Standards

ES 1b, 2a

What Is the Theory of Plate Tectonics?

As scientists learned more about sea-floor spreading and magnetic reversals, they formed a theory to explain how continents move. The theory of **plate tectonics** states that Earth's lithosphere is broken into many pieces—tectonic plates—that move slowly over the asthenosphere.

Tectonic plates move very slowly—only a few centimeters per year. Scientists can detect this motion only by using special equipment, such as global positioning systems (GPS). This equipment is sensitive enough to pick up even small changes in a continent's location.

STUDY TIP

Compare As you read, make a table showing the features of the three kinds of plate boundaries.

What Happens Where Tectonic Plates Touch?

The places where tectonic plates meet are called *boundaries*. Some features, such as earthquakes and volcanoes, are more common at tectonic plate boundaries than at other places on Earth. Other features, such as mid-ocean ridges and ocean trenches, form only at plate boundaries.

There are three types of plate boundaries:

- divergent boundaries, where plates move apart;
- convergent boundaries, where plates move together; and
- transform boundaries, where plates slide past each other.

The features that form at a plate boundary depend on what kind of plate boundary it is.

READING CHECK

1. Explain How do scientists detect tectonic plate motions?

_____

_____

_____

DIVERGENT BOUNDARIES

A **divergent boundary** forms where plates are moving apart. Most divergent boundaries are found beneath the oceans. Mid-ocean ridges form at these divergent boundaries. Because the plates are pulling away from each other, cracks form in the lithosphere. Melted rock can rise through these cracks. When the melted rock cools and hardens, it becomes new lithosphere.

READING CHECK

2. Describe What features are found at most divergent boundaries?

_____

_____

SECTION 3 The Theory of Plate Tectonics *continued*

CONVERGENT BOUNDARIES

A **convergent boundary** forms where plates are moving together. There are three different types of convergent boundaries:

- **Continent-Continent Boundaries** These form when continental lithosphere on one plate collides with continental lithosphere on another plate. Continent-continent convergent boundaries can produce very tall mountain ranges, such as the Himalayas.

- **Continent-Ocean Boundaries** These form when continental lithosphere on one plate collides with oceanic lithosphere on another plate. The denser oceanic lithosphere sinks underneath the continental lithosphere in a process called *subduction*. Subduction can cause a chain of mountains, such as the Andes, to form along the plate boundary.

- **Ocean-Ocean Boundaries** These form when oceanic lithosphere on one plate collides with oceanic lithosphere on another plate. One of the plates subducts beneath the other. A series of volcanic islands, called an *island arc*, can form along the plate boundary.

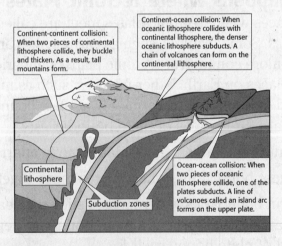

Continent-continent collision: When two pieces of continental lithosphere collide, they buckle and thicken. As a result, tall mountains form.

Continent-ocean collision: When oceanic lithosphere collides with continental lithosphere, the denser oceanic lithosphere subducts. A chain of volcanoes can form on the continental lithosphere.

Continental lithosphere

Ocean-ocean collision: When two pieces of oceanic lithosphere collide, one of the plates subducts. A line of volcanoes called an island arc forms on the upper plate.

Subduction zones

TRANSFORM BOUNDARIES

A **transform boundary** forms where plates slide past each other horizontally. Most transform boundaries are found near mid-ocean ridges. The ridges are broken into *segments*, or pieces. Transform boundaries separate the segments from one another.

One well-known transform boundary is the San Andreas fault system in California. It is located where the Pacific and North American plates slide past each other.

Critical Thinking

3. Infer Why do continent-continent convergent boundaries produce very tall mountain ranges?

_____

_____

_____

_____

_____

STANDARDS CHECK

ES 1b Lithospheric plates on the scales of continents and oceans constantly move in <u>response</u> to movements in the mantle. <u>Major</u> geological events, such as earthquakes, volcanic eruptions, and mountain building result from these plate motions.

Word Help: <u>response</u>
an action brought on by another action; a reaction

Word Help: <u>major</u>
of great importance or large scale

4. Identify Name two structures that can form at convergent plate boundaries.

_____

_____

Why Do Tectonic Plates Move?

Scientists do not know for sure what causes tectonic plates to move. They have three main hypotheses to explain plate movements: convection, slab pull, and ridge push.

Scientists used to think that convection in the mantle was the main force that caused plate motions. Remember that *convection* happens when matter carries heat from one place to another. Convection happens in the mantle as rock heats up and expands. As it expands, it becomes less dense and rises toward Earth's surface. ☑

As the hot material rises, cold, dense lithosphere sinks into the mantle at subduction zones. The rising hot material and the sinking cold material form *convection currents*. Until the 1990s, many scientists thought that these convection currents pulled the tectonic plates over Earth's surface.

Today, most scientists think that slab pull is the main force that causes plate motions. During subduction, oceanic lithosphere at the edge of a plate sinks into the mantle. The oceanic lithosphere sinks because it is colder and denser than the mantle. As the lithosphere at the edge of the plate sinks, it pulls the rest of the plate along with it. This process is called *slab pull*.

Another possible cause of plate motions is ridge push. At mid-ocean ridges, new oceanic lithosphere forms. This new lithosphere is warmer and less dense than the older lithosphere farther from the ridge. Therefore, it floats higher on the asthenosphere than the older lithosphere. As gravity pulls the new lithosphere down, the plate slides away from the mid-ocean ridge. This process is called *ridge push*.

☑ **READING CHECK**

5. Define What is convection?

_____

_____

_____

Critical Thinking

6. Compare How is slab pull different from ridge push?

_____

_____

_____

_____

_____

_____

_____

Driving Force	Description
Slab pull	Cold, sinking lithosphere at the edges of a tectonic plate pulls the rest of the plate across Earth's surface.
Ridge push	Gravity pulls newly formed lithosphere downward and away from the mid-ocean ridge. The rest of the plate moves because of this force.
Convection currents	Convection currents are produced when hot material in the mantle rises toward the surface and colder material sinks. The currents pull the plates over Earth's surface.

Name ______________________ Class ______________ Date ______________

Section 3 Review

NSES ES 1b, 2a

SECTION VOCABULARY

convergent boundary the boundary between tectonic plates that are colliding

divergent boundary the boundary between two tectonic plates that are moving away from each other

plate tectonics the theory that explains how large pieces of the Earth's outermost layer, called tectonic plates, move and change shape

transform boundary the boundary between tectonic plates that are sliding past each other horizontally

1. Define Write your own definition for plate tectonics.

2. Identify What are the three types of plate boundaries?

3. List Name three processes that may cause tectonic plates to move.

4. Describe How fast do tectonic plates move?

5. Identify Give two features that are found only at plate boundaries, and two features that are found most commonly at plate boundaries.

6. Explain Why does oceanic lithosphere sink beneath continental lithosphere at convergent boundaries?

CHAPTER 7 | Plate Tectonics

SECTION 4 Deforming the Earth's Crust

National Science Education Standards
ES 1b, 2a

BEFORE YOU READ

After you read this section, you should be able to answer these questions:

- What happens when rock is placed under stress?
- What are three kinds of faults?
- How do mountains form?

What Is Deformation?

In the left-hand figure below, the girl is bending the spaghetti slowly and gently. The spaghetti bends, but it doesn't break. In the right-hand figure, the girl is bending the spaghetti quickly and with a lot of force. Some of the pieces of spaghetti have broken.

STUDY TIP

Learn New Words As you read, underline words that you don't understand. When you learn what they mean, write the words and their definitions in your notebook.

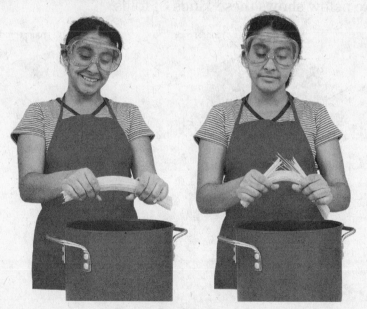

How can the same material bend in one situation but break in another? The answer is that the stress on the material is different in each case. *Stress* is the amount of force per unit area that is placed on an object.

TAKE A LOOK

1. Describe Circle the picture in which the girl is putting the most force on the spaghetti.

DEFORMATION

Like the spaghetti, rocks can bend or break under stress. When a rock is placed under stress, it *deforms*, or changes shape. When a small amount of stress is put on a rock slowly, the rock can bend. However, if the stress is very large or is applied quickly, the rock can break.

READING CHECK

2. Define What is stress?

_____

_____

SECTION 4 Deforming the Earth's Crust *continued*

What Happens When Rock Layers Bend?

Folding happens when rock layers bend under stress. Folding causes rock layers to look bent or buckled. The bends are called *folds*.

Most rock layers start out as horizontal layers. Therefore, when scientists see a fold, they know that deformation has happened. ☑

TYPES OF FOLDS

Three of the most common types of folds are synclines, anticlines, and monoclines. In a *syncline*, the oldest rocks are found on the outside of the fold. Most synclines are U-shaped. In an *anticline*, the youngest rocks are found on the outside of the fold. Most anticlines are ∩-shaped. In a *monocline*, rock layers are folded so that both ends of the fold are horizontal. The figure below shows these kinds of folds.

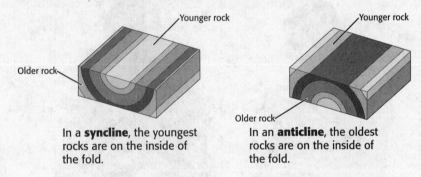

In a **syncline**, the youngest rocks are on the inside of the fold.

In an **anticline**, the oldest rocks are on the inside of the fold.

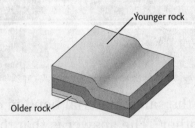

In a **monocline**, both sides of the fold are horizontal.

✔ **READING CHECK**

3. Explain How do folds indicate that deformation has happened?

_____

_____

_____

_____

_____

TAKE A LOOK
4. Identify Color the oldest rock layers in the figure blue. Color the youngest rock layers red.

What Happens When Rock Layers Break?

When rock is put under so much stress that it can no longer bend, it may break. The crack that forms when rocks break and move past each other is called a **fault**. The blocks of rock that are on either side of the fault are called *fault blocks*. When fault blocks move suddenly, they can cause earthquakes.

SECTION 4 Deforming the Earth's Crust *continued*

HANGING WALL AND FOOTWALL

When a fault forms at an angle, one fault block is called the *hanging wall* and the other is called the *footwall*. The figure below shows the difference between the hanging wall and the footwall.

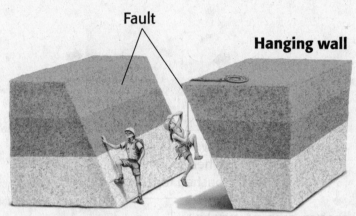

Fault

Hanging wall

Footwall

The footwall is the fault block that is below the fault. The hanging wall is the fault block that is above the fault.

Scientists classify faults by how the fault blocks have moved along the fault. There are three main kinds of faults: normal faults, reverse faults, and strike-slip faults.

NORMAL FAULTS

In a *normal fault*, the hanging wall moves down, or the footwall moves up, or both. Normal faults form when rock is under tension. **Tension** is stress that pulls rock apart. Therefore, normal faults are common along divergent boundaries, where Earth's crust stretches. ☑

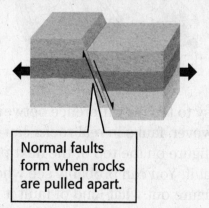

Normal faults form when rocks are pulled apart.

TAKE A LOOK

5. Compare How is the hanging wall different from the footwall?

_____

_____

✔ **READING CHECK**

6. Explain Why are normal faults common along divergent boundaries?

_____

_____

SECTION 4 Deforming the Earth's Crust *continued*

REVERSE FAULTS

In a *reverse fault*, the hanging wall moves up, or the footwall moves down, or both. Reverse faults form when rock is under compression. **Compression** is stress that pushes rock together. Therefore, reverse faults are common at convergent boundaries, where plates collide.

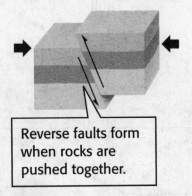

Reverse faults form when rocks are pushed together.

TAKE A LOOK
7. Identify Label the hanging walls and the footwalls on the normal and reverse faults.

STRIKE-SLIP FAULTS

In a *strike-slip fault*, the fault blocks move past each other horizontally. Strike-slip faults form when rock is under shear stress. *Shear stress* is stress that pushes different parts of the rock in different directions. Therefore, strike-slip faults are common along transform boundaries, where tectonic plates slide past each other.

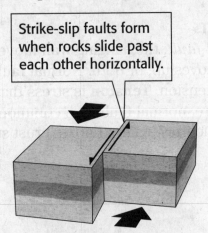

Strike-slip faults form when rocks slide past each other horizontally.

TAKE A LOOK
8. Describe How do strike-slip faults form?

_____

_____

_____

It can be easy to tell the difference between faults in a diagram. However, faults in real rocks can be harder to tell apart. The figure on the top of the next page shows an example of a fault. You can probably see where the fault is. How can you figure out what kind of fault it is? One way is to look at the rock layers around the fault. The dark rock layer in the hanging wall is lower than the same layer in the footwall. Therefore, this is a normal fault.

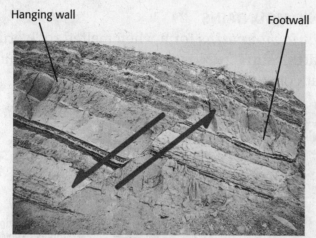

Hanging wall   Footwall

In these rocks, the hanging wall has moved down compared to the footwall. Therefore, this is a normal fault.

How Do Mountains Form?

As tectonic plates move over Earth's surface, the edges of the plates grind against each other. This produces a lot of stress in Earth's lithosphere. Over very long periods of time, the movements of the plates can form mountains. Mountains can form in three main ways: through folding, faulting, or volcanism.

FOLDED MOUNTAINS

Folded mountains form when rock layers are squeezed together and pushed upward. Folded mountains usually form at convergent boundaries, where continents collide. For example, the Appalachian Mountains formed hundreds of millions of years ago when North America collided with Europe and Africa.

FAULT-BLOCK MOUNTAINS

Fault-block mountains form when tension makes the lithosphere break into many normal faults. Along these faults, pieces of the lithosphere drop down compared with other pieces. This produces fault-block mountains. ☑

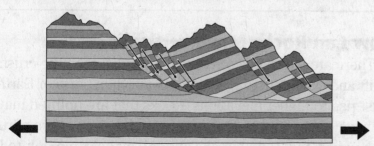

Fault-block mountains form when tension causes the crust to break into normal faults.

TAKE A LOOK

9. Explain How can you tell that this is a normal fault?

_____

_____

_____

Critical Thinking

10. Apply Concepts Why does it take a very long time for most mountains to form?

_____

_____

_____

_____

READING CHECK

11. Identify What kind of stress forms fault-block mountains?

_____

_____

SECTION 4 Deforming the Earth's Crust *continued*

The Andes mountains are examples of volcanic mountains. The Andes have formed where the Nazca plate is subducting beneath the South American plate.

Say It

Investigate Find out more about a volcanic mountain chain, such as the Andes, the Hawaiian islands, or Japan. Share what you learn with a small group.

TAKE A LOOK

12. Identify What kind of convergent boundary have the Andes mountains formed on?

_____

_____

VOLCANIC MOUNTAINS

Volcanic mountains form when melted rock erupts onto Earth's surface. Most major volcanic mountains are found at convergent boundaries.

Volcanic mountains can form on land or on the ocean floor. Volcanoes on the ocean floor can grow so tall that they rise above the surface of the ocean. These volcanoes form islands, such as the Hawaiian Islands.

Most of Earth's active volcanoes are concentrated around the edge of the Pacific Ocean. This area is known as the *Ring of Fire*.

TAKE A LOOK

13. Describe Fill in the table with the features of each kind of mountain. Include where the mountains form and what they are made of.

Type of Mountain	Description
Folded	
Fault-block	
Volcanic	

How Can Rocks Move Vertically?

There are two types of vertical movements in the crust: uplift and subsidence. **Uplift** happens when parts of Earth's crust rise to higher elevations. Rocks that are uplifted may or may not be deformed. **Subsidence** happens when parts of the crust sink to lower elevations. Unlike some uplifted rocks, rocks that subside do not deform.

SECTION 4 Deforming the Earth's Crust *continued*

CAUSES OF SUBSIDENCE AND UPLIFT

Temperature changes can cause uplift and subsidence. Hot rocks are less dense than cold rocks with the same composition. Therefore, as hot rocks cool, they may sink. If cold rocks are heated, they may rise. For example, the crust at mid-ocean ridges is hot. As it moves away from the ridge, it cools and subsides. Old, cold crust far from a ridge has a lower elevation than young, hot crust at the ridge.

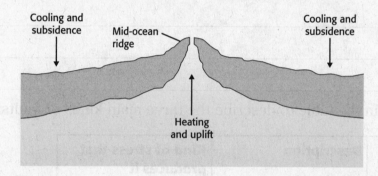

Cooling and subsidence

Mid-ocean ridge

Cooling and subsidence

Heating and uplift

TAKE A LOOK
14. Explain Why does ocean crust far from a mid-ocean ridge subside?

_____

_____

Changes in the weight on the crust can also cause uplift or subsidence. For example, glaciers are huge, heavy bodies of ice. When they form on the crust, they can push the crust down and cause subsidence. If the glaciers melt, the weight on the crust decreases. The crust slowly rises back to its original elevation in a process called *rebound*.

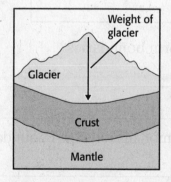

Weight of glacier

Glacier

Crust

Mantle

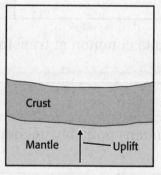

Crust

Mantle — Uplift

TAKE A LOOK
15. Identify What force caused the crust to subside in the left-hand figure?

_____

_____

Fault-block mountains are an example of a third way subsidence can happen. When the crust is under tension, rocks are stretched. They can break and form normal faults. The crust can sink along these faults, causing subsidence. This kind of subsidence is common in rift zones. A *rift zone* is a set of deep cracks that forms at a divergent boundary.

Name ____________________ Class ____________ Date ____________

Section 4 Review

NSES ES 1b, 2a

SECTION VOCABULARY

compression stress that occurs when forces act to squeeze an object

fault a break in a body of rock along which one block slides relative to another

folding the bending of rock layers due to stress

subsidence the sinking of regions of the Earth's crust to lower elevations

tension stress that occurs when forces act to stretch an object

uplift the rising of regions of the Earth's crust to higher elevations

1. Compare How are folding and faulting similar? How are they different?

2. Describe Fill in the spaces in the table to describe the three main kinds of faults.

Kind of fault	Description	Kind of stress that produces it
Normal		
	Hanging wall moves up; footwall moves down.	
		shear stress

3. Explain Why are strike-slip faults common at transform boundaries?

4. Infer Why are fault-block mountains probably uncommon at transform boundaries?

5. Define What is the Ring of Fire?

Name _____ Class _____ Date _____

What Are Earthquakes?

After you read this section, you should be able to answer these questions:

- Where do most earthquakes happen?
- What makes an earthquake happen?
- What are seismic waves?

National Science Education Standards
ES 1a, 1b

What Is an Earthquake?

Have you ever been in an earthquake? An *earthquake* is a movement or shaking of the ground. Earthquakes happen when huge pieces of Earth's crust move suddenly and give off energy. This energy travels through the ground and makes it move. **Seismology** is the study of earthquakes. Scientists who study earthquakes are called *seismologists*.

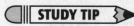

STUDY TIP

Learn New Words As you read this section, circle words that you don't understand. When you learn what they mean, write the words and their definitions in your notebook.

Where Do Most Earthquakes Happen?

Most earthquakes happen at places where two tectonic plates touch. Tectonic plates are always moving. In some places, they move away from each other. In some places, they move toward each other. And in some places, they grind past each other.

The movements of the plates cause Earth's rocky crust to break. A place where the crust is broken is called a *fault*. Earthquakes happen when rock breaks and slides along a fault. ☑

Earthquakes and Plate Boundaries

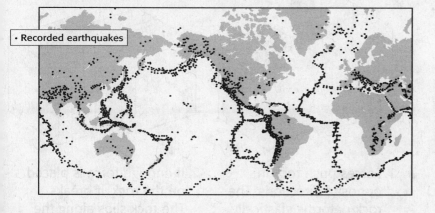

· Recorded earthquakes

READING CHECK

1. Define What is a fault?

TAKE A LOOK

2. Infer Use the earthquake locations to help you figure out where the tectonic plate boundaries are. Use a colored pen or marker to draw plate boundaries on the map.

Why Do Earthquakes Happen?

When tectonic plates move, pressure builds up on the rock near the edges of the plates. When rock is put under pressure, it changes shape, or deforms. This is called **deformation**.

Some rock can bend and fold like clay. When the pressure is taken away, the rock stays folded. When rock stays folded after the pressure is gone, the change is called *plastic deformation*.

TAKE A LOOK
3. Explain How do you know that the rock layers in the figure were once under a lot of pressure?

_____

_____

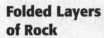
Folded Layers of Rock

Word Help: <u>response</u>
an action brought on by another action; a reaction

Word Help: <u>major</u>
of great importance or large scale

4. Explain How does the movement of tectonic plates cause earthquakes?

_____

_____

_____

_____

In some cases, rock acts more like a rubber band. It changes shape under pressure, but then it goes back to its original size and shape when the pressure goes away. This change is called *elastic deformation*.

Earthquakes happen when rock breaks under pressure. When the rock breaks, it snaps back to its original shape. This snap back is called **elastic rebound**. When the rock breaks and rebounds, it gives off energy. This energy creates faults and causes the ground to shake.

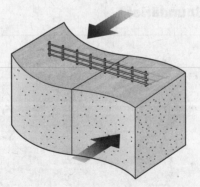

1. Forces push rock in opposite directions. The rock deforms elastically. It does not break.

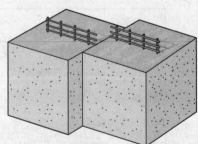

2. If enough force is placed on the rock, it breaks. The rock slips along the fault. Energy is released.

How Do Earthquakes Happen at Divergent Boundaries?

A *divergent boundary* is a place where two tectonic plates are moving away from each other. As the plates pull apart, the crust stretches. The crust breaks along faults. ☑

Most of the crust at divergent boundaries is thin and weak. Most earthquakes at divergent boundaries are small because only a little bit of pressure builds up before the rock breaks.

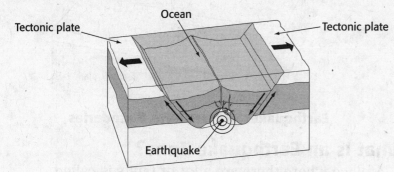

Earthquakes at Divergent Boundaries

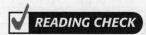

READING CHECK

5. Define What is a divergent boundary?

_____

_____

TAKE A LOOK

6. Identify Label the faults on the figure. Put a star where an earthquake is likely to happen.

How Do Earthquakes Happen at Convergent Boundaries?

A *convergent boundary* is a place where two tectonic plates collide. When two plates come together, the rock is put under a lot of pressure. The pressure grows and grows until the rock breaks.

The earthquakes that happen at convergent boundaries can be very strong because there is so much pressure. The strongest earthquakes ever recorded have all happened at convergent boundaries. ☑

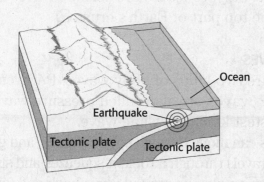

Earthquakes at Convergent Boundaries

READING CHECK

7. Explain Why are many earthquakes at convergent boundaries very strong?

_____

_____

TAKE A LOOK

8. Identify Draw arrows on the figure to show the directions that the two tectonic plates are moving.

SECTION 1 What Are Earthquakes? *continued*

How Do Earthquakes Happen at Transform Boundaries?

A *transform boundary* is a place where two tectonic plates slide past each other. As the plates move, pressure builds up on the rock. Eventually, the rock breaks and the plates slide past each other along a fault.

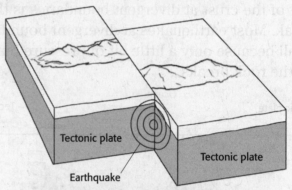

Earthquakes at Transform Boundaries

TAKE A LOOK
9. Identify Draw arrows showing the directions that the tectonic plates in the figure are moving.

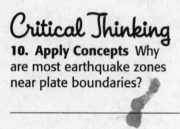

Critical Thinking

10. Apply Concepts Why are most earthquake zones near plate boundaries?

_____

_____

What Is an Earthquake Zone?

A place where there are a lot of faults is called an *earthquake zone*. The San Andreas Fault Zone in California is an example of an earthquake zone. Most earthquake zones are near plate boundaries, but some are in the middle of tectonic plates.

How Does Earthquake Energy Travel?

When an earthquake occurs, a lot of energy is given off. This energy travels through the Earth in the form of waves called **seismic waves**.

There are two kinds of seismic waves. *Body waves* are seismic waves that travel through the inside of Earth to the surface. *Surface waves* are seismic waves that travel through the top part of Earth's crust. ☑

READING CHECK

11. List What are the two kinds of seismic waves?

_____

_____

BODY WAVES

There are two kinds of body waves: P waves and S waves. **P waves** are also called pressure waves. They are the fastest kind of seismic wave.

P waves can move through solids, liquids, and gases. When a P wave travels through a rock, it squeezes and stretches the rock. P waves make the ground move back and forth.

S waves are also called shear waves. S waves move rock from side to side. They can travel only through solids. S waves travel more slowly than P waves.

SECTION 1 What Are Earthquakes? *continued*

SURFACE WAVES

Surface waves travel along the top of Earth's crust. Only the very top part of the crust moves when a surface wave passes.

Surface waves travel much more slowly than body waves. When an earthquake happens, surface waves are the last waves to be felt. Surface waves cause a lot more damage to buildings and landforms than body waves do. ☑

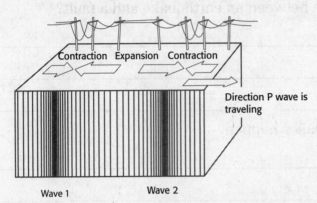

Wave 1   Wave 2

P waves are body waves that squeeze and stretch rock.

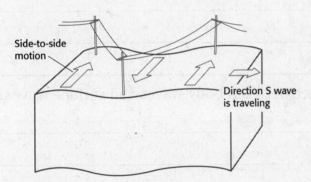

S waves are body waves that can move rock from side to side.

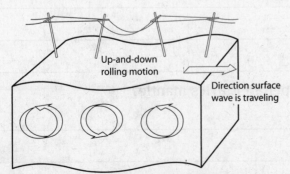

Surface waves can move the ground up and down in a circular motion.

✓ **READING CHECK**

12. Compare Which kind of seismic wave travels the most slowly?

_____

_____

TAKE A LOOK

13. Compare How are the motions of P waves and S waves different?

_____

_____

_____

_____

Critical Thinking

14. Infer What do you think is the reason surface waves usually cause the most damage?

_____

_____

_____

_____

Section 1 Review

SECTION VOCABULARY

deformation the bending, tilting, and breaking of the Earth's crust; the change in the shape of rock in response to stress	**S wave** a seismic wave that causes particles of rock to move in a side-to-side direction
elastic rebound the sudden return of elastically deformed rock to its undeformed shape	**seismic wave** a wave of energy that travels through the Earth, away from an earthquake in all directions
P wave a seismic wave that causes particles of rock to move in a back-and-forth direction	**seismology** the study of earthquakes

1. **Compare** What is the difference between an earthquake and a fault?

_____

_____

_____

2. **Identify** Where do most earthquakes happen?

_____

_____

3. **Describe** What causes earthquakes?

_____

_____

_____

4. **Compare** What is the main difference between body waves and surface waves?

_____

_____

_____

5. **Apply Concepts** Why are some earthquakes stronger than others?

_____

_____

_____

6. **Infer** Why do few earthquakes happen in Earth's mantle?

_____

_____

_____

CHAPTER 8 Earthquakes

SECTION 2 # Earthquake Measurement

National Science
Education Standards
ES 1b

BEFORE YOU READ

After you read this section, you should be able to answer these questions:

• How do scientists know exactly where and when an earthquake happened?

• How are earthquakes measured?

How Do Scientists Study Earthquakes?

Scientists who study earthquakes use an important tool called a seismograph. A **seismograph** records vibrations that are caused by seismic waves. When the waves from an earthquake reach a seismograph, it records them as lines on a chart called a **seismogram**.

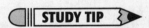

STUDY TIP

Ask Questions Read this section quietly to yourself. As you read, write down questions that you have. Discuss your questions in a small group.

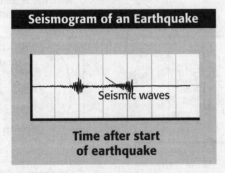

Seismogram of an Earthquake

Seismic waves

Time after start of earthquake

Remember that earthquakes happen when rock in Earth's crust breaks. The rock might break in one small area, but the earthquake can be felt many miles away.

The place inside the Earth where the rock first breaks is called the earthquake's **focus**. The place on Earth's surface that is right above the focus is called the **epicenter**. Seismologists can use seismograms to find the epicenter of an earthquake. ☑

READING CHECK

1. Explain What is the difference between the epicenter and the focus of an earthquake?

_____

_____

_____

_____

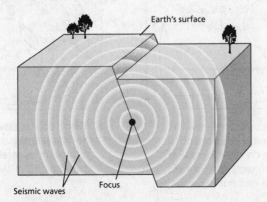

Earth's surface

Seismic waves

Focus

TAKE A LOOK

2. Identify On the figure, mark the epicenter of the earthquake with a star.

How Do Seismologists Know When and Where an Earthquake Happened?

Seismograms help us learn when an earthquake happened. They can also help seismologists find the epicenter of an earthquake. The easiest way to do this is to use the S-P time method. This is how the S-P time method works:

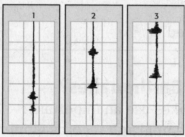

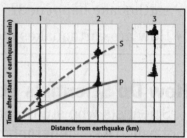

1. The seismologist uses seismograms of the earthquake made at three different places.

2. The seismologist lines up the P waves and S waves on each seismogram with the curves on a graph of time versus distance. The curves on the graph were made using information from earthquakes that happened in the past.

Math Focus

3. Read a Graph Look at the middle seismogram in step 3. What is the difference between the time the P waves arrived and the time the S waves arrived?

_____

4. Read a Graph Look at step 3. How far away from the epicenter is the farthest seismograph station?

_____

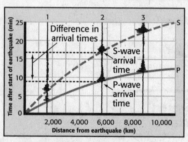

3. Then, the seismologist uses the graph to figure out the difference in arrival times of the P and S waves at each location. The seismologist can use the difference in arrival times to figure out when the earthquake happened. The seismologist can also determine how far away each station is from the epicenter of the earthquake.

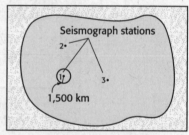

4. On a map, a circle is drawn around a seismograph station. The radius of the circle equals the distance from the seismograph to the epicenter. (This distance is taken from the time-distance graph.)

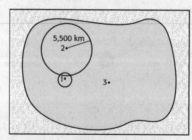

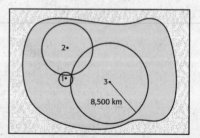

TAKE A LOOK

5. Identify On the map in step 6, draw a star at the earthquake's epicenter.

5. When a second circle is drawn around another seismograph station, the circle overlaps the first circle in two spots. One of these spots is the earthquake's epicenter.

6. When a circle is drawn around the third seismograph station, all three circles meet in one spot—the earthquake's epicenter.

SECTION 2 Earthquake Measurement *continued*

What Is the Magnitude of an Earthquake?

Scientists study seismograms to find out how much the ground moved during an earthquake. They can use the seismograms to figure out how strong the earthquake was.

Have you ever heard someone say that an earthquake was 6.8 or 7.4 "on the Richter scale"? The Richter scale is used to describe the strength, or *magnitude*, of an earthquake. The higher the number, the stronger the earthquake.

Richter magnitude	Effects
2.0	can be detected only by a seismograph
3.0	can be felt at the epicenter
4.0	can be felt by most people in the area
5.0	causes damage at the epicenter
6.0	can cause widespread damage
7.0	can cause great, widespread damage

The Richter scale can be used to compare the magnitudes of different earthquakes. When the Richter magnitude of an earthquake goes up by one unit, the amount of ground shaking caused by the earthquake goes up 10 times. For example, an earthquake with a magnitude of 5.0 is 10 times stronger than an earthquake with a magnitude of 4.0.

What Is the Intensity of an Earthquake?

The *intensity* of an earthquake describes how much damage the earthquake caused and how much it was felt by people. Seismologists in the United States use the Modified Mercalli Intensity Scale to compare the intensity of different earthquakes.

The effects of an earthquake can be very different from place to place. An earthquake can have many different intensity numbers, even though it has only one magnitude.

Mercalli intensity (from I to XII)	Effects
I	shaking felt by only a few people
IV	shaking felt indoors by many, no damage
VIII	damage to some buildings
XII	total damage

Critical Thinking

6. Infer How do you think scientists use seismograms to determine the magnitude of an earthquake?

_____

_____

_____

_____

Math Focus

7. Calculate How many times stronger is a magnitude 6.0 earthquake than a magnitude 4.0 earthquake? Explain your answer.

_____

_____

_____

_____

Name ______________________ Class ______________ Date ______________

Section 2 Review

NSES ES 1b

SECTION VOCABULARY

epicenter the point on Earth's surface directly above an earthquake's starting point, or focus **focus** the point along a fault at which the first motion of an earthquake occurs	**seismogram** a tracing of earthquake motion that is created by a seismograph **seismograph** an instrument that records vibrations in the ground and determines the location and strength of an earthquake

1. Compare What is the difference between a seismograph and a seismogram?

____

____

____

2. Apply Concepts Which city would more likely be the epicenter of an earthquake: San Francisco, California, or St. Paul, Minnesota? Explain your answer.

____

____

____

3. Explain How does a seismologist use the graph of time versus distance for seismic waves to find the location of an earthquake's epicenter?

____

____

____

4. Analyze Methods What can you learn from only one seismogram? What can't you learn?

____

____

____

____

5. Infer How can an earthquake with a moderate magnitude have a high intensity?

____

____

____

CHAPTER 8 | Earthquakes
SECTION 3 **Earthquakes and Society**

**National Science
Education Standards**
ES 1b

BEFORE YOU READ

After you read this section, you should be able to answer these questions:

- Can scientists predict when earthquakes will happen?
- Why do some buildings survive earthquakes better than others?
- How can you prepare for an earthquake?

What Is Earthquake Hazard?

Earthquake hazard tells how likely it is that a place will have a damaging earthquake in the future. Scientists look to the past to figure out earthquake-hazard levels. A place that has had a lot of strong earthquakes in the past has a high earthquake-hazard level. A place that has had few or no earthquakes has a much lower level.

STUDY TIP

Be Prepared As you read, underline important safety information that can help you to prepare for an earthquake.

Earthquake Hazard Map of the Continental United States

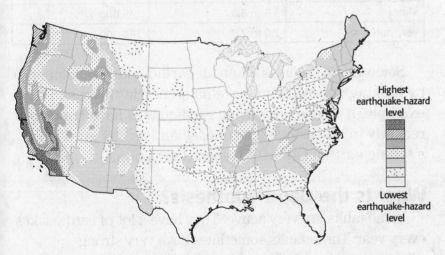

Highest earthquake-hazard level

Lowest earthquake-hazard level

TAKE A LOOK
1. **Identify** On the map, find the place where you live. What is its earthquake-hazard level?

Look at the map above. Notice that California has the highest earthquake-hazard level in the country. The San Andreas Fault Zone runs through most of California, and a lot of earthquakes happen there. Minnesota has a very low earthquake-hazard level. Very few strong earthquakes have been recorded in Minnesota.

SECTION 3 Earthquakes and Society *continued*

Can Scientists Predict Earthquakes?

You know that earthquakes have different magnitudes. You can probably guess that earthquakes don't happen on a set schedule. But what you may not know is that the strength of earthquakes is related to how often they happen.

Scientists can't predict earthquakes. However, by looking at how often earthquakes have happened in the past, they can estimate where and when an earthquake is likely to happen.

Look at the table below. It shows the number of earthquakes of different sizes that happen every year. There are many more weak earthquakes than strong earthquakes every year.

Math Focus

2. Calculate About how many times more light earthquakes than strong earthquakes happen every year?

_____

_____

Description	Magnitude on the Richter scale	Average number per year
Great	8.0 and higher	1
Major	7.0 to 7.9	18
Strong	6.0 to 6.9	120
Moderate	5.0 to 5.9	800
Light	4.0 to 4.9	6,200
Minor	3.0 to 3.9	49,000
Very minor	2.0 to 2.9	365,000

Scientists can guess when an earthquake will happen by looking at how many have happened in the past. For example, if only a few strong earthquakes have happened recently in an earthquake zone, scientists can guess that a strong earthquake will happen there soon.

What Is the Gap Hypothesis?

Some faults are very active. They have a lot of earthquakes every year. These faults sometimes have very strong earthquakes. A part of an active fault that hasn't had a strong earthquake in a long time is called a **seismic gap**.

The **gap hypothesis** says that if an active fault hasn't had a strong earthquake in a long time, it is likely to have one soon. In other words, it says that strong earthquakes are more likely to happen in seismic gaps.

Critical Thinking

3. Apply Concepts What do you think makes strong earthquakes more likely to happen in seismic gaps?

_____

_____

_____

_____

_____

How Do Earthquakes Affect Buildings?

Have you ever seen pictures of a city after a strong earthquake has hit? You may have noticed that some buildings don't have very much damage. Other buildings, however, are totally destroyed. Engineers can study the damage to learn how to make buildings that are stronger and safer.

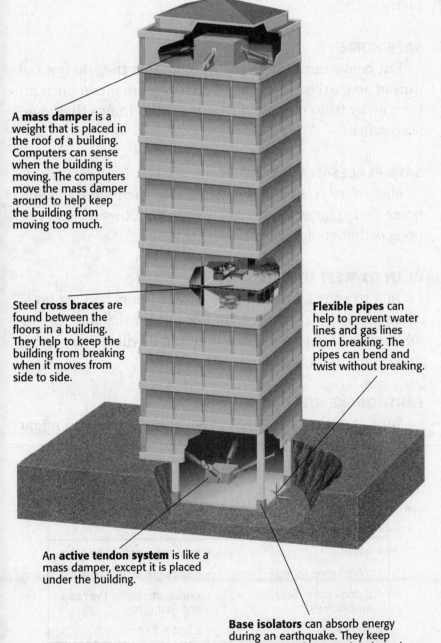

A **mass damper** is a weight that is placed in the roof of a building. Computers can sense when the building is moving. The computers move the mass damper around to help keep the building from moving too much.

Steel **cross braces** are found between the floors in a building. They help to keep the building from breaking when it moves from side to side.

Flexible pipes can help to prevent water lines and gas lines from breaking. The pipes can bend and twist without breaking.

An **active tendon system** is like a mass damper, except it is placed under the building.

Base isolators can absorb energy during an earthquake. They keep seismic waves from moving through the building. Base isolators are made of rubber, steel, and lead.

Critical Thinking

4. List Give three factors that can affect how much a building will be damaged by an earthquake.

_____

_____

_____

TAKE A LOOK

5. Compare How is a mass damper different from an active tendon system?

_____

_____

_____

How Can You Prepare for an Earthquake?

If you live in a place where earthquakes happen often, you and your family should have an earthquake plan. You should practice your plan so you will be prepared if an earthquake happens. ☑

How Can You Make an Earthquake Plan?

There are several things to include in your earthquake plan.

SAFE HOME

Put heavy things near the floor so that they do not fall during an earthquake. Make sure things that can burn are kept away from electric wires and other things that can start a fire.

SAFE PLACES IN YOUR HOME

Make sure you know a safe place in each room in your home. Safe places are areas far from windows or heavy objects that could fall or break. ☑

PLAN TO MEET OTHERS

Talk to your family, friends, or neighbors and set up a place where you all will meet after an earthquake. If you all know where to meet one another, it will be easy to make sure that everyone is safe.

EARTHQUAKE KIT

Your earthquake kit should have things that you might need after an earthquake. Remember that you may not have electricity or running water after an earthquake.

READING CHECK

6. Explain Why is it important to make and practice an earthquake plan?

_____

_____

_____

_____

READING CHECK

7. Identify Think about your bedroom. Write down a safe place in your bedroom that you can go during an earthquake.

_____

_____

_____

TAKE A LOOK

8. List List four foods that would be useful to have in an earthquake kit.

_____

_____

_____

_____

What Should Be in an Earthquake Kit	
• water	• food that won't go bad
• a fire extinguisher	• a flashlight with batteries
• a small radio that runs on batteries	• extra batteries for the radio and flashlight
• medicines	• a first-aid kit

SECTION 3 Earthquakes and Society *continued*

What Should You Do During an Earthquake?

If you are inside when an earthquake happens, crouch or lie facedown under a table or a desk. Make sure you are far away from windows or heavy objects that might fall. Cover your head with your hands. ☑

If you are outside during an earthquake, lie face down on the ground. Make sure you are far from buildings, power lines, and trees. Cover your head with your hands.

If you are in a car or bus, you should ask the driver to stop. Everyone should stay inside the car or bus until the earthquake is over.

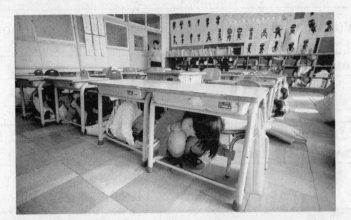

What Should You Do After an Earthquake?

Being in an earthquake can be scary. After an earthquake happens, people are often confused about what happened. They may not know what to do or where to go.

After an earthquake, try to stay calm. Look around you. If you are near something dangerous, like a power line or broken glass, get away as quickly as you can. Never go into a building after an earthquake until your parent, a teacher, a police officer, or a firefighter tells you it is safe. ☑

Always remember that there could be aftershocks. Aftershocks are weaker earthquakes that can happen after a large earthquake. Even though they are weaker than the main earthquake, aftershocks can still be very strong and damaging.

Stick to your earthquake plan. Stay together with your family or friends so that they know you are safe.

READING CHECK

9. List Look around your classroom. List two places that you could go in case of an earthquake.

_____

_____

Say It

Share Experiences Have you ever been in an earthquake? In a small group, talk about what it was like.

READING CHECK

10. Identify Who should you ask if you want to know whether it is safe to go back into a building after an earthquake?

_____

_____

Section 3 Review

SECTION VOCABULARY

gap hypothesis a hypothesis that is based on the idea that a major earthquake is more likely to occur along the part of an active fault where no earthquakes have occurred for a certain period of time	**seismic gap** an area along a fault where relatively few earthquakes have occurred recently but where strong earthquakes have occurred in the past

1. Identify Why are seismologists interested in seismic gaps?

_____

_____

_____

2. Describe Fill in the chart below to show what you should do during an earthquake.

If you are...	Then you should...
...inside a building	
	...lie face down on the ground with your hands on your head, far from power lines or fire hazards.
...in a car or bus	

3. Identify What do engineers do to learn how to make a building more likely to survive an earthquake?

_____

_____

4. Identify Relationships What is the relationship between the strength of an earthquake and how often it occurs?

_____

_____

5. Infer In most cases, you should stay inside a car or a bus in an earthquake. When might it be best to leave a car or a bus during an earthquake?

_____

_____

_____

CHAPTER 9 | Volcanoes

Volcanic Eruptions

National Science
Education Standards
ES 1c

BEFORE YOU READ

After you read this section, you should be able to answer these questions:

• What are two kinds of volcanic eruptions?

• How does the composition of magma affect eruptions?

• What are two ways that magma can erupt from a volcano?

What Is a Volcano?

When you think of a volcano, what comes into your mind? Most people think of a steep mountain with smoke coming out. In fact, a **volcano** is any place where gases and *magma*, or melted rock, come out of the ground. A volcano can be a tall mountain or a small hole in the ground.

THE PARTS OF A VOLCANO

If you could look inside an erupting volcano, it would look similar to the figure below. Below the volcano is a body of magma called a **magma chamber**. The magma from the magma chamber rises to the surface and erupts at the volcano. Magma escapes from the volcano through openings in the Earth's crust called **vents**. When magma flows onto the Earth's surface, it is called *lava*.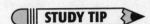

Lava runs out of the vents and down the side of the volcano. The rivers of lava are called lava flows. As they cool and harden, they make the volcano bigger.

When the magma reaches the surface, it erupts out of vents.

When the magma chamber is full, magma rises through the crust and erupts out of the volcano.

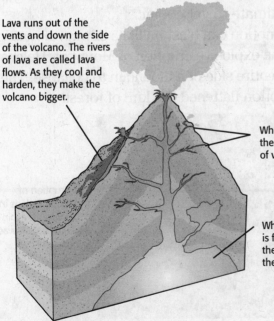

STUDY TIP

Compare After you read this section, make a chart that describes the features of each kind of lava and pyroclastic material.

READING CHECK

1. Define What is a magma chamber?

_____

_____

TAKE A LOOK

2. Describe What makes volcanoes grow larger?

_____

_____

What Happens When Volcanoes Erupt?

Many people think that all volcanic eruptions are alike. However, this is not the case. Scientists put volcanic eruptions into two groups: nonexplosive eruptions and explosive eruptions.

NONEXPLOSIVE ERUPTIONS

Nonexplosive volcanic eruptions are the most common type of eruption. These eruptions produce fairly calm flows of lava. The lava flows over the Earth's surface. Nonexplosive eruptions do not produce very much ash or dust, but they can release huge amounts of lava. For example, most of the rock of the ocean floor was produced by nonexplosive eruptions. ☑

EXPLOSIVE ERUPTIONS

Explosive eruptions are much less common than nonexplosive eruptions. However, explosive eruptions can be more destructive than nonexplosive eruptions. During an explosive eruption, clouds of hot ash, gas, and rock fragments shoot rapidly out of a volcano.

Most explosive eruptions do not produce lava flows. Instead of flowing calmly over the Earth's surface, magma sprays into the air in tiny droplets. The droplets harden to form particles called *ash*. The ash from an explosive eruption can reach the upper parts of the Earth's atmosphere. It can stay there for years, blocking sunlight and causing the climate to get cooler.

An explosive eruption can blast millions of tons of material from a volcano. The explosive eruption of Mount St. Helens in 1980 caused an entire side of a mountain to collapse. The blast from the eruption flattened 600 km^2 of forest.

✓ READING CHECK

3. Identify What is the most common type of volcanic eruption?

_____

_____

Critical Thinking

4. Compare How are nonexplosive eruptions different from explosive eruptions? Give two ways.

_____

_____

_____

_____

📢 Say It

Investigate Find out more information about the eruptions of Mount St. Helens. Share your findings with a small group.

The eruption of Mount St. Helens in 1980 was an explosive eruption. It was very destructive.

SECTION 1 Volcanic Eruptions *continued*

Why Do Volcanoes Erupt?

By comparing magma from different eruptions, scientists have been able to figure out why volcanoes erupt in different ways. The main factor affecting an eruption is the composition of the magma. The amounts of water, silica, and gas in the magma determine the type of eruption. ☑

WATER CONTENT

If magma contains a lot of water, an explosive eruption is more likely. Beneath the surface, magma is under high pressure. The high pressure allows water to dissolve into the magma. If the magma rises quickly, the pressure suddenly decreases and the water turns to bubbles of gas. As the gases expand, they cause an explosion.

This is similar to what happens when you shake a can of soda and open it. When you shake the can, the gas dissolved in the soda forms bubbles. Pressure builds up inside the can. When you open the can, the pressure causes the soda to shoot out.

SILICA AND GAS CONTENT

The amount of silica in magma also affects how explosive an eruption is. *Silica* is a compound made of the elements silicon and oxygen. Magma that contains a lot of silica is very thick and stiff. It flows slowly and may harden inside a volcano's vents, blocking them. As more magma pushes up from below, the pressure increases. If enough pressure builds up, the volcano can explode. ☑

Silica-rich magma may be so stiff that water vapor and other gases cannot move out of the magma. Trapped bubbles of gas may expand until they explode. When they explode, the magma shatters and ash is blasted from the vent. Magma with less silica is thinner and runnier. Therefore, gases can move out of the magma easily, and explosive eruptions are less likely.

Material	How it affects eruptions
Water	
Silica	

☑ **READING CHECK**

5. Identify What is the main factor that determines how a volcano erupts?

_____

_____

☑ **READING CHECK**

6. Describe How can magma that contains a lot of silica cause an explosive eruption?

_____

_____

_____

TAKE A LOOK

7. Identify Relationships Fill in the blank spaces in the table.

How Can Magma Erupt from a Volcano?

There are two main ways that magma can erupt from a volcano: as lava or as pyroclastic material. *Pyroclastic material* is hardened magma that is blasted into the air. Nonexplosive eruptions produce mostly lava. Explosive eruptions produce mostly pyroclastic material. ☑

Most eruptions produce either lava or pyroclastic material, but not both. However, a single volcano may erupt many times. It may produce lava during some eruptions and pyroclastic material during others.

TYPES OF LAVA

Geologists classify lava by the shapes it forms when it cools. Some kinds of lava form smooth surfaces. Others form sharp, jagged edges as they cool. The figure below shows four kinds of lava flows.

Aa is lava that forms a thick, brittle crust as it cools. The crust is torn into sharp pieces as lava moves underneath it.

Pahoehoe is lava that forms a thin, flexible crust as it cools. The crust wrinkles as the lava moves underneath it.

Blocky lava is cool, stiff lava that does not travel very far from the volcano. Blocky lava usually oozes from a volcano and forms piles of rocks with sharp edges.

Pillow lava is lava that erupts under water. As it cools, it forms rounded lumps that look like pillows.

TAKE A LOOK

9. Compare How are aa and blocky lava similar?

_____

_____

TYPES OF PYROCLASTIC MATERIAL

Pyroclastic material forms when magma explodes from a volcano. The magma solidifies in the air. Pyroclastic material also forms when powerful eruptions shatter existing rock.

Geologists classify pyroclastic material by the size of its pieces. Pieces of pyroclastic material can be the size of houses or as small as dust particles. The figure on the top of the next page shows four kinds of pyroclastic materials.

Volcanic bombs are large blobs of lava that harden in the air.

Lapilli are small bits of lava that harden before they hit the ground. Lapilli are usually about the size of pebbles."

Volcanic ash forms when gases trapped in magma or lava form bubbles. When the bubbles explode, they create millions of tiny pieces.

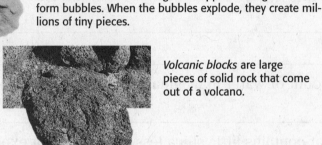

Volcanic blocks are large pieces of solid rock that come out of a volcano.

TAKE A LOOK
10. Describe How do lapilli form?

_____

_____

_____

PYROCLASTIC FLOWS

A *pyroclastic flow* is a dangerous type of volcanic flow. Pyroclastic flows form when ash and dust race down the side of a volcano like a river. Pyroclastic flows are very dangerous. They can be as hot as 700°C and can move at 200 km/h. A pyroclastic flow can bury or destroy everything in its path. A pyroclastic flow from the eruption of Mount Pinatubo is shown in the figure below.

Math Focus
11. Convert How fast can pyroclastic flows move? Give your answer in miles per hour.
1 km = 0.62 mi

_____

This pyroclastic flow formed during the 1991 eruption of Mount Pinatubo, in the Philippines.

Name _____ Class _____ Date _____

Section 1 Review

NSES ES 1c

SECTION VOCABULARY

magma chamber the body of molten rock that feeds a volcano **vent** an opening at the surface of the Earth through which volcanic material passes	**volcano** a vent or fissure in the Earth's surface through which magma and gases are expelled

1. Compare How is lava different from magma?

_____

2. Identify What are the two kinds of volcanic eruptions?

_____

_____

3. Explain How does the amount of water in magma affect how a volcano erupts?

_____

4. Explain Why is magma that contains little silica less likely to erupt explosively?

_____

_____

5. Compare How is pahoehoe lava different from pillow lava? How are they similar?

_____

_____

_____

6. Describe How do volcanic bombs form?

_____

_____

7. Describe How does volcanic ash form?

_____

_____

_____

8. Define What is a pyroclastic flow?

_____

_____

_____

9. Infer Do pyroclastic flows form during explosive or nonexplosive eruptions?

_____

Copyright © by Holt, Rinehart and Winston. All rights reserved.
Interactive Textbook 156 Volcanoes

CHAPTER 9 | Volcanoes

SECTION
2 **Effects of Volcanic Eruptions**

BEFORE YOU READ

After you read this section, you should be able to answer these questions:

• How can volcanoes affect climate?

• What are three kinds of volcanoes?

• What are three structures that volcanic eruptions can form?

National Science Education Standards
ES 1c

How Can Volcanoes Affect Climate?

In 1815, a huge volcanic explosion happened on Mount Tambora in Indonesia. Historians estimate that the explosion killed 12,000 people. As many as 80,000 people died from hunger and disease following the explosion. However, the explosion did not affect only the people living in Indonesia. It also affected the climate worldwide.

Ash and dust from the explosion flew into the upper atmosphere. There, they spread across the Earth. They blocked sunlight from reaching the Earth's surface. As a result, global temperatures dropped. In 1816, there was a snowstorm in June! The colder temperatures caused food shortages in North America and Europe.

STUDY TIP
Compare After you read this section, make a chart comparing the three kinds of volcanoes. Describe how each type of volcano forms and what it looks like.

TAKE A LOOK
1. Identify What causes global temperatures to drop after a large explosive eruption?

_____

_____

_____

A large explosive eruption	→ produces →	a lot of ash and dust

which ↓

Global cooling	← causing ←	block sunlight from reaching Earth's surface

In 1991, an explosive eruption on Mount Pinatubo caused global temperatures to drop. Explosive eruptions may cause global temperatures to decrease by 0.5°C to 1°C. This may seem like a small change, but even small temperature changes can disrupt world climates.

How Can Volcanoes Affect the Earth's Surface?

In addition to affecting climate, volcanoes can have important effects on the Earth's surface. Volcanoes produce many unique *landforms*, or surface features.

The most well-known volcanic landforms are the volcanoes themselves. There are three main kinds of volcanoes: shield volcanoes, cinder cone volcanoes, and composite volcanoes.

SECTION 2 Effects of Volcanic Eruptions *continued*

SHIELD VOLCANOES

Shield volcanoes form when layers of lava from many nonexplosive eruptions build up. The lava that forms shield volcanoes is thin and runny. Therefore, it spreads out in thin layers over a wide area. This produces a volcano with a wide base and gently sloping sides.

Shield volcanoes can be very large. For example, Mauna Kea in Hawaii is a shield volcano. Measured from the base on the ocean floor, Mauna Kea is taller than Mount Everest!

STANDARDS CHECK

ES 1c Land forms are the result of a combination of constructive and destructive forces. Constructive forces include crustal deformation, volcanic eruption, and deposition of sediment, while destructive forces include weathering and erosion.

2. Describe How do shield volcanoes form?

_____

_____

_____

_____

Lava flows

Shield volcanoes form when many layers of lava build up over time.

CINDER CONE VOLCANOES

Cinder cone volcanoes are made of pyroclastic material. The pyroclastic material is produced from explosive eruptions. As it piles up, it forms a mountain with steep slopes. Cinder cones are small. Most of them erupt for only a short time. For example, Paricutín is a cinder cone volcano in Mexico. In 1943, Paricutín appeared in a cornfield. It erupted for only nine years. ☑

Most cinder cone volcanoes are found in clusters. They may be found on the sides of other volcanoes. They erode quickly because the pyroclastic material is loose and not stuck together.

READING CHECK

3. Identify What are cinder cone volcanoes made of?

_____

_____

TAKE A LOOK

4. Identify Which type of volcanic eruption produces cinder cone volcanoes?

_____

_____

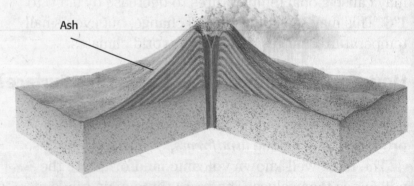

Ash

Cinder cone volcanoes form when ash from explosive eruptions piles up. Most cinder cones are small.

| SECTION 2 | Effects of Volcanic Eruptions *continued* |

COMPOSITE VOLCANOES

Composite volcanoes are the most common type of volcano. They form when a volcano erupts both explosively and nonexplosively. They have layers of lava flows and pyroclastic material. They usually have a broad base and sides that get steeper toward the top. Mount St. Helens is a composite volcano.

5. Infer The word stratum means "layer." Why are composite volcanoes sometimes also called stratovolcanoes?

_____

_____

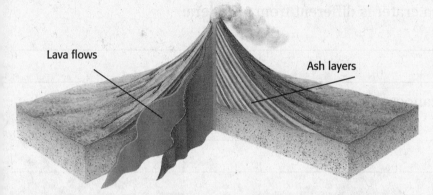

Lava flows

Ash layers

Composite volcanoes form from layers of ash and lava. Most have steep sides.

TAKE A LOOK
6. Identify What two materials are composite volcanoes made of?

_____

_____

What Are Other Types of Volcanic Landforms?

In addition to volcanoes, other landforms are created by volcanic activity. The landforms include craters, calderas, and lava plateaus.

CRATERS

A **crater** is a funnel-shaped pit around the central vent at the top of a volcano. Lava and pyroclastic material can pile up around the vent. This produces a crater in the middle of the cone. ☑

✔ **READING CHECK**
7. Define What is a crater?

_____

_____

CALDERAS

A **caldera** is a large *depression*, or pit, that forms when a magma chamber collapses. The ground over the magma chamber sinks, forming a caldera. Calderas can look similar to craters, but calderas are much larger.

LAVA PLATEAUS

A **lava plateau** is a large area of land covered by a huge volume of lava. Lava plateaus are the largest volcanic landforms. They do not form at tall volcanoes. Instead, lava plateaus form when a large volume of lava erupts from a crack in the crust. Most of the lava on the Earth's surface is found in lava plateaus.

Section 2 Review

SECTION VOCABULARY

caldera a large, circular depression that forms when the magma chamber below a volcano partially empties and causes the ground above to sink	**crater** a bowl-shaped depression that forms on the surface of an object when a falling body strikes the object's surface or when an explosion occurs
	lava plateau a wide, flat landform that results from repeated nonexplosive eruptions of lava that spread over a large area

1. Compare Explain how a crater is different from a caldera.

_____

_____

2. Describe How can volcanoes affect climate?

_____

_____

_____

3. Identify What are the three main types of volcanoes?

_____

_____

_____

4. Explain Why do shield volcanoes have wide bases?

_____

_____

5. Explain Why do cinder cone volcanoes erode quickly?

_____

6. Identify What is the largest kind of volcanic landform?

_____

7. Apply Concepts Does the lava that forms shield volcanoes probably have a lot of silica or water in it? Explain your answer.

_____

_____

_____

CHAPTER 9 | Volcanoes

SECTION 3 **Causes of Volcanic Eruptions**

BEFORE YOU READ

After you read this section, you should be able to answer these questions:

• How does magma form?

• Where do volcanoes form?

• How can geologists predict volcanic eruptions?

How Does Magma Form?

Magma forms deep in the Earth's crust and in the upper parts of the mantle. In these areas, the temperature and pressure are very high. Changes in pressure and temperature can cause magma to form.

Part of the upper mantle is made of very hot, solid rock. The rock is so hot that it can flow, like soft chewing gum, even though it is solid. If rock of this temperature were found at the Earth's surface, it would be *molten*, or melted. The rock in the mantle does not melt because it is under high pressure. This pressure is produced by the weight of the rock above the mantle. ☑

In the figure below, the curved line shows the melting point of a rock. The *melting point* is the temperature at which the rock melts for a certain pressure.

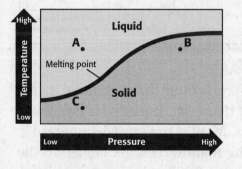

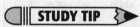

STUDY TIP

Describe After you read this section, make flowcharts showing how magma forms at divergent boundaries and at convergent boundaries.

READING CHECK

1. Explain Why doesn't the hot rock in the mantle melt?

_____

_____

_____

Math Focus

2. Describe A rock starts out at point C. Then, its temperature increases. What will happen to the rock if its temperature continues to rise?

_____

_____

The curved line shows the melting point of the rock. Rock with the temperature and pressure of point A is liquid. Rock at the same temperature but higher pressure (B) is solid. Rock at the same pressure but lower temperature (C) is also solid.

MAGMA FORMATION IN THE MANTLE

The temperature of the mantle is fairly constant. Magma usually forms because of a decrease in pressure. Therefore, a lot of magma forms at the boundary between separating tectonic plates, where pressure decreases. Magma is less dense than the solid rock it forms from. Therefore, it rises toward the surface and erupts.

SECTION 3 Causes of Volcanic Eruptions *continued*

Where Do Volcanoes Form?

The locations of volcanoes give clues about how volcanoes form. The figure below shows the locations of some of the world's major active volcanoes. The map also shows the boundaries between tectonic plates. Most volcanoes are found at tectonic plate boundaries. For example, there are many volcanoes on the plate boundaries surrounding the Pacific Ocean. Therefore, the area is sometimes called the *Ring of Fire*.

Remember that tectonic plate boundaries are areas where plates collide, separate, or slide past one another. Most volcanoes are found where plates move together or apart. About 15% of active volcanoes on land form where plates separate, and about 80% form where plates collide. The remaining few volcanoes on land are found far from tectonic plate boundaries.

Volcanoes and Tectonic Plate Boundaries

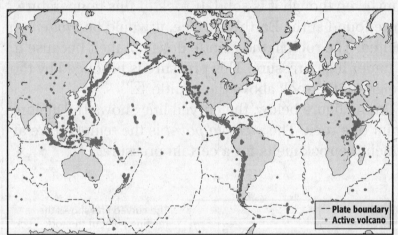

- - Plate boundary
- • Active volcano

WHERE PLATES MOVE APART

At a *divergent boundary*, tectonic plates move away from each other. A set of deep cracks called a **rift zone** forms between the plates. Mantle rock moves upward to fill in the gap. When the mantle rock gets close to the surface, the pressure decreases. The decrease in pressure causes the mantle rock to melt, forming magma. The magma rises through the rift zones and erupts. ☑

Most divergent boundaries are on the ocean floor. Lava that flows from undersea rift zones produces volcanoes and mountain chains. These volcanoes and mountain chains are called *mid-ocean ridges*. The mid-ocean ridges circle the ocean floor.

3. Describe Where are most volcanoes found?

_____

_____

 **READING CHECK**

4. Identify What causes magma to melt at divergent boundaries?

_____

_____

SECTION 3 Causes of Volcanic Eruptions *continued*

How Magma Forms at a Divergent Boundary

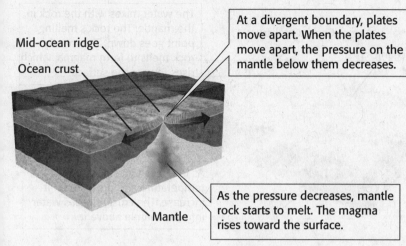

Mid-ocean ridge

Ocean crust

Mantle

At a divergent boundary, plates move apart. When the plates move apart, the pressure on the mantle below them decreases.

As the pressure decreases, mantle rock starts to melt. The magma rises toward the surface.

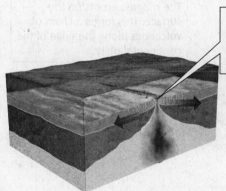

When the magma reaches the surface, it erupts onto the sea floor. When it cools and hardens, new ocean crust forms.

TAKE A LOOK
5. Explain How does new ocean crust form?

_____

_____

_____

WHERE PLATES MOVE TOGETHER

At a *convergent boundary*, the tectonic plates collide. When an oceanic plate collides with a continental plate, the oceanic plate slides under the continental plate. This is called *subduction*. The oceanic crust sinks into the mantle because it is more dense than the continental crust. ☑

As the ocean crust sinks, the temperature and pressure on it increase. Because the ocean crust forms below the ocean, the rock contains a lot of water. The heat and pressure on the ocean crust cause this water to be released.

The water mixes with the mantle rock above the oceanic plate. When the mantle rock mixes with water, it can melt at a lower temperature. The mantle rock begins to melt at the subduction zone. The magma rises to the surface and erupts as a volcano.

READING CHECK

6. Explain Why does oceanic crust sink below continental crust?

_____

_____

| SECTION 3 | Causes of Volcanic Eruptions *continued* |

How Magma Forms at a Convergent Boundary

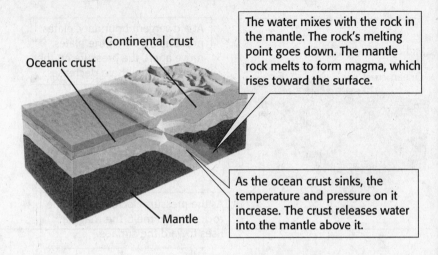

The water mixes with the rock in the mantle. The rock's melting point goes down. The mantle rock melts to form magma, which rises toward the surface.

Continental crust

Oceanic crust

As the ocean crust sinks, the temperature and pressure on it increase. The crust releases water into the mantle above it.

Mantle

The magma erupts on the surface. This forms a chain of volcanoes along the edge of the continental plate.

TAKE A LOOK

7. Explain How does subduction produce magma?

_____

_____

_____

_____

IN THE MIDDLE OF PLATES

Although most volcanoes form at plate boundaries, not all volcanoes form there. Some volcanoes, such as the Hawaiian Islands, form at hot spots. **Hot spots** are places on the Earth's surface where volcanoes form far from plate boundaries. Most scientists think that hot spots form above hot columns of mantle rock called *mantle plumes*. Some scientists think that hot spots form where magma rises through cracks in the Earth's crust. ☑

Long chains of volcanoes are common at hot spots. One theory to explain this is that a mantle plume stays in one place while the plate moves over it. Another theory states that hot-spot volcanoes occur in long chains because they form along cracks in the Earth's crust. Scientists are not sure which of these theories is correct. It is possible that some hot spots form over plumes, but others form over cracks.

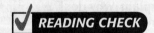 **READING CHECK**

8. Define What is a
hot spot?

_____

_____

_____

How Can We Predict Volcanic Eruptions?

Scientists cannot always predict when and where a volcano will erupt. However, by studying ancient and modern volcanoes, scientists have been able to identify some signs that an eruption may happen.

One feature that scientists use to predict whether an eruption will happen is the state of the volcano. Geologists put volcanoes into three groups based on how active they are.

- *Extinct* volcanoes have not erupted in recorded history and probably will not erupt again.
- *Dormant* volcanoes are currently not erupting, but they may erupt again.
- *Active* volcanoes are currently erupting or show signs of erupting in the near future.

SMALL QUAKES AND VOLCANIC GASES

Most active volcanoes produce small earthquakes as the magma within them moves upward. This happens because the magma pushes on the rocks as it rises. In many cases, the number and strength of these earthquakes increases before a volcanic eruption. Therefore, monitoring earthquakes is one of the best ways to predict an eruption.

Scientists also study the volume and composition of gases given off by the volcano. Just before an eruption, many volcanoes give off more gas. The composition of the gas may also change before an eruption. By monitoring the gases, scientists can predict when an eruption may happen.

MEASURING SLOPE AND TEMPERATURE

As magma rises before an eruption, it can cause the Earth's surface to swell. The side of a volcano may even bulge. Scientists can use an instrument called a *tiltmeter* to measure the slope of the volcano's sides. Changes in the slope can indicate that an eruption is likely. ☑

One of the newest methods for predicting volcanic eruptions involves using satellite images. Satellites can record the surface temperatures at and around volcanoes. As magma rises, the surface temperature of the volcano may increase. Therefore, an increase in surface temperature can indicate that an eruption is likely.

Critical Thinking

9. Compare What is the difference between dormant volcanoes and extinct volcanoes?

_____

_____

_____

Critical Thinking

10. Infer Why may a volcano that is about to erupt give off more gas?
(Hint: Why are some eruptions explosive?)

_____

_____

_____

_____

☑ **READING CHECK**

11. Explain Why may the Earth's surface swell before an eruption?

_____

_____

Section 3 Review

SECTION VOCABULARY

hot spot a volcanically active area of Earth's surface, commonly far from a tectonic plate boundary	**rift zone** an area of deep cracks that forms between two tectonic plates that are pulling away from each other

1. Identify Where do rift zones form?

_____

2. Apply Concepts The map below shows the locations of many volcanoes. On the map, circle three volcanoes that are probably found at hot spots.

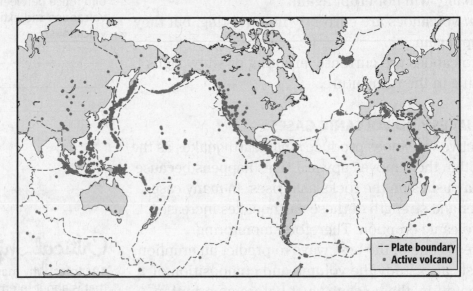

Plate boundary
Active volcano

3. Identify What is the most common cause of magma formation in the mantle?

4. Describe How does magma form at divergent boundaries?

_____

_____

_____

5. List Give four signs that a volcanic eruption is likely.

_____

_____

_____

_____

CHAPTER 10 Weathering and Soil Formation

Weathering

National Science Education Standards
ES 1c, 1d, 1k

What Is Weathering?

How do large rocks turn into smaller rocks? **Weathering** is the process in which rocks break down. There are two main kinds of weathering: mechanical weathering and chemical weathering.

What Is Mechanical Weathering?

Mechanical weathering happens when rocks are broken into pieces by physical means. There are many *agents*, or causes, of mechanical weathering. ☑

ICE

Ice is one agent of mechanical weathering. Cycles of freezing and thawing can cause *ice wedging*, which can break rock into pieces.

The cycle of ice wedging starts when water seeps into cracks in a rock. When the water freezes, it expands. The ice pushes against the cracks. This causes the cracks to widen. When the ice melts, the water seeps further into the cracks. As the cycle repeats, the cracks get bigger. Finally, the rock breaks apart.

Ice Wedging

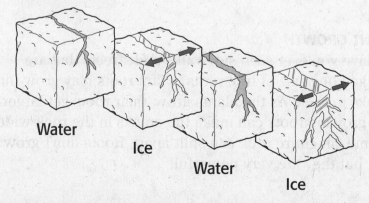

Water
Ice
Water
Ice

STUDY TIP

Compare Make a chart showing the ways that mechanical weathering and chemical weathering can happen.

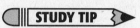

READING CHECK

1. Define What is mechanical weathering?

_____

_____

Critical Thinking

2. Infer Would ice wedging happen if water did not expand as it froze? Explain your answer.

_____

_____

_____

WIND, WATER, AND GRAVITY

As you scrape a large block of chalk against a board, tiny pieces of the chalk rub off on the board. The large piece of chalk wears down and becomes smaller. The same process happens with rocks. **Abrasion** is a kind of mechanical weathering that happens when rocks are worn away by contact with other rocks. Abrasion happens whenever one rock hits another. Water, wind, and gravity can cause abrasion.

Water can cause abrasion by moving rocks and making them hit each other. The rocks in this river are rounded because of abrasion.

Wind can cause abrasion when it blows sand against rocks. This rock has been shaped by blowing sand.

Gravity can cause abrasion by making rocks rub against each other as they slide downhill. As the rocks grind against each other, they are broken into smaller pieces.

Say It

Discuss In a small group, talk about some different environments in which abrasion may happen.

TAKE A LOOK

3. Explain How does running water cause abrasion?

_____

_____

_____

_____

_____

PLANT GROWTH

Have you ever seen sidewalks and streets that are cracked because of tree roots? Plant roots may grow into cracks in rock. As the plants grow, their roots get larger. The growing roots can make the cracks in the rock wider. In time, an entire rock can split apart. Roots don't grow fast, but they are very powerful!

ANIMALS

Did you know that earthworms cause a lot of weathering? They tunnel through the soil and move pieces of rock around. This motion breaks some of the rocks into smaller pieces. It also exposes more rock surfaces to other agents of weathering.

Any animal that burrows in the soil causes mechanical weathering. Ants, worms, mice, coyotes, and rabbits are just a few of the animals that can cause weathering. The mixing and digging that animals do can also cause chemical weathering, another kind of weathering.

What Is Chemical Weathering?

In addition to physical weathering, rocks can be broken down by chemical means. **Chemical weathering** happens when rocks break down because of chemical reactions.

Water, acids, and air are all agents of chemical weathering. They react with the chemicals in the rock. The reactions can break the bonds in the minerals that make up the rock. When the bonds in the minerals are broken, the rock can be worn away. ☑

WATER

If you drop a sugar cube into a glass of water, the sugar cube will dissolve after a few minutes. In a similar way, water can dissolve some of the chemicals that make up rocks. Even very hard rocks, such as granite, can be broken down by water. However, this process may take thousands of years or more.

Chemical Weathering in Granite

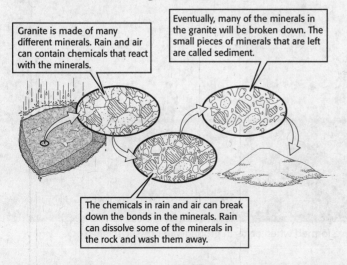

Granite is made of many different minerals. Rain and air can contain chemicals that react with the minerals.

Eventually, many of the minerals in the granite will be broken down. The small pieces of minerals that are left are called sediment.

The chemicals in rain and air can break down the bonds in the minerals. Rain can dissolve some of the minerals in the rock and wash them away.

STANDARDS CHECK

ES 1k Living organisms have played many roles in the Earth system, including affecting the composition of the atmosphere, producing some types of rocks, and contributing to the weathering of rocks.

Word Help: role
a part or function; purpose

Word Help: affect
to change; to act upon

4. Describe How can earthworms cause weathering?

_____

_____

_____

_____

_____

_____

☑ **READING CHECK**

5. List What are three agents of chemical weathering?

_____

_____

_____

TAKE A LOOK

6. Infer What do you think is the reason it takes a very long time for granite to break down?

_____

_____

_____

_____

_____

ACID PRECIPITATION

Precipitation, such as rain and snow, always contains a little bit of acid. However, sometimes precipitation contains more acid than normal. Rain, sleet, or snow that contains more acid than normal is called **acid precipitation**.

Acid precipitation forms when small amounts of certain gases mix with water in the atmosphere. The gases come from natural sources, such as active volcanoes. They are also produced when people burn fossil fuels, such as coal and oil. ☑

The acids in the atmosphere fall back to the ground in rain and snow. Acids can dissolve materials faster than plain water can. Therefore, acid precipitation can cause very rapid weathering of rock.

ACIDS IN GROUNDWATER

In some places, water flows through rock underground. This water, called *groundwater*, may contain weak acids. When the groundwater touches some kinds of rock, a chemical reaction happens. The chemical reaction dissolves the rock. Over a long period of time, huge caves can form where rock has been dissolved.

This cave formed when acids in groundwater dissolved the rock.

TAKE A LOOK

8. Explain Caves like the one in the picture are not found everywhere. What do you think controls where a cave forms?

_____

_____

_____

ACIDS FROM LIVING THINGS

All living things make weak acids in their bodies. When the living things touch rock, some of these acids are transferred to the surface of the rock.

The acids react with chemicals in the rock and weaken it. The different kinds of mechanical weathering can more easily remove rock in these weakened areas. ☑

The rock may also crack in the weakened areas. Even the smallest crack can expose more of the rock to both mechanical weathering and chemical weathering.

AIR

Have you ever seen a rusted car or building? Rusty metal is an example of chemical weathering. Metal reacted with something to produce rust. What did the metal react with? In most cases, the answer is air.

The oxygen in the air can react with many metals. These reactions are a kind of chemical weathering called *oxidation*. Rust is a common example of oxidation. Rocks can rust if they have a lot of iron in them.

Many people think that rust forms only when metal gets wet. In fact, oxidation can happen even without any water around. However, when water is present, oxidation happens much more quickly.

Oxidation can cause rocks to weaken. Oxidation changes the metals in rocks into different chemicals. These chemicals can be broken down more easily than the metals that were there before.

☑ **READING CHECK**

9. Explain How can acids from living things cause weathering?

_____
_____
_____
_____
_____

Factor	How does it cause chemical weathering?
Water	
Acid precipitation	
Acids in groundwater	
Acids from living things	
Air	

TAKE A LOOK

10. Describe Fill in the blank spaces in the table to describe how different factors cause chemical weathering.

Section 1 Review

NSES ES 1c, 1d, 1k

SECTION VOCABULARY

abrasion the grinding and wearing away of rock surfaces through the mechanical action of other rock or sand particles	**mechanical weathering** the process by which rocks break down into smaller pieces by physical means
acid precipitation rain, sleet, or snow that contains a high concentration of acids	**weathering** the natural process by which atmospheric and environmental agents, such as wind, rain, and temperature changes, disintegrate and decompose rocks
chemical weathering the process by which rocks break down as a result of chemical reactions	

1. List What are three things that can cause abrasion?

_____

_____

_____

2. Explain Fill in the spaces to show the steps in the cycle of ice wedging.

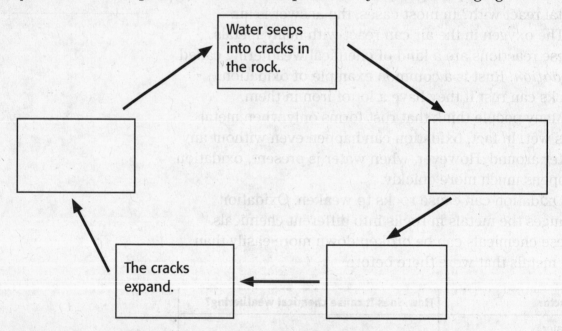

Water seeps into cracks in the rock.

The cracks expand.

3. Identify How can acids cause chemical weathering?

_____

_____

_____

4. Compare How is mechanical weathering caused by ice wedging similar to mechanical weathering caused by plant roots?

_____

_____

CHAPTER 10 Weathering and Soil Formation
SECTION
2 **Rates of Weathering**

National Science Education Standards
ES 1c, 1d

What Is Differential Weathering?

Hard rocks, such as granite, weather more slowly than softer rocks, such as limestone. **Differential weathering** happens when softer rock weathers away and leaves harder, more resistant rock behind. The figures below show an example of how differential weathering can shape the landscape.

📝 **STUDY TIP**

Apply Concepts After you read this section, think about how the different factors that control weathering can affect objects that you see every day. Discuss your ideas in a small group.

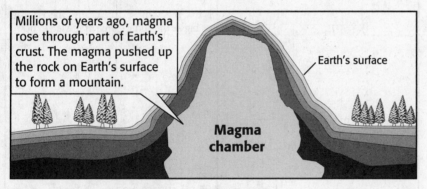

Millions of years ago, magma rose through part of Earth's crust. The magma pushed up the rock on Earth's surface to form a mountain.

Earth's surface

Magma chamber

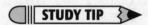

📣 **Say It**

Discuss In a small group, share your ideas about how some different landscape features formed because of differential weathering.

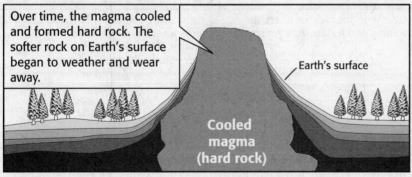

Over time, the magma cooled and formed hard rock. The softer rock on Earth's surface began to weather and wear away.

Earth's surface

Cooled magma (hard rock)

TAKE A LOOK
1. **Infer** Imagine that the rock on the outside of the mountain was much harder than the cooled magma. How might Devils Tower look today? Explain your answer.

Today, all of the soft rock of the mountain has weathered away. The only thing that is left is the hard rock from inside the mountain. This hard rock forms the unusual structure called Devils Tower, in Wyoming.

_____

_____

_____

_____

How Does a Rock's Surface Area Affect Weathering?

Most chemical weathering takes place only on the outer surface of a rock. Therefore, rocks with a lot of surface area weather faster than rocks with little surface area. However, if a rock has a large volume as well as a large surface area, it takes longer for the rock to wear down. The figure below shows how the surface area and volume of a rock affect how fast it wears down.

Critical Thinking

2. Explain Why does most chemical weathering happen only on the outer surface of a rock?

_____

_____

_____

_____

Math Focus

3. Calculate Determine the surface areas of the large cube and the eight smaller cubes. Write these values in the blank lines on the figure.

TAKE A LOOK

4. Apply Concepts Why do the edges and corners of the cubes weather faster than the faces? (Hint: remember that objects with large surface areas weather quickly.)

_____

_____

_____

Imagine a rock in the shape of a cube. Each side of the rock is 4 m long. The volume of the cube is 4 m × 4 m × 4 m = 64 m³. Each face of the rock has an area of 4 m × 4 m = 16 m². Because there are six faces on the rock, it has a surface area of 6 × 16 m² = _____ m². 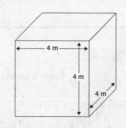 Cube with 4 m sides: volume = 64 m³, surface area = _____ m²	Now imagine eight smaller rocks that are shaped like cubes. Each small rock's side is 2 m long. The volume of each small rock is 2 m × 2 m × 2 m = 8 m³. The total volume of all 8 rocks is 8 × 8 m³ = 64 m³, the same as the large rock. Each face of each small rock has an area of 2 m × 2 m = 4 m². Each small rock has six faces, and there are eight rocks total. Therefore, the total surface area of all eight small rocks is 8 × 6 × 4 m² = _____ m². This is twice as big as the surface area of the large rock. 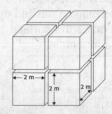 Eight cubes with 2 m sides: volume = 64 m³, surface area = _____ m²
Over time, the rock weathers. Its volume and surface area get smaller. 	In the same amount of time, the smaller rocks weather more and become much smaller. They lose a larger fraction of their volume than the larger rock.
More time passes. The large rock is weathered even more. It is now much smaller than it was before it was weathered. 	In the same amount of time, the small rocks have completely worn away. They took much less time to wear away than the large rock. Even though their volume was the same, they had more surface area than the large rock. The large surface area allowed them to wear away more quickly.

How Does Climate Affect Weathering?

The rate of weathering in an area is affected by the climate of that area. *Climate* is the average weather conditions of an area over a long period of time. Some features of climate that affect weathering are temperature, moisture, elevation, and slope.

TEMPERATURE

Temperature is a major factor in both chemical and mechanical weathering. Cycles of freezing and thawing increase the chance that ice wedging will take place. Areas in the world that have many freezes and thaws have faster mechanical weathering than other regions do.

High temperatures can also speed up weathering. Many chemical reactions happen faster at higher temperatures. These reactions can break down rock quickly in warm areas. ☑

MOISTURE

Water can interact with rock as precipitation, as running water, or as water vapor in the air. Water can speed up many chemical reactions. For example, oxidation can happen faster when water is present.

Water is also important in many kinds of mechanical weathering. For example, ice wedging cannot happen without water. Abrasion also happens faster when water is present.

ELEVATION AND SLOPE

Elevation and slope can affect how fast weathering occurs. *Elevation* is a measure of how high a place is above sea level. *Slope* is a measure of how steep the sides of a mountain or hill are. The table below shows how elevation and slope affect weathering.

Factor	How the factor affects weathering
Elevation	Rocks at high elevations are exposed to low temperatures and high winds. They can weather very quickly. Rocks at sea level can be weathered by ocean waves.
Slope	The steep sides of mountains and hills make water flow down them faster. Fast-moving water has more energy to break down rock than slow-moving water. Therefore, rocks on steep slopes can weather faster than rocks on level ground.

✓ **READING CHECK**

5. Identify What is one way that temperature affects mechanical weathering?

_____

_____

_____

TAKE A LOOK

6. Explain Why do rocks on the sides of mountains weather faster than rocks on level ground?

_____

_____

_____

_____

Name ______________________ Class ______________ Date ______________

Section 2 Review

NSES ES 1c, 1d

SECTION VOCABULARY

differential weathering the process by which softer, less weather resistant rocks wear away at a faster rate than harder, more weather resistant rocks do	

1. Identify What are four factors that affect how fast weathering happens?

____

____

____

____

2. Explain Why does it take less time for small rocks to wear away than it does for large rocks to wear away?

____

____

____

____

3. Describe Imagine a rock on a beach and a rock on the side of a mountain. How would the factors that control weathering be different for these two rocks?

____

____

____

____

4. Apply Concepts Two rivers run into the ocean. One river is very long. The other is very short. Which river probably drops the smallest rock pieces near the ocean? Explain your answer.

____

____

____

____

SECTION
3 **From Bedrock to Soil**

BEFORE YOU READ

After you read this section, you should be able to answer these questions:

• What is soil?

• How do the features of soil affect the plants that grow in it?

• What is the effect of climate on soil?

National Science Education Standards
ES 1c, 1e, 1g, 1k

Where Does Soil Come From?

What do you think of when you think of soil? Most people think of dirt. However, soil is more than just dirt. **Soil** is a loose mixture of small mineral pieces, organic material, water, and air. All of these things help to make soil a good place for plants to grow.

Soil is made from weathered rocks. The rock that breaks down and forms a soil is called the soil's **parent rock**. Different parent rocks are made of different chemicals. Therefore, the soils that form from these rocks are also made of different chemicals. ☑

Bedrock is the layer of rock beneath soil. Because the material in soil is easily moved, the bedrock may not be the same as the soil's parent rock. Soil that has been moved away from its parent rock is called *transported soil*.

In some cases, the bedrock is the same as the parent rock. In these cases, the soil remains in place above its parent rock. Soil that remains above its parent rock is called *residual soil*.

STUDY TIP

Summarize in Pairs Read this section quietly to yourself. Then, talk about the material with a partner. Together, try to figure out the parts that you didn't understand.

READING CHECK

1. Explain Why are different soils made of different chemicals?

_____

_____

The soil is weathered from bedrock.	The soil is carried in from another place.
The bedrock is the same as the parent rock.	The bedrock is not the same as the parent rock.

_____ _____

TAKE A LOOK
2. Identify Fill in the blanks with the terms *residual soil* and *transported soil*.

SECTION 3 From Bedrock to Soil *continued*

What Are the Properties of Soil?

Some soils are great for growing plants. However, plants cannot grow in some other soils. Why is this? To better understand how plants can grow in soil, you must know about the properties of soil. These properties include soil texture, soil structure, and soil fertility.

SOIL TEXTURE

Soil is made of particles of different sizes. Some particles, such as sand, are fairly large. Other particles are so small that they cannot be seen without a microscope. **Soil texture** describes the amounts of soil particles of different sizes that a soil contains. ☑

Soil texture affects the consistency of soil and how easily water can move into the soil. *Soil consistency* describes how easily a soil can be broken up for farming. For example, soil that contains a lot of clay can be hard, which makes breaking up the soil difficult. Most plants grow best in soils that can be broken up easily.

READING CHECK

3. Define What is soil texture?

_____

_____

_____

Math Focus

4. Calculate How many times larger is the biggest silt particle than the biggest clay particle?

_____

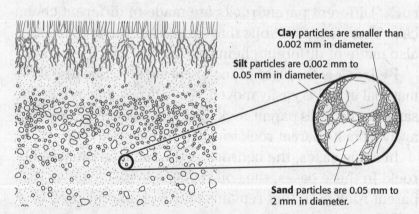

Clay particles are smaller than 0.002 mm in diameter.

Silt particles are 0.002 mm to 0.05 mm in diameter.

Sand particles are 0.05 mm to 2 mm in diameter.

Soil contains particles of many different sizes. However, all of the particles are smaller than 2 mm in diameter.

SOIL STRUCTURE

The particles in soil are not always evenly mixed. Sometimes, particles of a certain kind of material will form clumps in the soil. **Soil structure** describes the arrangement of particles in a soil.

SECTION 3 From Bedrock to Soil *continued*

SOIL FERTILITY

Plants need to get nutrients from soil in order to grow. Some soils are rich in nutrients. Other soils may have few nutrients or may be unable to give the nutrients to plants. The ability of soil to hold nutrients and to supply nutrients to plants is called *soil fertility*.

Some of the nutrients in a soil come from its parent rock. Other nutrients come from **humus**. Humus is the organic material that forms in soil from the remains of plants and animals. These remains are broken down into nutrients by decomposers, such as bacteria and fungi. It is humus that gives dark-colored soils their color. ☑

What Are the Different Layers in Soil?

Most soil forms in layers. The layers are horizontal, so soil scientists call them *horizons*.

✔ **READING CHECK**

5. Define What is humus?

_____

_____

_____

Horizon name	Description
O	The O horizon is made of decaying material from dead organisms. It is found in some areas, such as forests, but not in others.
A	The A horizon is made of topsoil. Topsoil contains more humus than any other soil horizon does.
E	The E horizon is a layer of sediment with very few nutrients in it. The nutrients in the E horizon have been removed by water.
B	The B horizon is very rich in nutrients. The nutrients that were washed out of other horizons collect in the B horizon.
C	The C horizon is made of partly weathered bedrock or of sediments from other locations.
R	The R horizon is made of bedrock that has not been weathered very much.

Water dissolves and removes nutrients as it passes through the soil. This is called **leaching**.

STANDARDS CHECK

ES 1e Soil consists of weathered rocks and decomposed organic material from dead plants, animals, and bacteria. Soils are often found in layers, with each having a different chemical composition and texture.

Word Help: layer
a separate or distinct portion of matter that has thickness

6. Identify Which three soil horizons probably contain the most nutrients?

_____

_____

_____

SECTION 3 From Bedrock to Soil *continued*

Why Is the pH of a Soil Important?

The *pH scale* is used to measure how acidic or basic something is. The scale ranges from 0 to 14. A pH of 7 is a *neutral* pH. Soil that has a pH below 7 is *acidic*. Soil that has a pH above 7 is *basic*.

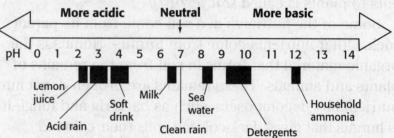

TAKE A LOOK

7. Identify Which is more acidic, lemon juice or a soft drink?

_____

_____

The pH of a soil affects how nutrients dissolve in the soil. Many plants are unable to get certain nutrients from soils that are very acidic or basic. The pH of a soil therefore has a strong effect on soil fertility. Most plants grow best in soil with a pH of 5.5 to 7.0. A few plants grow best in soils with higher or lower pH.

Soil pH is determined partly by the soil's parent rock. Soil pH is also affected by the acidity of rainwater, the use of fertilizers, and the amount of chemical weathering. ☑

 **READING CHECK**

8. List What are three things that affect soil pH?

_____

_____

_____

How Does Climate Affect Soil?

Soil types vary from place to place. The kinds of soils that develop in an area depend on its climate. The different features of these soils affect the number and kinds of organisms that can survive in different areas.

TROPICAL CLIMATES

Tropical rain forests receive a lot of direct sunlight and rain. Because of these factors, plants grow year-round. The heat and moisture also cause dead organisms to decay easily. This decay produces a lot of rich humus in the soil.

Even though a lot of humus can be produced in tropical rain forests, their soils are often poor in nutrients. One reason for this is that tropical rain forests have heavy rains. The heavy rains in this climate zone can leach nutrients from the topsoil. The rainwater carries the nutrients deep into the soil, where the plants can't reach them. In addition, the many plants that grow in tropical climates can use up the nutrients in the soil. ☑

READING CHECK

9. Explain Why are many tropical soils poor in nutrients?

_____

_____

_____

DESERTS AND ARCTIC CLIMATES

Deserts and arctic climates receive little rainfall. Therefore, the nutrients in the soil are not leached by rainwater. However, the small amount of rain in these climates makes weathering happen more slowly. As a result, soil forms slowly.

Few plants and animals live in deserts and arctic climates. Therefore, most soils there contain very little humus.

Sometimes, desert soils can become harmful, even to desert plants. Groundwater can seep into the desert soil. The groundwater often contains salt. When the water evaporates, the salt is left in the soil. The salt can build up in the soil and harm plants.

TEMPERATE FORESTS AND GRASSLANDS

Most of the continental United States has a temperate climate. Because the temperature changes often, mechanical weathering happens quickly in temperate climates. Thick layers of soil can build up.

Temperate areas get a medium amount of rain. The rain is enough to weather rock quickly, but it is not enough to leach many nutrients from the soil.

Many different kinds of plants can grow in temperate soils. Therefore, they contain a lot of humus. The large amount of humus makes the soils very rich in nutrients. The most fertile soils in the world are found in temperate climates. For example, the Midwestern part of the United States is often called the United States' "breadbasket" because of the many crops that grow there.

Critical Thinking

10. Apply Concepts As in deserts, groundwater in arctic climates can contain salt. Salt does not build up in arctic soils as quickly as in desert soils. What do you think is the reason for this?

_____

_____

_____

Type of climate	Description of climate	Features of the soil in this climate
Tropical climates	warm temperatures a lot of rain many living things	
Deserts and arctic climates		has little humus poor in nutrients
	medium amount of rain temperature changes often	

TAKE A LOOK

11. Describe Fill in the chart to show the features of soils in different climates.

Section 3 Review

SECTION VOCABULARY

bedrock the layer of rock beneath soil **humus** dark, organic material formed in soil from the decayed remains of plants and animals **leaching** the removal of substances that can be dissolved from rock, ore, or layers of soil due to the passing of water **parent rock** a rock formation that is the source of soil	**soil** a loose mixture of rock fragments, organic material, water, and air that can support the growth of vegetation **soil structure** the arrangement of soil particles **soil texture** the soil quality that is based on the proportions of soil particles

1. Summarize What are three properties of soil?

_____

_____

_____

2. Compare What climate feature do arctic climates and desert climates share that makes their soils similar?

_____

_____

_____

3. Analyze How can flowing water affect the fertility of soils?

_____

_____

_____

4. Identify How does soil pH affect plant growth?

_____

_____

_____

5. Explain What determines a soil's texture?

_____

_____

_____

6. Identify Name three things that are found in soils.

_____

_____

_____

CHAPTER 10 Weathering and Soil Formation

SECTION
4 Soil Conservation

BEFORE YOU READ

After you read this section, you should be able to answer these questions:

• Why is soil important?

• How can farmers conserve soil?

Why Is Soil Important?

You have probably heard about endangered plants and animals. Did you know that soil can be endangered, too? Soil can take many years to form. It is not easy to replace. Therefore, soil is considered a nonrenewable resource.

Soil is important for many reasons. Soil provides nutrients for plants. If the soil loses its nutrients, plants will not be able to grow. Soil also helps to support plant roots so the plants can grow well.

Animals get their energy from plants. The animals get energy either by eating plants or by eating animals that have eaten plants. If plants are unhealthy because the soil has few nutrients, then animals will be unhealthy, too.

Soil also provides a home, or *habitat*, for many living things. Bacteria, insects, mushrooms, and many other organisms live in soil. If the soil disappears, so does the habitat for these living things. ☑

Soil is also very important for storing water. It holds water that plants can use. Soil also helps to prevent floods. When rain falls, the soil can soak it up. The water is less likely to cause floods.

What does soil provide?	Why is it important?
Nutrients	
Habitat	
Water storage	

If we do not take care of soils, they could become unusable. In order to keep our soils usable, we need to conserve them. **Soil conservation** means protecting soils from erosion and nutrient loss. Soil conservation can help to keep soils fertile and healthy.

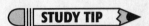

STUDY TIP

Compare Create a chart that shows the similarities and differences in the ways that farmers can help conserve soil.

READING CHECK

1. Explain Why is soil important for animals?

_____

_____

_____

TAKE A LOOK
2. Identify In the table, fill in the reasons that nutrients, habitat, and water storage are important.

SECTION 4 Soil Conservation *continued*

How Can Soil Be Lost?

Soil loss is a major problem around the world. One cause of soil loss is soil damage. Soil can be damaged if it is overused. Overused soil can lose its nutrients and become infertile. Plants can't grow in infertile soil.

Plants help to hold water in the soil. If plants can't grow somewhere because the soil is infertile, the area can become a desert. This process is known as *desertification*.

EROSION

Another cause of soil loss is erosion. **Erosion** happens when wind, water, or gravity transports soil and sediment from one place to another. If soil is not protected, it can be exposed to erosion.

Plant roots help to keep soil in place. They prevent water and wind from carrying the soil away. If there are no plants, soil can be carried away through erosion. ☑

How Can Farmers Help to Conserve Soil?

Farming can cause soil damage. However, farmers can prevent soil damage if they use certain methods when they plow, plant, and harvest their fields.

CONTOUR PLOWING

Water that runs straight down a hill can carry away a lot of soil. Farmers can plow their fields in special ways to help slow the water down. When the water moves more slowly down a hill, it carries away less soil. *Contour plowing* means plowing a field in rows that run across the slope of a hill.

☑ **READING CHECK**

3. Describe How do plant roots prevent erosion?

_____
_____
_____
_____

TAKE A LOOK

4. Identify Fill in the blank line in the figure to explain how contour plowing reduces erosion.

Contour plowing helps water to run more slowly down hills. This reduces erosion because _____

SECTION 4 Soil Conservation *continued*

TERRACES

On very steep hills, farmers can use terraces to prevent soil erosion. *Terraces* change one very steep field into many smaller, flatter fields.

Terraces keep water from running downhill very quickly.

NO-TILL FARMING

In *no-till farming*, farmers leave the stalks from old crops lying on the field while the newer crops grow. The old stalks protect the soil from rain and help reduce erosion.

The stalks left behind in no-till farming reduce erosion by protecting the soil from rain.

COVER CROPS

Cover crops are crops that are planted between harvests of a main crop. Cover crops can help to replace nutrients in the soil. They can also prevent erosion by providing cover from wind and rain.

CROP ROTATION

If the same crop is grown year after year in the same field, the soil can lose certain nutrients. To slow this process, a farmer can plant different crops in the field every year. Different crops use different nutrients from the soil. Some crops used in crop rotation can replace soil nutrients.

Critical Thinking

5. Infer What do you think is the reason farmers use terraces only on very steep hills?

_____

_____

_____

Critical Thinking

6. Apply Concepts How can crop rotation affect the number of plants that soil can support?

_____

_____

_____

_____

_____

_____

Section 4 Review

SECTION VOCABULARY

erosion the process by which wind, water, ice, or gravity transports soil and sediment from one location to another	**soil conservation** a method to maintain the fertility of soil by protecting the soil from erosion and nutrient loss

1. Define Write your own definition for soil conservation.

_____

_____

2. Identify Give three things that soil provides to living things.

_____

_____

_____

3. Compare How is weathering different from erosion?

_____

_____

_____

4. Identify What are two causes of soil loss?

_____

_____

5. List What are five ways that farmers can help to conserve soil?

_____

_____

_____

_____

6. Explain How does no-till farming help to reduce erosion?

_____

_____

_____

CHAPTER 11 The Flow of Fresh Water

SECTION 1

The Active River

National Science Education Standards
ES 1c, 1f

BEFORE YOU READ

After you read this section, you should be able to answer these questions:

- How does moving water change the surface of Earth?
- What is the water cycle?
- What factors affect the rate of stream erosion?

What Is Erosion?

Six million years ago, the Colorado River began carving through rock to form the Grand Canyon. Today, the river has carved through 1.6 km (about 1 mi) of rock!

Before the Grand Canyon was formed, the land was flat. Then the rock in the area began to lift upward because of plate tectonics. As Earth's crust lifted upward, water began to run downhill. The moving water cut into the rock and started forming the Grand Canyon.

Over millions of years, water cut into rock through the process of erosion. During **erosion**, wind, water, ice, and gravity move soil and rock from one place to another. Water is the main force in forming the Grand Canyon and in changing the Earth's landscape. ☑

STUDY TIP

Describe As you read, make a list of the different ways in which water can change the landscape of Earth.

READING CHECK

1. Identify What formed the Grand Canyon?

_____

_____

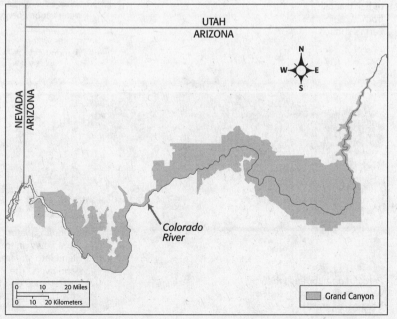

UTAH
ARIZONA

N
W—E
S

NEVADA
ARIZONA

Colorado River

0   10   20 Miles
0   10   20 Kilometers

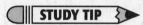 Grand Canyon

Six million years ago, the Colorado River started flowing through northern Arizona. Today, it has carved the Grand Canyon, which is about 1.6 km deep and 446 km long.

Math Focus

2. Calculate How long is the Grand Canyon in miles? Show your work.

1 km = 0.62 mi

_____

_____

How Does the Water Cycle Work?

Have you ever wondered where the water in rivers comes from? It is part of the water cycle. The **water cycle** is the nonstop movement of water between the air, the land, and the oceans. The major source of energy that drives the water cycle is the sun.

In the water cycle, water comes to Earth's surface from the clouds as rain, snow, sleet, or hail. The water moves downward through the soil or flows over the land. Water that flows over the land collects in streams and rivers and flows to the oceans.

Energy from the sun changes the water on Earth's surface into a gas that rises up to form clouds. The gas is called *water vapor*. The water vapor in clouds moves through the atmosphere until it falls to Earth's surface again.

STANDARDS CHECK

ES 1f Water, which covers the majority of the earth's surface, circulates through the crust, oceans, and atmosphere in what is known as the "water cycle." Water evaporates from the earth's surface, rises and cools as it moves to higher elevations, condenses as rain or snow, and falls to the surface where it collects in lakes, oceans, soil, and in rocks underground.

3. Define What is the water cycle?

_____

_____

_____

_____

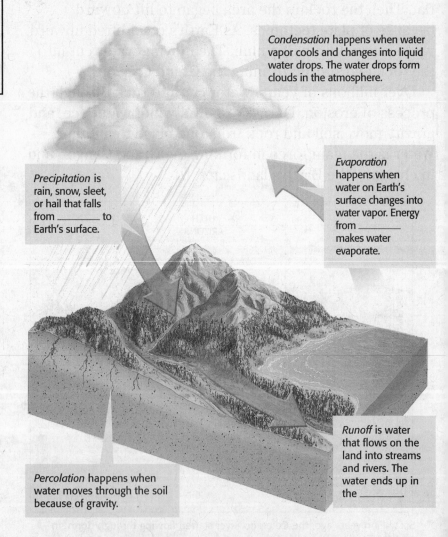

Condensation happens when water vapor cools and changes into liquid water drops. The water drops form clouds in the atmosphere.

Precipitation is rain, snow, sleet, or hail that falls from _____ to Earth's surface.

Evaporation happens when water on Earth's surface changes into water vapor. Energy from _____ makes water evaporate.

Runoff is water that flows on the land into streams and rivers. The water ends up in the _____.

Percolation happens when water moves through the soil because of gravity.

TAKE A LOOK
4. Identify In the figure, fill in the blank lines with the correct words.

What Is a River System?

What happens when you turn on the shower in your family's bathroom? When water hits the shower floor, the individual drops of water join together to form small streams. The small streams join together to form larger streams. The larger streams carry the water down the drain.

The water in your shower is like the water in a river system. A *river* is a stream that has many tributaries. A *river system* is a group of streams and rivers that drain an area of land. A **tributary** is a stream that flows into a lake or a larger stream. ☑

How Do River Systems Work?

River systems are divided into areas called watersheds. A **watershed** is the land that is drained by a river system. Many tributaries join together to form the rivers in a watershed.

The largest watershed in the United States is the Mississippi River watershed. It covers over one-third of the United States. It has hundreds of tributaries. The Mississippi River watershed drains into the Gulf of Mexico.

Watersheds are separated from each other by an area of higher ground called a **divide**. All of the rivers on one side of a divide flow away from it in one direction. All of the rivers on the other side of the divide flow away from it in the opposite direction. The Continental Divide separates the Mississippi River watershed from the watersheds in the western United States.

READING CHECK

5. Define What is a river system?

_____

_____

_____

TAKE A LOOK

6. List Name three rivers that are tributaries to the Mississippi River.

_____

_____

_____

What Is a Stream Channel?

A stream forms as water wears away soil and rock to make a channel. A **channel** is the path that a stream follows. As more soil and rock are washed away, the channel gets wider and deeper.

Over time, tributaries flow into the main channel of a river. The main channel has more water in it than the tributaries. The larger amount of water makes the main channel longer and wider. ☑

What Causes Stream Erosion?

Gradient is a measure of the change in the height of a stream over a certain distance. Gradient can be used to measure how steep a stream is. The left-hand picture below shows a stream with a high gradient. The water in this stream is moving very fast. The fast-moving water easily washes away rock and soil.

A river that is flat, as in the right-hand picture, has a low gradient and flows slowly. The slow water washes away less rock and soil.

7. Explain Why is the main channel of a river longer and wider than the channels of its tributaries?

_____

_____

_____

TAKE A LOOK

8. Explain How does the gradient of a stream affect how much erosion it causes?

_____

_____

_____

_____

This stream has a large gradient. It flows very fast.

This stream has a small gradient. It flows very slowly.

The amount of water that flows in a stream during a certain amount of time is called the *discharge*. The discharge of a stream can change. A large rainfall or a lot of melted snow can increase the stream's discharge.

When the discharge of a stream gets bigger, the stream can carry more sediment. The larger amount of water will flow fast and erode more land.

SECTION 1 The Active River *continued*

How Does a Stream Carry Sediment?

A stream's **load** is the material carried in the stream's water. A fast-moving stream can carry large rocks. The large rocks can cause rapid erosion by knocking away more rock and soil.

A slow-moving stream carries smaller rocks in its load. The smaller particles erode less rock and soil. The stream also carries material that is dissolved in the water.

Large rocks that bounce along the bottom of the stream are called *bed load*.

Materials that are floating in the water are called *suspended load*. They often make the stream look muddy or cloudy.

Tiny particles that are dissolved in the water are called *dissolved load*.

9. Identify How does a stream carry material from one place to another? Give three ways.

_____

_____

_____

How Do Scientists Describe Rivers?

All rivers have different features. These features can change with time. Many factors, such as weather, surroundings, gradient, and load, control the changes in a river. Scientists use special terms to describe rivers with certain features.

SECTION 1 The Active River *continued*

YOUTHFUL RIVERS

Youthful rivers are fast-flowing rivers with high gradients. Many of them flow over rapids and waterfalls. Youthful rivers make narrow, deep channels for the water to flow in. The picture below shows a youthful river.

TAKE A LOOK

10. Describe What features of this river tell you that it is a youthful river?

_____

_____

_____

MATURE RIVERS

Mature rivers erode rock and soil to make wide channels. Many tributaries flow into a mature river, so mature rivers carry large amounts of water. The picture below shows a mature river bending and curving through the land. The curves and bends are called *meanders*. ☑

✔ **READING CHECK**

11. Explain Why do mature rivers carry a lot of water?

_____

_____

TAKE A LOOK

12. Identify Label the meanders on this picture of a mature river.

SECTION 1 The Active River *continued*

REJUVENATED RIVERS

Rejuvenated rivers form where land has been raised up by plate tectonics. This gives a river a steep gradient. Therefore, rejuvenated rivers flow fast and have deep channels. As shown in the picture below, steplike gradients called *terraces* may form along the sides of rejuvenated rivers.

TAKE A LOOK
13. Identify Label the terraces on this picture of a rejuvenated river.

OLD RIVERS

Old rivers have very low gradients. Instead of widening and deepening its channel, an old river deposits soil and rock along its channel. Since very few tributaries flow into an old river, the river does not quickly erode land. Old rivers have wide, flat floodplains and many meanders. In the picture below, a bend in an old river's channel has eroded into a lake. This is called an *oxbow lake*.

Critical Thinking

14. Infer Very few tributaries flow into an old river. Do you think it will have a large or a small discharge?

_____

_____

Interactive Textbook   **193**   The Flow of Fresh Water

Name _____ Class _____ Date _____

Section 1 Review

NSES ES 1c, 1f

SECTION VOCABULARY

channel the path that a stream follows	**load** the materials carried by a stream
divide the boundary between drainage areas that have streams that flow in opposite directions	**tributary** a stream that flows into a lake or into a larger stream
erosion the process by which wind, water, ice, or gravity transports soil and sediment from one location to another	**water cycle** the continuous movement of water between the atmosphere, the land, and the oceans
	watershed the area of land that is drained by a river system

1. Explain Why do most rivers have wider channels than most streams?

_____

_____

2. Show a Sequence Fill in the Process Chart to show what happens in the water cycle.

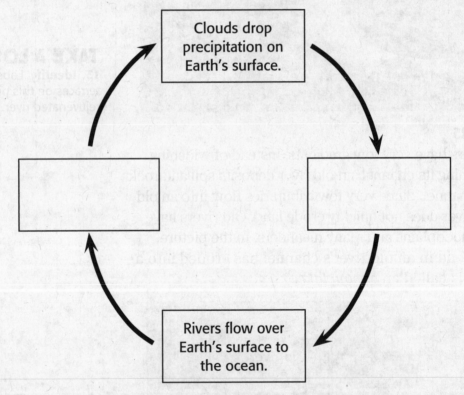

Clouds drop precipitation on Earth's surface.

Rivers flow over Earth's surface to the ocean.

3. Identify What is the main source of energy for the water cycle?

_____

4. Describe How do rivers change Earth's surface?

_____

_____

_____

CHAPTER 11 The Flow of Fresh Water
SECTION
2 **Stream and River Deposits**

National Science
Education Standards
ES 1c

BEFORE YOU READ

After you read this section, you should be able to answer these questions:

• What types of deposits are caused by streams?

• Why do floods happen?

• How do floods affect humans?

How Do Rivers and Streams Rebuild Land?

Earlier, you learned that rivers and streams erode rock and soil. However, rivers also carry this loose rock and soil to new places. Rivers can help to form new land through deposition. **Deposition** is the process in which material is laid down, or dropped.

The rock and soil that a river erodes from the land move downstream in the river's load. When the flow of water slows down, the river deposits, or drops, some of its load. Material, such as rock and soil, that is deposited by rivers and streams is called *sediment*. ☑

In a river, erosion happens on the outside of a bend, where the water flows quickly. Deposition happens on the inside of a bend, where the water flows more slowly.

Erosion happens along the outside of a bend, where water is flowing _____.

Deposition happens along the inside of a bend, where water is flowing _____.

STUDY TIP

Compare In your notebook, create a table to compare the different landforms created by stream deposits.

READING CHECK

1. **Define** What is sediment?

_____

_____

_____

TAKE A LOOK

2. **Identify** Fill in the blanks in the figure with the correct words.

PLACER DEPOSITS

Heavy minerals, such as gold, can be carried by very fast-moving rivers. In places where the rivers slow down, the heavy minerals may be deposited. This kind of sediment is called a *placer deposit*.

What Are Deltas and Alluvial Fans?

Rivers deposit their loads of rocks and soil when their flow of water slows down. When a river enters the ocean, it flows much more slowly. Therefore, it deposits its load into the ocean.

When the river enters the ocean, it deposits its load under the water in a fan-shaped pattern called a **delta**. The river deposits can build up above the water's surface to form new land and build new coastline.

Rivers and streams can also deposit their loads on dry land. When a fast-moving stream flows from a mountain onto flat land, the stream slows down quickly. As it slows down, the stream deposits its rocks and soil in a fan-shaped pattern known as an **alluvial fan**.

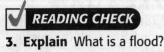
Say It

Investigate Learn about a place in the world where people live on a delta. Give a short talk to your class about the place you studied.

Ocean

Delta

River

Mountains

River

Alluvial fan

Why Do Rivers and Streams Flood?

Rivers and streams are always changing. They may have different amounts of water in them in different seasons. If there is a lot of rain or melting snow, a stream will have a lot of water in it.

Floods are natural events that can happen with the change of seasons. A *flood* happens when there is too much water for the channel to hold. The stream flows over the sides of its channel. ☑

During a flood, the land along the sides of the stream is covered in water. The stream drops its sediment on this land. This area is called a **floodplain**. Floodplains are good areas for farming because flooding brings new soil to the land.

✓ READING CHECK

3. Explain What is a flood?

_____

_____

_____

_____

_____

What Are the Effects of Floods?

Floods are powerful and can cause a lot of damage. They can ruin homes and buildings. They can also wash away land where animals and people live.

In 1993, when the Mississippi River flooded, farms and towns in nine states were damaged. Floods can cover roads and carry cars and people downstream. They can also drown people and animals. ☑

Floods sometimes happen very fast. After a very bad rainstorm, water can rush over the land and cause a *flash flood*. Flash floods can be hard to predict and are dangerous.

Floodwater can flow strongly enough to move cars. These people were trapped in their car in a flash flood.

✓ READING CHECK

4. List List three ways that floods can be harmful to people and animals.

_____

_____

_____

TAKE A LOOK

5. Explain Why can a flash flood be dangerous to people driving in cars?

_____

_____

_____

_____

Humans try to control flooding by building barriers around streams. One kind of barrier is called a dam. A *dam* is a barrier that can guide floodwater to a reservoir, such as a lake or pond. A dam can prevent flooding in one area and create an artificial lake in another area. The water in the artificial lake can be used for farming, drinking, or producing electricity.

Another barrier that people build is called a levee. A *levee* is a wall of sediment on the side of a river. The barrier helps keep the river from flooding the nearby land. Many levees form naturally from sediment that is deposited by the river. People can use sandbags to create artificial levees. ☑

✓ READING CHECK

6. Identify What do people do to try to protect themselves from a flood?

_____

_____

_____

_____

Section 2 Review

NSES ES 1c

SECTION VOCABULARY

alluvial fan a fan-shaped mass of material deposited by a stream when the slope of the land decreases sharply	**deposition** the process in which material is laid down
delta a fan-shaped mass of material deposited at the mouth of a stream	**floodplain** an area along a river that forms from sediments deposited when the river overflows its banks

1. Explain Why do floods happen?

_____

_____

_____

2. Compare Complete the table to describe the features of different kinds of stream deposits.

Type of Deposit	How is it formed?	Where is it formed?
Alluvial fan		
Delta		
Floodplain		

3. Explain How can a flood be both helpful and harmful to people?

_____

_____

_____

_____

4. List Give two kinds of barriers that people use to control floodwater.

_____

_____

_____

CHAPTER 11 The Flow of Fresh Water

SECTION
3 **Water Underground**

National Science
Education Standards
ES 1c

BEFORE YOU READ

After you read this section, you should be able to answer
these questions:

• What is a water table?

• What is an aquifer?

• What is the difference between a spring and a well?

Where Is Fresh Water Found?

Some of the Earth's fresh water is found in streams
and lakes. However, a large amount of water is also
found underground. Rainwater and water from streams
move through the soil and into the spaces between rocks
underground. This underground area is divided into two
zones. The *zone of aeration* is the area that rainwater
passes through. The spaces between particles in the zone
of aeration contain both water and air.

The *zone of saturation* is the area where water collects. The spaces between particles in the zone of saturation are filled with only water. The water found inside
underground rocks is called *groundwater*. ☑

The zone of aeration and the zone of saturation meet
at a boundary called the **water table**. The depth of the
water table is not the same all the time or in all places.
The water table can move closer to the surface during
wet seasons and farther from the surface during dry seasons. In wet regions, the water table may be just below
the surface. In dry regions, the water table may be hundreds of meters below the surface.

STUDY TIP
Summarize As you read,
underline the important ideas
in this section. When you are
finished reading, write a
one- or two-paragraph
summary of the section, using
the underlined ideas.

READING CHECK
1. Define What is
groundwater?

_____

_____

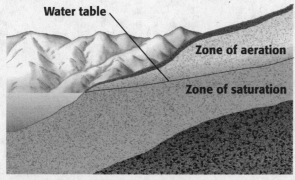

Water table

Zone of aeration

Zone of saturation

The water table is the boundary between the zone of
aeration and the zone of saturation.

TAKE A LOOK
2. Infer If a region receives
a lot of rainfall, will the water
table in the region probably
rise or fall?

_____

_____

SECTION 3 Water Underground *continued*

How Can Water Pass Through Rock?

A layer of rock or sediment that stores groundwater is called an **aquifer**. Most aquifers are made of sedimentary rock. There can be many *pores*, or open spaces, between the particles in an aquifer. The more open spaces there are, the more water the aquifer can hold. The fraction of a rock's volume that is taken up by pores is called the rock's **porosity**. ☑

Imagine filling a jar with large pebbles. The pebbles cannot fill all of the space in the jar, so there will be many open spaces. In other words, the jar has a high porosity. Now, imagine pouring sand over the pebbles in the jar. The sand can fill the spaces between the pebbles, leaving little open space in the jar. The jar has a low porosity.

Like the jar, the sizes of the particles in a rock affect the rock's porosity. Rocks made of the same-sized particles tend to have high porosity, like the jar of pebbles. Rocks made of different-sized particles tend to have low porosity, like the jar with sand and pebbles.

If the open spaces in the rock layer are connected, water can move through the rock. A rock's ability to let water pass through is called **permeability**. Rock that water can not flow through is called *impermeable*.

The size of rock particles also affects permeability. Rock made of large particles tends to have a high permeability. This is because the large particles produce less friction on the water moving through them. *Friction* is a force that slows down moving objects. Rock particles produce friction on water when the water touches the rock particles.

3. Define What is porosity?

_____

_____

_____

Critical Thinking

4. Apply Concepts Shale has a very high porosity. However, shale does not form many aquifers, because water cannot move through it easily. Explain why this might be the case.

(Hint: What is permeability?)

_____

_____

_____

_____

_____

_____

TAKE A LOOK
5. Explain Why do small particles produce more friction on the water than large particles?

_____

_____

The large particles touch the water in only a few places. They produce little friction on the water, so they have a high permeability.

The small particles touch the water in many places. They produce a lot of friction on the water, so they have a low permeability.

SECTION 3 Water Underground *continued*

RECHARGE ZONES

Like rivers, aquifers depend on precipitation to keep their water level constant. Precipitation that falls onto land can flow through the ground and into an aquifer. The ground surface where water enters an aquifer is called the aquifer's **recharge zone**. Recharge zones are found where the soil and rock above an aquifer are permeable. ☑

Some aquifers are small, but many cover large underground areas. Many cities and farms depend on aquifers for fresh water.

People's actions can affect the amount and quality of water in an aquifer. If people build roads or buildings in a recharge zone, less water can enter the aquifer. If people dump chemicals in a recharge zone, the chemicals can enter the aquifer and pollute the water in it.

What Is a Spring?

Like all water, groundwater tends to move downhill. Remember that the water table can be at different depths in different places. Groundwater tends to move to follow the slope of the water table. When the water table meets the Earth's surface, water flows out and forms a *spring*. Springs are important sources of drinking water. ☑

In some places, an aquifer is found between two layers of impermeable rock. This is called an *artesian formation*. The top layer of impermeable rock is called the *cap rock*. If the water in the aquifer flows through a crack in the cap rock, it forms an **artesian spring**.

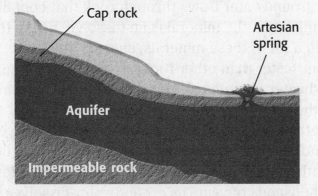

The water from most springs is cool. However, rock far below Earth's surface can be very hot. Therefore, water that flows through deep aquifers may be very hot. When this water reaches the surface, it can form a *hot spring*.

☑ **READING CHECK**

6. Identify Where does the water in aquifers come from?

_____

_____

☑ **READING CHECK**

7. Describe How does a spring form?

_____

_____

_____

TAKE A LOOK

8. Explain What causes an artesian spring to form?

_____

_____

_____

How Do People Get Water Out of the Ground?

When people need a supply of water, they often dig a well. A *well* is a hole dug by people that is deeper than the water table. If the well does not reach below the water table, the well will not produce water. In addition, too many wells may remove water from an aquifer more quickly than the aquifer can refill. Then, the water table can drop and all the wells can go dry.

9. Identify Two people drilled wells to try to get water out of the ground. The white bars in the figure show where the two people drilled their wells. Which of the wells will probably produce water? Explain your answer.

_____

_____

_____

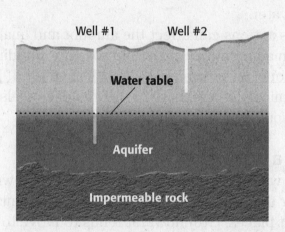

Well #1   Well #2

Water table

Aquifer

Impermeable rock

How Can Groundwater Cause Weathering?

Remember that streams can cause physical weathering when particles in the stream hit other particles and cause them to break. Groundwater can also cause weathering. However, instead of physical weathering, groundwater causes chemical weathering by dissolving rock.

Many minerals, such as calcite, can dissolve in water. When groundwater flows through rock that contains these minerals, the minerals can dissolve. Rocks that contain a lot of these minerals, such as limestone, can weather faster than other rocks. In addition, some groundwater contains weak acids. These acids can dissolve the rock more quickly than pure water can.

Say It

Investigate Learn more about an area that has large underground caves. What kind of rock did the caves form in? How did the caves form? Share your findings with a small group.

Weathering by groundwater can form large caves. In fact, most of the caves in the world have formed because of weathering by groundwater. The caves form slowly over thousands of years as groundwater dissolves limestone. The figure on the next page shows how large some of these caves can become.

SECTION 3 Water Underground *continued*

This cave in New Mexico formed when groundwater dissolved huge volumes of limestone.

What Structures Can Groundwater Form?

Many caves show signs of deposition as well as weathering. Water flowing through caves can have many minerals dissolved in it. If the water drips from a crack in the cave's ceiling, it can deposit these minerals. These deposits form icicle-shaped structures that hang from the ceilings. They are called *stalactites*.

Water that falls to the cave floor can build cone-shaped structures called *stalagmites*. Sometimes, a stalactite and a stalagmite join together to form a *dripstone column*.

Sometimes, the roof of a cave can collapse. This forms a circular *depression*, or pit, on the Earth's surface called a *sinkhole*. Sinkholes can damage buildings, roads, and other structures on the surface. Streams can "disappear" into sinkholes and flow through the cave underground. In areas where the water table is high, lakes can form inside sinkholes.

This sinkhole formed in Florida when a cave collapsed.

TAKE A LOOK
10. Identify Name one kind of rock that can be dissolved by groundwater.

_____

_____

Critical Thinking

11. Compare How is a stalactite different from a stalagmite?

_____

_____

_____

TAKE A LOOK
12. Identify Relationships How are caves and sinkholes related?

_____

_____

_____

Section 3 Review

SECTION VOCABULARY

aquifer a body of rock or sediment that stores groundwater and allows the flow of groundwater	**porosity** the percentage of the total volume of a rock or sediment that consists of open spaces
artesian spring a spring whose water flows from a crack in the cap rock over an aquifer	**recharge zone** an area in which water travels downward to become part of an aquifer
permeability the ability of a rock or sediment to let fluids pass through its open spaces, or pores	**water table** the upper surface of underground water; the upper boundary of the zone of saturation

1. Explain Why is it important to know where the water table is located?

_____

_____

_____

2. Describe How does particle size affect the porosity of an aquifer?

_____

_____

_____

3. Infer How could building new roads affect the recharge zone of an aquifer?

_____

_____

_____

4. Compare What is the difference between a spring and a well?

_____

_____

_____

5. List What are two features that are formed by underground weathering?

_____

_____

6. Describe How does a dripstone column form?

_____

_____

_____

CHAPTER 11 The Flow of Fresh Water

SECTION 4 Using Water Wisely

BEFORE YOU READ

After you read this section, you should be able to answer these questions:

• What are two forms of water pollution?

• How is wastewater cleaned?

• How is water conserved?

Is Water A Limited Resource?

Although the Earth is covered with oceans, lakes, and rivers, only 3% of the Earth's water is fresh water. Most of that 3% is frozen in the polar icecaps. Therefore, people must take care of and protect their water resources.

Cities, factories, and farms can pollute water. Water can become so polluted that it is no longer safe to use. Two types of water pollution are point-source pollution and nonpoint-source pollution.

Where Does Water Pollution Come From?

Pollution that comes from one specific site is called **point-source pollution**. For example, a leak from a sewer pipe is point-source pollution. Because point-source pollution comes from a single place, it is easier to control than nonpoint-source pollution. ☑

Pollution that comes from many sources is called **nonpoint-source pollution**. Most nonpoint-source pollution gets into water by runoff. *Runoff* is water that flows over the ground into rivers, streams, or oceans. As runoff flows over the ground, it can pick up chemicals and other pollutants. These pollutants are carried to clean bodies of water.

People use chemicals, such as fertilizers, on the land. Runoff can carry these chemicals to clean bodies of water.

STUDY TIP

Identify As you learn about different types of pollution, think about where you live. Make a list of some sources of pollution in your community.

READING CHECK

1. Define What is point-source pollution?

_____

_____

TAKE A LOOK

2. Explain How can chemicals that are spilled on land end up in oceans or other water bodies?

_____

_____

Why Does Water Have To Be Clean?

Water is important to many organisms. If the water is not clean, the organisms using it will not be healthy. Three important properties of water that affect water quality are dissolved oxygen, nitrates, and pH.

DISSOLVED OXYGEN

Fish and other organisms that live in water need oxygen to survive. The oxygen that is dissolved in water is called *dissolved oxygen*, or *DO*. If the DO in water is too low, many organisms can become sick or die. ☑

Pollution can cause the DO level in water to decrease. An increase in water temperature can also cause the DO level to decrease. Many energy facilities, such as nuclear power plants, release hot water into the environment. This can increase the temperature of water in natural water bodies. This *thermal pollution* can decrease DO levels.

NITRATES

Nitrates are naturally formed compounds of nitrogen and oxygen. All water contains some nitrates. However, too much nitrate in the water can be harmful to organisms. An increase in nitrate levels can also cause the DO level to decrease. Some kinds of pollution, such as animal wastes, increase the level of nitrates in water.

Animal wastes contain a lot of nitrates. Runoff can carry these nitrates to water bodies, causing water pollution.

pH

The *pH* of water is a measure of how acidic the water is. Most organisms cannot live in very acidic water. Acid rain and some kinds of wastes can make water bodies more acidic. Water with a high *alkalinity*, or ability to react with acids, can protect organisms from acid rain and other pollution.

✓ READING CHECK

3. Identify What is dissolved oxygen?

_____

_____

TAKE A LOOK

4. Identify What is one source of nitrates in water?

_____

_____

SECTION 4 Using Water Wisely *continued*

How Can Dirty Water Be Cleaned?

What happens to water that you flush down the toilet or wash down the drain? If you live in a city or a large town, the water probably flows through sewer pipes to a sewage treatment plant. **Sewage treatment plants** are facilities that clean waste materials out of water. After water has passed through a sewage treatment plant, it can safely be released into the environment. ☑

A sewage treatment plant cleans water in two ways. The first steps are called the *primary treatment*. First, the dirty water is passed through a large screen. This screen catches solid objects, such as paper, rags, and bottle caps.

The water is then placed in a large tank. As the water sits, small pieces of material sink to the bottom of the tank. These small pieces, such as food or soil, are filtered out. Any material that floats on the surface is also removed.

After going through primary treatment, the water is ready for *secondary treatment*. During secondary treatment, the water is placed in an *aeration tank*. There, the water mixes with oxygen and bacteria. The bacteria use the oxygen to consume the wastes dissolved in the water.

The water is then placed in a settling tank. Any dirt in the water sinks to the bottom of the tank and is removed. The water is then mixed with chlorine to disinfect it. Finally, it is sent to a river, lake, or ocean.

☑ **READING CHECK**

5. Identify What do sewage treatment plants do?

_____

_____

Critical Thinking

6. Infer Why is secondary treatment necessary?

_____

_____

_____

_____

_____

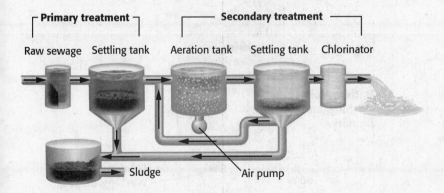

Primary treatment — Secondary treatment

Raw sewage   Settling tank   Aeration tank   Settling tank   Chlorinator

Sludge   Air pump

TAKE A LOOK

7. Identify What is the purpose of the chlorinator?

_____

_____

SECTION 4 Using Water Wisely *continued*

SEPTIC TANKS

If you live in an area without a sewage treatment plant, your waste probably goes into a septic tank. A **septic tank** is a large underground tank that cleans the wastewater from one household. The wastewater flows into the tank, where the solids sink to the bottom. Bacteria break down these solids. The water then flows into pipes buried underground. The pipes take the water to nearby ground called a *drain field*.

TAKE A LOOK
8. Identify Why do people use septic tanks?

_____

_____

_____

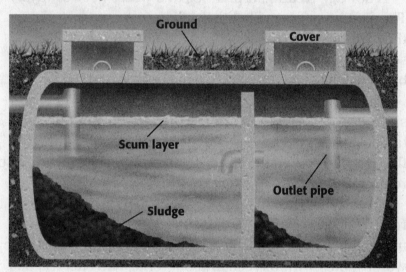

How Do People Use Water?

The average household in the United States uses 100 gallons of water each day. The graph below shows how an average household uses water.

Math Focus
9. Read a Graph What does an average United States household use most of its water for?

_____

_____

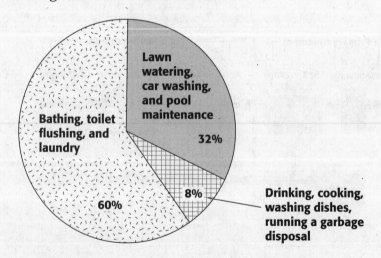

The average U.S. household uses 100 gallons of water per day.

SECTION 4 Using Water Wisely *continued*

INDUSTRY AND AGRICULTURE

About 19% of the water used in the world is used by industries. Water is used in manufacturing, mining minerals, and electricity generation. Most industries recycle and reuse water. Recycling helps keep more water in the environment.

Many farmers get their water from aquifers. When farmers use too much water from an aquifer, less water may be available for farming. For example, the Ogallala aquifer provides water for about one-fifth of the cropland in the United States. Recently, the water level in the aquifer has dropped a great deal. Scientists estimate that it would take at least 1,000 years for the water level to get back to normal if no more water is taken from the aquifer.

When farmers water their crops, a large amount of water is lost through evaporation and runoff. Drip irrigation, in which water is placed directly on the plant's roots, wastes less water.

How Can You Conserve Water at Home?

Individual families can conserve water also. Toilets and shower heads that use less water are good choices for conservation. If you have to water your lawn, water it at night and use a drip watering system.

Each person can do his or her part to conserve water. Simple choices, such as taking shorter showers and turning off the water when you brush your teeth, are helpful. If everyone tries to conserve water, we can make a big difference.

Math Focus

10. Calculate What fraction of the water used in the world is NOT used by industry?

_____

Say It

Investigate Find out more about ways that farmers can help conserve water. Share what you learn with a small group.

Things You Can Do to Conserve Water

• Use water-saving toilets and showerheads.
• Water the lawn at night or don't water it at all.
•
•
•
•

TAKE A LOOK

11. Brainstorm With a partner, come up with other ways that you can help conserve water. Write your ideas in the table.

The Flow of Fresh Water

Section 4 Review

SECTION VOCABULARY

nonpoint-source pollution pollution that comes from many sources rather than from a single, specific site	**septic tank** a tank that separates solid waste from liquids and that has bacteria that break down solid waste
point-source pollution pollution that comes from a specific site	**sewage treatment plant** a facility that cleans the waste materials found in water that comes from sewers or drains

1. Explain How can pollution affect the level of oxygen in water? Why is this important?

_____

_____

_____

2. Compare Complete the Venn Diagram to compare how a sewage treatment plant and a septic tank clean wastewater.

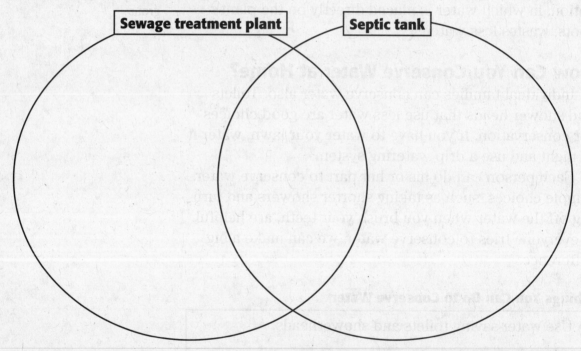

Sewage treatment plant

Septic tank

3. Apply Concepts Many farms are found along the banks of the Mississippi River. Describe what kind of pollution you might find in this river. Is the pollution point-source pollution or nonpoint-source pollution? Explain your answer.

_____

_____

_____

SECTION 1

Shoreline Erosion and Deposition

BEFORE YOU READ

After you read this section, you should be able to answer these questions:

- What is a shoreline?
- How do waves shape shorelines?

National Science Education Standards
ES 1c

How Do Waves Form?

Waves form when wind blows over the surface of the ocean. Strong winds produce large waves. The waves move toward land. When waves crash into the land over a long time, they can break rock down into smaller pieces. These pieces are called *sand*.

A **shoreline** is a place where the land and the water meet. Most shorelines contain sand. The motion of waves helps to shape shorelines. During *erosion*, waves remove sand from shorelines. During *deposition*, waves add sand to shorelines. ☑

WAVE TRAINS

Waves move in groups called *wave trains*. The waves in a wave train are separated by a period of time called the *wave period*. You can measure the wave period by counting the seconds between waves breaking on the shore. Most wave periods are 10 to 20 s long.

When a wave reaches shallow water, the bottom of the wave drags against the sea floor. As the water gets shallower, the wave gets taller. Soon, it can't support itself. The bottom slows down. The top of the wave begins to curl, fall over, and break. Breaking waves are called *surf*.

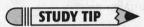
STUDY TIP

Summarize Read this section quietly to yourself. Talk about what you learned with a partner. Together, try to figure out the answers to any questions that you have.

☑ **READING CHECK**

1. Compare How is wave erosion different from wave deposition?

_____

_____

_____

_____

Waves travel in groups called wave trains. The time between one wave and the next is the wave period.

Math Focus

2. Calculate A certain wave train contains 6 waves. The time between the first wave and the last wave is 72 seconds. What is the wave period?

_____

POUNDING SURF

The energy in waves is constantly breaking rock into smaller and smaller pieces. Crashing waves can break solid rock and throw the pieces back toward the shore. Breaking waves can enter cracks in the rock and break off large boulders. Waves also pick up fine grains of sand. The loose sand wears down other rocks on the shore through abrasion. ☑

What Are the Effects of Wave Erosion?

Wave erosion can produce many features along a shoreline. For example, *sea cliffs* form when waves erode rock to form steep slopes. As waves strike the bottom of the cliffs, the waves wear away soil and rock and make the cliffs steeper.

How fast sea cliffs erode depends on how hard the rock is and how strong the waves are. Cliffs made of hard rock, such as granite, erode slowly. Cliffs made of soft rock, such as shale, erode more quickly.

During storms, large, high-energy waves can erode the shore very quickly. These waves can break off large chunks of rock. Many of the features of shorelines are shaped by storm waves. The figures below and on the next page show some features that form because of wave erosion.

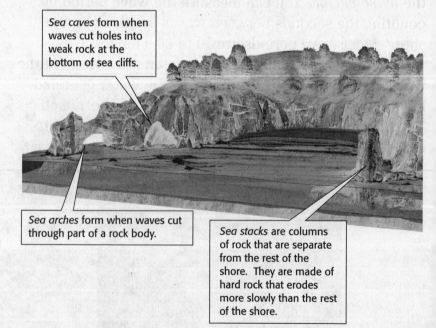

Sea caves form when waves cut holes into weak rock at the bottom of sea cliffs.

Sea arches form when waves cut through part of a rock body.

Sea stacks are columns of rock that are separate from the rest of the shore. They are made of hard rock that erodes more slowly than the rest of the shore.

READING CHECK

3. Identify Give two ways that waves can break rock into smaller pieces.

_____

_____

_____

_____

Critical Thinking

4. Identify Relationships When may a storm not produce high-energy waves?

_____

_____

_____

TAKE A LOOK

5. Compare How is a sea stack different from a sea arch?

_____

_____

_____

SECTION 1 Shoreline Erosion and Deposition *continued*

Headlands are finger-shaped bodies of rock that stick out into the sea. They are made of harder rock than the rest of the shore.

Wave-cut terraces form when sea cliffs are worn back from the shore. This produces a nearly flat platform beneath the water at the base of the cliff.

6. Define What is a headland?

_____

_____

_____

What Are the Effects of Wave Deposition?

Waves carry many materials, such as sand, shells, and small rocks. When the waves deposit these materials on the shoreline, a beach forms. A **beach** is any area of shoreline that is made of material deposited by waves. Some beach material is deposited by rivers and moves down the shoreline by the action of waves. ☑

Many people think that all beaches are made of sand. However, beaches may be made of many materials, not just sand. The size and shape of beach material depend on how far the material traveled before it was deposited. They also depend on how the material is eroded. For example, beaches in stormy areas may be made of large rocks because smaller particles are removed by the waves.

The color of a beach can vary, too. A beach's color depends on what particles make up the beach. Light-colored sand is the most common beach material. Most light-colored sand is made of the mineral quartz. Many Florida beaches are made of quartz sand. On many tropical beaches, the sand is white. It is made of finely ground white coral. ☑

Beaches can also be black or dark-colored. Black-sand beaches are found in Hawaii. Their sands are made of eroded lava from volcanoes. This lava is rich in dark-colored minerals, so the sand is also dark-colored. The figures on the next page show some examples of beaches.

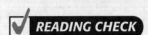 **READING CHECK**

7. Define Write your own definition for *beach*.

_____

_____

_____

_____

_____

✓ **READING CHECK**

8. Identify What mineral is most light-colored sand made of?

_____

_____

SECTION 1 Shoreline Erosion and Deposition *continued*

This beach in New England is made of large rocks. Smaller sand particles are washed away during storms.

This beach in Florida is made of light-colored quartz sand.

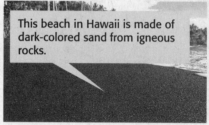

This beach in Hawaii is made of dark-colored sand from igneous rocks.

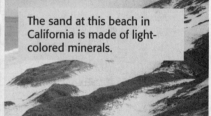

The sand at this beach in California is made of light-colored minerals.

TAKE A LOOK
9. Explain Why are some beaches made mostly of larger rock pieces, instead of sand?

_____

_____

WAVE ANGLE AND SAND MOVEMENT

Waves can move sand along a beach. The movement of the sand depends on the angle at which the waves hit the shore. *Longshore currents* form when waves hit the shore at an angle. The waves wash sand onto the shore at the same angle that the waves are moving. However, when the waves wash back into the ocean, they move sand directly down the slope of the beach. This causes the sand to move in a zigzag pattern, as shown in the figure below.

TAKE A LOOK
10. Infer Why don't longshore currents form in places where waves hit the shore head-on?

_____

_____

_____

_____

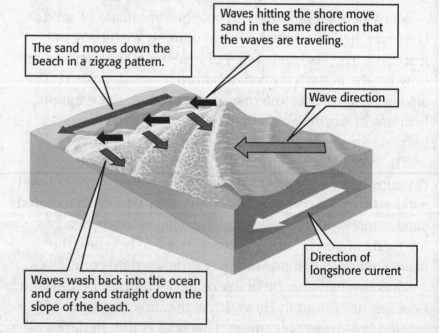

The sand moves down the beach in a zigzag pattern.

Waves hitting the shore move sand in the same direction that the waves are traveling.

Wave direction

Direction of longshore current

Waves wash back into the ocean and carry sand straight down the slope of the beach.

SECTION 1 Shoreline Erosion and Deposition *continued*

OFFSHORE DEPOSITS

Longshore currents can carry beach material offshore. This process can produce landforms in open water. These landforms include sandbars, barrier spits, and barrier islands.

A *sandbar* is a ridge of sand, gravel, or broken shells that is found in open water. Sandbars may be completely under water or they may stick up above the water. ☑

A *barrier spit* is a sandbar that sticks up above the water and is connected to the shoreline. Cape Cod, Massachusetts, is an example of a barrier spit. It is shown in the figure below.

READING CHECK

11. Define What is a sandbar?

_____

_____

_____

Cape Cod, Massachusetts, is an example of a barrier spit. Barrier spits form when sandbars are connected to the shoreline.

TAKE A LOOK

12. Identify What is a barrier spit?

_____

_____

A *barrier island* is a long, narrow island that forms parallel to the shoreline. Most barrier islands are made of sand.

Santa Rosa Island in Florida is an example of a barrier island.

TAKE A LOOK

13. Compare What is the difference between a barrier island and a barrier spit?

_____

_____

_____

Name ______________________ Class ______________ Date ______________

Section 1 Review

NSES ES 1c

SECTION VOCABULARY

beach an area of the shoreline that is made up of deposited sediment	**shoreline** the boundary between land and a body of water

1. Compare How is a shoreline different from a beach?

____

2. Explain Where does the energy to change the shoreline come from? Explain your answer.

____

____

____

____

3. Identify Give two examples of different-colored beach sand and explain why each kind is a certain color.

____

____

____

____

4. Explain How do longshore currents move sand?

____

____

____

5. List Give five landforms that are produced by wave erosion.

____

____

____

____

____

CHAPTER 12 Agents of Erosion and Deposition

SECTION
2
Wind Erosion and Deposition

National Science
Education Standards
ES 1c, 2a

BEFORE YOU READ

After you read this section, you should be able to answer these questions:

• How can wind erosion shape the landscape?
• How can wind deposition shape the landscape?

How Can Wind Erosion Affect Rocks?

Wind can move soil, sand, and small pieces of rock. Therefore, wind can cause erosion. However, some areas are more likely to have wind erosion than other areas. For example, plant roots help to hold soil and rock in place. Therefore, areas with few plants, such as deserts and coastlines, are more likely to be eroded by wind. These areas also may be made of small, loose rock particles. Wind can move these particles easily. ☑

Wind can shape rock pieces in three ways: saltation, deflation, and abrasion.

SALTATION

Wind moves large grains of soil, sand, and rock by saltation. **Saltation** happens when sand-sized particles skip and bounce along in the direction that the wind is moving. When moving sand grains hit one another, some of the grains bounce up into the air. These grains fall back to the ground and bump other grains. These other grains can then move forward.

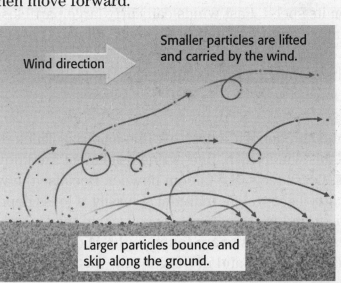

Wind direction

Smaller particles are lifted and carried by the wind.

Larger particles bounce and skip along the ground.

STUDY TIP

Learn New Words As you read this section, underline words you don't understand. When you figure out what they mean, write the words and their definitions in your notebook.

READING CHECK

1. Explain How do plant roots help to prevent wind erosion?

_____

_____

TAKE A LOOK

2. Apply Concepts Why can't the wind lift and carry large particles?

_____

_____

_____

_____

DEFLATION

Wind can blow tiny particles away from larger rock pieces during deflation. **Deflation** happens when wind removes the top layers of fine sediment or soil and leaves behind larger rock pieces. ☑

Deflation can form certain land features. It can produce *desert pavement*, which is a surface made of pebbles and small, broken rocks. In some places, the wind can scoop out small, bowl-shaped areas in sediment on the ground. These areas are called *deflation hollows*.

ABRASION

Wind can grind and wear down rocks by abrasion. **Abrasion** happens when rock or sand wears down larger pieces of rock. Abrasion happens in areas where there are strong winds, loose sand, and soft rocks. The wind blows the loose sand against the rocks. The sand acts like sandpaper to erode, smooth, and polish the rocks.

Process	Description
	Large particles bounce and skip along the ground.
Deflation	
Abrasion	

What Landforms Are Produced by Wind Deposition?

Wind can carry material over long distances. The wind can carry different amounts and sizes of particles depending on its speed. Fast winds can carry large particles and may move a lot of material. However, all winds eventually slow down and drop their material. The heaviest particles fall first, while light material travels the farthest.

LOESS

Wind can deposit extremely fine material. Thick deposits of this windblown, fine-grained sediment are known as **loess**. Loess feels like talcum powder. Because the wind can carry light-weight material so easily, a loess deposit can be found far away from its source. In the United States, loess deposits are found in the Midwest, the Mississippi Valley, and in Oregon and Washington states.

✔ READING CHECK

3. Define What is deflation?

_____

_____

_____

_____

TAKE A LOOK
4. Complete Fill in the blank spaces in the table.

Critical Thinking

5. Infer What do you think is the reason that fast winds can carry larger particles than slower winds?

_____

_____

_____

DUNES

Barriers, such as plants and rocks, can cause wind to slow down. As it slows, the wind deposits particles on top of the barrier. As the dropped material builds up, the barrier gets larger. The barrier causes the wind to slow down even more. More and more material builds up on the barrier until a mound forms.

A mound of wind-deposited sand is called a **dune**. Dunes are common in sandy deserts and along sandy shores of lakes and oceans.

THE MOVEMENT OF DUNES

Wind conditions affect a dune's shape and size. As the wind blows sand through a desert, it is removed from some places and deposited in others. This can cause dunes to seem to move across the desert.

In general, dunes move in the same direction the wind is blowing. A dune has one gently sloped side and one steep side. The gently sloped side faces the wind. It is called the *windward slope*. The wind constantly moves sand up this side. As sand moves over the top of the dune, the sand slides down the steep side. The steep side is called the *slip face*. ☑

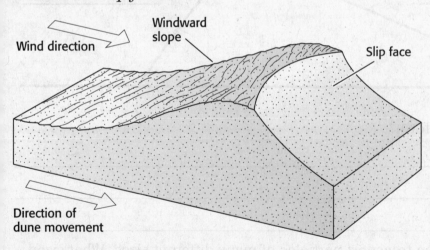

Wind direction

Windward slope

Slip face

Direction of dune movement

The wind blows sand up the windward slope of the dune. The sand moves over the top of the dune and falls down the steep slip face. In this way, dunes move across the land in the direction that the wind blows.

STANDARDS CHECK

ES 1c Land forms are the result of a combination of constructive and destructive forces. Constructive forces include crustal deformation, volcanic eruption, and deposition of sediment, while destructive forces include weathering and erosion.

6. Define What is a dune?

_____

_____

✓ **READING CHECK**

7. Identify In what direction do dunes generally move?

_____

_____

TAKE A LOOK

8. Compare How is the windward slope of a dune different from the slip face?

_____

_____

Section 2 Review

SECTION VOCABULARY

abrasion the grinding and wearing away of rock surfaces through the mechanical action of other rock or sand particles	**loess** fine-grained sediments of quartz, feldspar, hornblende, mica, and clay deposited by the wind
deflation a form of wind erosion in which fine, dry soil particles are blown away	**saltation** the movement of sand or other sediments by short jumps and bounces that is caused by wind or water
dune a mound of wind-deposited sand that moves as a result of the action of wind	

1. Identify Give two land features that can form because of deflation.

_____

_____

2. Describe What areas are most likely to be affected by wind erosion? Give two examples.

_____

_____

3. Identify The figure shows a drawing of a sand dune. Label the windward slope and the slip face. Draw an arrow to show the direction of the wind.

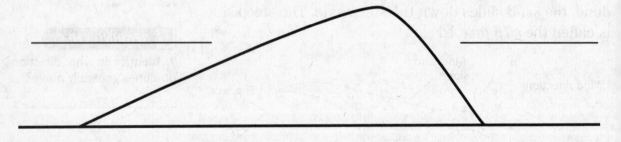

4. Explain How do dunes form?

_____

_____

_____

5. Apply Concepts Wind can transport particles of many different sizes. What sized particles are probably carried the farthest by the wind? Explain your answer.

_____

_____

_____

_____

CHAPTER 12 Agents of Erosion and Deposition

SECTION 3 Erosion and Deposition by Ice

National Science
Education Standards
ES 1c, 2a

BEFORE YOU READ

After you read this section, you should be able to answer these questions:

• What are glaciers?

• How do glaciers affect the landscape?

What Are Glaciers?

A **glacier** is a huge piece of moving ice. The ice in glaciers contains most of the fresh water on Earth. Glaciers are found on every continent except Australia.

There are two kinds of glaciers: continental and alpine. *Continental glaciers* are ice sheets that can spread across entire continents. *Alpine glaciers* are found on the tops of mountains. Both continental and alpine glaciers can greatly affect the landscape. ☑

Glaciers form in areas that are so cold that snow stays on the ground all year round. For example, glaciers are common in polar areas and on top of high mountains. In these areas, layers of snow build up year after year. Over time, the weight of the top layers pushes down on the lower layers. The lower layers change from snow to ice.

HOW GLACIERS MOVE

Glaciers can move in two ways: by sliding and by flowing. As more ice builds up on a slope, the glacier becomes heavier. The glacier can start to slide downhill, the way a skier slides downhill. Glaciers can also move by flowing. The solid ice in glaciers can move slowly, like soft putty or chewing gum.

Thick glaciers move faster than thin glaciers. Glaciers on steep slopes move faster than those on gentler slopes.

STUDY TIP

Compare As you read, make a table comparing the landforms that glaciers can produce.

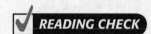

READING CHECK

1. Identify What are the two kinds of glaciers?

_____

_____

_____

This is McBride Glacier in Alaska.

TAKE A LOOK

2. Define What is a glacier?

_____

_____

SECTION 3 Erosion and Deposition by Ice *continued*

How Do Glaciers Affect the Landscape?

Glaciers can produce many different features as they move over Earth's surface. As a glacier moves, it can pick up and carry the rocks in its path. Glaciers can carry rocks of many different sizes, from dust all the way up to boulders. These rocks can scrape grooves into the land below the glacier as the glacier moves.

Continental glaciers tend to flatten the land that they pass over. However, alpine glaciers can produce sharp, rugged landscapes. The figure below shows some of the features that alpine glaciers can form.

Critical Thinking

3. Identify Relationships How is erosion by glaciers an example of water shaping the landscape?

_____

_____

_____

TAKE A LOOK
4. Explain How are horns, cirques, and arêtes related?

_____

_____

_____

_____

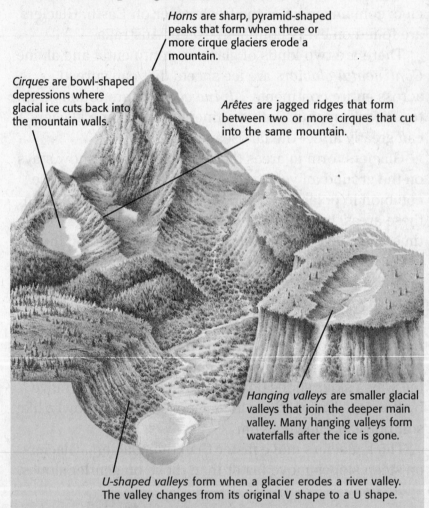

Horns are sharp, pyramid-shaped peaks that form when three or more cirque glaciers erode a mountain.

Cirques are bowl-shaped depressions where glacial ice cuts back into the mountain walls.

Arêtes are jagged ridges that form between two or more cirques that cut into the same mountain.

Hanging valleys are smaller glacial valleys that join the deeper main valley. Many hanging valleys form waterfalls after the ice is gone.

U-shaped valleys form when a glacier erodes a river valley. The valley changes from its original V shape to a U shape.

GLACIAL DEPOSITS

As a glacier melts, it drops all of the material that it is carrying. The material that is carried and deposited by glaciers is called **glacial drift**. There are two kinds of glacial drift: till and stratified drift.

SECTION 3 Erosion and Deposition by Ice *continued*

TILL DEPOSITS

Till is unsorted rock material that is deposited by melting glacial ice. It is called "unsorted" because the rocks are of all different sizes. Till contains fine sediment as well as large boulders. When the ice melts, it deposits this material onto the ground. ☑

The most common till deposits are *moraines*. Moraines form ridges along the edges of glaciers. There are many types of moraines. They are shown in the figure below.

READING CHECK

5. Explain Why is till considered unsorted?

_____

_____

Lateral moraines form along each side of a glacier.

Medial moraines form when valley glaciers that have lateral moraines meet.

Ground moraines form from unsorted materials left beneath a glacier.

Terminal moraines form when sediment is dropped at the front of the glacier.

🔈 **Say It**

Learn New Words Look up the words *lateral, medial,* and *terminal* in a dictionary. In a group, talk about why these words are used to describe different kinds of moraines.

STRATIFIED DRIFT

When a glacier melts, the water forms streams that carry rock material away from the glacier. The streams deposit the rocks in different places depending on their size. Larger rocks are deposited closer to the glacier. The rocks form a sorted deposit called **stratified drift**. The large area where the stratified drift is deposited is called an *outwash plain*. ☑

In some cases, a block of ice is left in the outwash plain as the glacier melts. As the ice melts, sediment builds up around it. The sediment forms a bowl-shaped feature called a *kettle*. Kettles can fill with water and become ponds or lakes.

READING CHECK

6. Define Write your own definition for *stratified drift*.

_____

_____

_____

_____

Section 3 Review

SECTION VOCABULARY

glacial drift the rock material carried and deposited by glaciers **glacier** a large mass of moving ice	**stratified drift** a glacial deposit that has been sorted and layered by the action of streams or meltwater **till** unsorted rock material that is deposited directly by a melting glacier

1. List Give two kinds of glacial drift.

_____

_____

2. Identify What are four kinds of moraines?

_____

_____

_____

_____

3. Compare How are continental glaciers different from alpine glaciers?

_____

_____

_____

_____

4. Explain How do glaciers form?

_____

_____

_____

5. Describe How does a kettle form?

_____

_____

_____

6. Infer How can a glacier deposit both unsorted and sorted material?

_____

_____

_____

_____

CHAPTER 12 Agents of Erosion and Deposition

SECTION 4 **The Effect of Gravity on Erosion and Deposition**

BEFORE YOU READ

After you read this section, you should be able to answer these questions:

• What is mass movement?

• How does mass movement shape Earth's surface?

• How can mass movement affect living things?

National Science Education Standards

ES 1c, 2a

What Is Mass Movement?

Gravity can cause erosion and deposition. Gravity makes water and ice move. It also causes rock, soil, snow, or other material to move downhill in a process called **mass movement**.

STUDY TIP

Ask Questions As you read this section, write down any questions you have. Talk about your questions in a small group.

ANGLE OF REPOSE

Particles in a steep sand pile move downhill. They stop when the slope of the pile becomes stable. The *angle of repose* is the steepest angle, or slope, at which the loose material no longer moves downhill. If the slope of a pile of material is larger than the angle of repose, mass movement happens. ☑

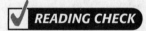

The slope of this pile of sand is equal to the sand's angle of repose. The sand pile is stable. The sand particles are not moving.

READING CHECK

1. Define What is the angle of repose?

_____

_____

_____

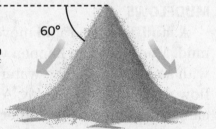

The slope of this pile of sand is larger than the angle of repose. Therefore, particles of sand move down the slope of the pile.

TAKE A LOOK

2. Explain Why are sand particles moving downhill in the bottom picture?

_____

_____

_____

The angle of repose can be different in different situations. The composition, size, weight, and shape of the particles in a material affect its angle of repose. The amount of water in a material can also change the material's angle of repose.

What Are the Kinds of Mass Movement?

Mass movement can happen suddenly and quickly. Rapid mass movement can be very dangerous. It can destroy or bury everything in its path.

LANDSLIDES

A **landslide** happens when a large amount of rock and soil moves suddenly and rapidly downhill. Landslides can carry away or bury plants and animals and destroy their habitats. Several factors can make landslides more likely. ☑

- Heavy rains can make soil wet and heavy, which makes the soil more likely to move downhill.
- Tree roots help to keep land from moving. Therefore, *deforestation*, or cutting down trees, can make landslides more likely.
- Earthquakes can cause rock and soil to start moving.
- People may build houses and other buildings on unstable hillsides. The extra weight of the buildings can cause a landslide. ☑

The most common kind of landslide is a *slump*. Slumps happen when a block of material moves downhill along a curved surface.

ROCK FALLS

A **rock fall** happens when loose rocks fall down a steep slope. Many such slopes are found on the sides of roads that run through mountains. Gravity can cause the loose and broken rocks above the road to fall. The rocks in a rock fall may be many different sizes.

MUDFLOWS

A **mudflow** is a rapid movement of a large amount of mud. Mudflows can happen when a lot of water mixes with soil and rock. The water makes the slippery mud flow downhill very quickly. A mudflow can carry away cars, trees, houses, and other objects that are in its path.

Mudflows are common in mountain regions when a long dry season is followed by heavy rain. Mudflows may also happen when trees and other plants are cut down. Without plant roots to hold soil in place and help water drain away, large amounts of mud can quickly form.

✔ **READING CHECK**

3. Describe How can landslides affect wildlife habitats?

_____
_____
_____

✔ **READING CHECK**

4. Identify Give three factors that can make landslides more likely.

_____
_____
_____
_____
_____

Critical Thinking

5. Infer Does water probably increase or decrease the angle of repose of soil? Explain your answer.

_____
_____
_____
_____

SECTION 4 The Effect of Gravity on Erosion and Deposition *continued*

LAHARS

Volcanic eruptions can produce dangerous mudflows called *lahars*. A volcanic eruption on a snowy peak can suddenly melt a great amount of snow and ice. The water mixes with soil and ash to produce a hot flow that rushes downhill. Lahars can travel faster than 80 km/h.

CREEP

Not all mass movement is fast. In fact, very slow mass movement is happening on almost all slopes. **Creep** is the name given to this very slow movement of material downhill. Even though creep happens very slowly, it can move large amounts of material over a long period of time. ☑

Many factors can affect creep. Water can loosen soil and rock so that they move more easily. Plant roots can cause rocks to crack and can push soil particles apart. Burrowing animals, such as moles and gophers, can loosen rock and soil particles. All of these factors may make creep more likely.

☑ **READING CHECK**

6. Compare How is creep different from the other kinds of mass movement that are discussed in this section?

_____

_____

Type of Mass Movement	Description
Landslide	Material moves suddenly and rapidly down a slope.
Rock fall	
Mudflow	
	Water mixes with volcanic ash to produce a fast-moving, dangerous mudflow.
	Material moves downhill very slowly.

TAKE A LOOK

7. Describe Fill in the blank spaces in the table.

Name ______________________ Class ____________ Date ____________

Section 4 Review

NSES ES 1c, 2a

SECTION VOCABULARY

creep the slow downhill movement of weathered rock material

landslide the sudden movement of rock and soil down a slope

mass movement the movement of a large mass of sediment or a section of land down a slope

mudflow the flow of a mass of mud or rock and soil mixed with a large amount of water

rock fall the rapid mass movement of rock down a steep slope or cliff

1. List What are four kinds of mass movement?

2. Infer Why is it important for people to think about mass movement when they decide how to use land?

3. Identify Relationships How is mass movement related to the angle of repose?

4. Identify What force causes mass movements?

5. Compare How are landslides different from mudflows?

6. List Give four things that can affect a material's angle of repose.

Name _____ Class _____ Date _____

Earth's Oceans

National Science Education Standards
ES 1b, 1f, 1g, 1h, 1j, 2a

BEFORE YOU READ

After you read this section, you should be able to answer these questions:

- What affects the salinity of ocean water?
- What affects the temperature of ocean water?
- How does the ocean affect air temperatures?

What Is the Global Ocean?

Earth has more liquid water on its surface than any other planet in the solar system. In fact, 71% of Earth's surface is covered by liquid water. Most of Earth's water is found in its oceans. There are five main oceans on Earth. However, the oceans are all connected to each other. Therefore, scientists often refer to all the oceans on Earth as the *global ocean*.

The continents divide the global ocean into the five main oceans. The largest ocean is the *Pacific Ocean*. The *Atlantic Ocean* is the second largest ocean. It has half the volume of the *Pacific Ocean*. The *Indian Ocean* is the third largest ocean. The *Southern Ocean* extends from the coast of Antarctica to 60°S latitude. The *Arctic Ocean* is the smallest ocean. Much of its surface is covered by ice. The figure below shows where these oceans are found. ☑

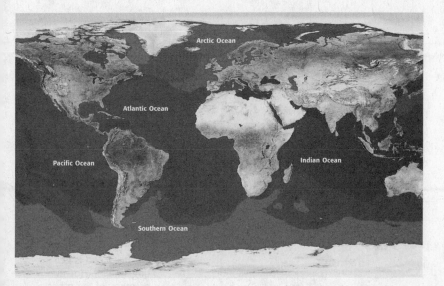

STUDY TIP
Ask Questions As you read this section, write down any questions you have. After you read, discuss your questions in a small group.

READING CHECK
1. Describe What divides the global ocean into five parts?

_____

_____

TAKE A LOOK
2. Identify What are the five main oceans?

_____

_____

_____

_____

SECTION 1 Earth's Oceans *continued*

How Did The Oceans Form?

Soon after the Earth formed, it was very different than it is today. There were no oceans. Volcanoes produced large amounts of ash, dust, and gases, including water vapor. These gases began to form the atmosphere. Slowly, Earth cooled. By about 4 billion years ago, the temperature was low enough for the water vapor in the atmosphere to condense. The liquid water fell as rain. The rain filled the basins in Earth's surface, and the first oceans formed. ☑

✔ **READING CHECK**

3. Identify Where did some of the gases that formed Earth's atmosphere come from?

_____

_____

Plate tectonics has caused the shapes and locations of Earth's oceans to change over time.

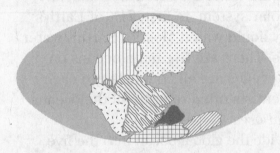

About 245 million years ago, the continents were joined into one large land mass. The oceans were one large body of water.

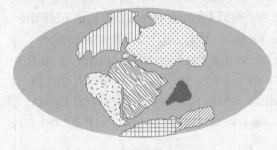

By about 180 million years ago, the continents had started to break apart. The North Atlantic Ocean and the Indian Ocean began to form.

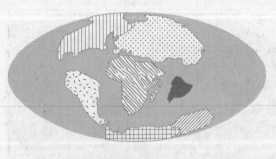

By about 65 million years ago, most of the Atlantic Ocean had formed. However, the South Atlantic Ocean was much narrower than it is today.

Today, the continents are still moving slowly, about 1 to 10 cm per year. The Pacific Ocean is getting smaller. However, many of the other oceans are growing.

STANDARDS CHECK

ES 1b Lithospheric plates on the scales of continents and oceans constantly move at rates of centimeters per year in <u>response</u> to movements in the mantle. <u>Major</u> geological events, such as earthquakes, volcanic eruptions, and mountain building result from these plate motions.

Word Help: <u>response</u>
an action brought on by another action; a reaction

Word Help: <u>major</u>
of great importance or large scale

4. Explain Why have the shapes and locations of the oceans changed with time?

_____

_____

_____

Why Is Ocean Water Salty?

Ocean water is different from the water that we drink. People cannot use ocean water for drinking because it is salty.

Most of the salt in the ocean is the same kind of salt we use on our food. This type of salt is called sodium chloride. It is a compound made from the elements sodium, Na, and chlorine, Cl. Ocean water also contains other dissolved solids, including magnesium and calcium. These dissolved solids make the water taste salty.

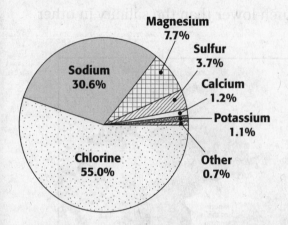

Magnesium 7.7%
Sulfur 3.7%
Calcium 1.2%
Sodium 30.6%
Potassium 1.1%
Chlorine 55.0%
Other 0.7%

This graph shows the amounts of different kinds of solids in ocean water.

Math Focus

5. Read a Graph Which two elements make up most of the dissolved solids in sea water?

_____

_____

As rivers and streams flow toward the ocean, they dissolve minerals from rocks and soil. These minerals include halite (sodium chloride) and calcite (calcium carbonate). The rivers carry the dissolved minerals to the ocean. At the same time, liquid water in the oceans evaporates to form water vapor. As the water evaporates, it leaves behind the minerals that were dissolved in it. ☑

What Is Salinity?

Salinity is a measure of the amount of solid material that is dissolved in a certain amount of liquid. It is usually measured as grams of dissolved solids per kilogram. On average, ocean water has 35 g/kg of dissolved solids in it. This means that 1 kg of ocean water has about 35 g of solids dissolved in it. If you evaporated 1 kg of ocean water, 35 g of solids would remain.

✓ **READING CHECK**

6. Identify Where does most of the salt in the ocean come from?

_____

_____

EFFECTS OF LOCATION ON SALINITY

Some parts of the ocean are saltier than others. Most oceans in hot, dry climates have high salinities. In these areas, the hot weather causes water to evaporate quickly. Salt is left behind. For example, the Red Sea in the Middle East is very salty. The climate there is very hot and dry. ☑

Some parts of the ocean are less salty than others. Along the coastlines, fresh water from streams and rivers runs into the ocean. As fresh water mixes with ocean water, the salinity of the ocean water decreases. For example, the salinity of the ocean waters near the Amazon River is much lower than the salinity in other parts of the ocean.

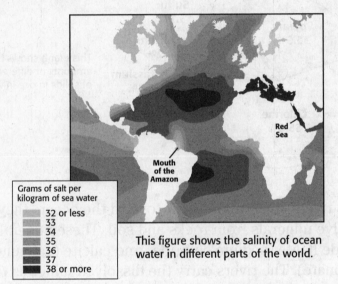

Red Sea

Mouth of the Amazon

Grams of salt per kilogram of sea water
32 or less
33
34
35
36
37
38 or more

This figure shows the salinity of ocean water in different parts of the world.

READING CHECK

7. Explain Why do oceans in hot, dry climates tend to have high salinity?

_____

_____

_____

TAKE A LOOK

8. Infer The Gulf of Mexico is located between Mexico and Florida. Why is the ocean water in the Gulf of Mexico less salty than in other places?

_____

_____

_____

EFFECTS OF WATER MOVEMENT ON SALINITY

The movement of water also affects salinity. Slow-moving ocean water tends to have higher salinity than fast-moving water. Parts of the ocean with slow-moving water, such as gulfs, bays, and seas, often have high salinities. Parts of the ocean without currents are likely have higher salinities as well. ☑

What Affects Ocean Temperatures?

The temperature of ocean water decreases as depth increases. However, the temperature change is not uniform. Scientists can divide the water in the ocean into three layers based on how temperature changes. These three layers are the surface zone, the thermocline, and the deep zone.

READING CHECK

9. Describe How does water movement affect salinity?

_____

_____

Ocean Temperature and Depth

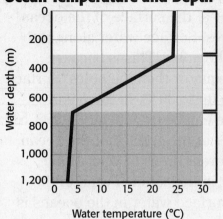

Surface zone The surface zone is the warm, top layer of ocean water. It extends to about 300 m below the surface. Sunlight heats the top 100 m of the surface zone. Convection currents mix the heated water with the cooler water below.

Thermocline The thermocline is the layer of water below the surface zone. It extends to between 700 m and 1,000 m below the surface. In the thermocline, temperature decreases quickly as depth increases.

Deep zone The deep zone is the deepest layer of the ocean. The temperature in the deep zone is only about 2°C.

TAKE A LOOK
10. Identify How does warm water mix with cool water in the surface zone?

_____

_____

SURFACE ZONE

The *surface zone* is the top layer of ocean water. It is heated by energy from the sun. As the ocean water is heated, it becomes less dense and rises above denser, cooler water. Convection currents form as the water moves. These currents can move heat within the surface zone to a depth of 100 m to 300 m. Therefore, the water temperature within the surface zone is fairly uniform.

THERMOCLINE

The *thermocline* is the layer of the ocean just beneath the surface zone. Within the thermocline, the temperature of the water decreases a lot as depth increases. This is because the sun cannot heat the water below the surface zone. In addition, the warm water of the surface zone cannot mix easily with the water below.

The depth of the thermocline is different in different places. It can extend from 100 m to almost 1,000 m below the surface of the ocean.

DEEP ZONE

The *deep zone* is the layer below the thermocline. In the deep zone, the temperature of the water is about 2°C. This very cold water is very dense. It moves slowly across the ocean floor and forms the deep ocean currents.

Critical Thinking

11. Predict Consequences What would happen to the water temperature in the surface zone if convection currents did not form there? Explain your answer.

_____

_____

_____

_____

_____

SECTION 1 Earth's Oceans *continued*

CHANGES IN SURFACE TEMPERATURE

The temperature of water at the surface of the ocean is different in different places. Surface water along the equator is warmer than at the poles. This is because more sunlight reaches the equator than the poles. Surface water at the equator can be up to 30°C. In the polar oceans, water at the surface can be as cold as −1.9°C. ☑

The temperature of water at the surface of the ocean can also change during different times of year. Many areas receive more sunlight in the summer than in the winter. In these areas, the surface water in the oceans is warmer in the summer.

How Does Water Move on Earth?

Imagine that you could see Earth from outer space. You would see green and brown landmasses, blue oceans, and white clouds. These parts are involved in the water cycle, which is shown below. The **water cycle** is the movement of water between the atmosphere, the ocean, and the land. The ocean is important to the water cycle because it holds nearly all of the Earth's water. ☑

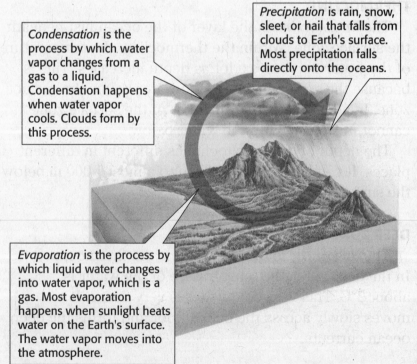

Condensation is the process by which water vapor changes from a gas to a liquid. Condensation happens when water vapor cools. Clouds form by this process.

Precipitation is rain, snow, sleet, or hail that falls from clouds to Earth's surface. Most precipitation falls directly onto the oceans.

Evaporation is the process by which liquid water changes into water vapor, which is a gas. Most evaporation happens when sunlight heats water on the Earth's surface. The water vapor moves into the atmosphere.

READING CHECK

12. Explain Why is surface water warmer at the equator than at the poles?

_____

_____

READING CHECK

13. Define What is the water cycle?

_____

_____

_____

TAKE A LOOK

14. Identify Where does most precipitation fall?

_____

How Do the Oceans Affect Air Temperatures?

You may know that many living things depend on the oceans for food and shelter. However, the oceans also help to keep the rest of the planet suitable for living things. This is because the oceans absorb and hold energy from sunlight. This helps to keep temperatures in the atmosphere from changing too much.

The ocean absorbs and releases heat much more slowly than the land does. Air can absorb heat from the oceans. Therefore, the oceans help to keep air temperatures steady, as shown below.

STANDARDS CHECK

ES 1j Global patterns of atmospheric movement influence local weather. Oceans have a <u>major</u> effect on climate, because water in the oceans holds a large amount of heat.

Word Help: <u>major</u>
of great importance or large scale

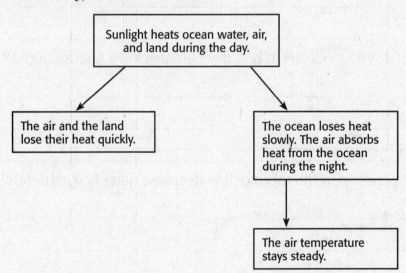

Sunlight heats ocean water, air, and land during the day.

The air and the land lose their heat quickly.

The ocean loses heat slowly. The air absorbs heat from the ocean during the night.

The air temperature stays steady.

15. Explain How do the oceans affect air temperatures?

_____

_____

_____

_____

If there were no oceans on Earth, the air temperature could vary from above 100°C to below −100°C in a single day! Such large temperature changes could cause a lot of severe weather. Life as we know it could not exist in these conditions.

The ocean can also affect the climate of different areas. Remember that ocean water at the equator is warmer than ocean water at the poles. Currents in the ocean move the ocean water from place to place. This helps to distribute heat throughout the Earth. ☑

The warm water from the equator can flow past land at high latitudes. The air absorbs heat from the warm ocean water. As a result, the land can have milder temperatures than other areas at the same latitude. For example, the islands of the United Kingdom are at about the same latitude as parts of Canada. However, a warm ocean current flows past the United Kingdom. As a result, its climate is much warmer than much of Canada's.

✓ READING CHECK

16. Identify What helps to distribute the heat in ocean water?

_____

Name _____ Class _____ Date _____

Section 1 Review

NSES   ES 1b, 1f, 1g, 1h, 1j, 2a

SECTION VOCABULARY

salinity a measure of the amount of dissolved salts in a given amount of liquid	**water cycle** the continuous movement of water between the atmosphere, the land, and the oceans

1. Explain Why do scientists call the ocean water on Earth the global ocean?

_____

_____

2. Identify What are two factors that can affect salinity?

_____

_____

3. Identify Relationships Why do most oceans in hot, dry climates have high salinities?

_____

_____

_____

4. Explain Why does the temperature in the thermocline decrease quickly with depth?

_____

_____

_____

5. Explain In the space below, make a drawing of the water cycle. Explain what is happening in each part of your drawing.

CHAPTER 13 Exploring the Oceans
SECTION
2 **The Ocean Floor**

BEFORE YOU READ

After you read this section, you should be able to answer these questions:

• What are the features of the ocean floor?

• How do the features of the ocean floor form?

How Do Scientists Study the Ocean Floor?

The ocean floor can be as far as 11,000 m below the surface of the ocean. People cannot survive at such great depths without a lot of special equipment. Scientists sometimes use such equipment to travel to the ocean floor. However, scientists can also study the ocean floor from the surface of the ocean. They can even study the ocean floor from outer space!

STUDYING THE OCEAN FLOOR WITH SHIPS

People cannot survive the high pressures on the ocean floor. Therefore, in order to explore the deep zone, scientists had to build an underwater vessel that can survive under high pressures. One such vessel is called *Deep Flight*. Future ships may be able to carry scientists to 11,000 m below the ocean's surface. ☑

Scientists can also use Remotely Operated Vehicles, or ROVs, to explore the ocean floor directly. These ROVs carry special equipment deep below the ocean surface. Scientists control ROVs using radio signals from the surface. Because ROVs do not carry people, they can explore parts of the ocean floor that are too dangerous for people to travel to.

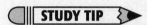
STUDY TIP

Compare As you read, make a chart showing the similarities and differences between the different ways that scientists study the ocean floor.

READING CHECK

1. Explain Why did scientists have to build a special ship to explore the deep zone?

_____

_____

_____

Critical Thinking

2. Infer *Deep Flight* and other deep-ocean vessels have very thick windows and walls. Why do you think this is?

_____

_____

_____

The *Deep Flight* is a vessel that can carry scientists deep beneath the ocean surface.

SECTION 2 The Ocean Floor *continued*

STUDYING THE OCEAN FLOOR WITH SONAR

Sonar stands for *sound navigation and ranging*. Sonar instruments on a ship send pulses of sound down into the ocean. The sound bounces off the ocean floor and returns to the ship. The deeper the water, the longer it takes for the sound to return to the ship. ☑

Scientists use sonar to determine how deep the ocean floor is. They know that sound travels about 1,500 m/s in ocean water. They measure how long it takes for the sound to return to the ship. Then, they use this information to figure out how deep the ocean floor is.

☑ **READING CHECK**

3. Identify What is sonar?

_____

_____

_____

_____

Math Focus

4. Calculate The ship sends out a sonar signal. It picks up the return signal 4 s later. About how deep is the ocean floor? Show your work.

_____

_____

_____

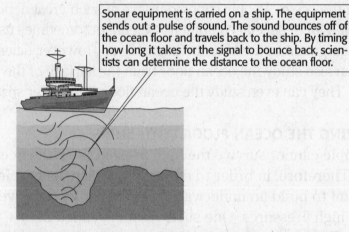

Sonar equipment is carried on a ship. The equipment sends out a pulse of sound. The sound bounces off of the ocean floor and travels back to the ship. By timing how long it takes for the signal to bounce back, scientists can determine the distance to the ocean floor.

STUDYING THE OCEAN FLOOR USING SATELLITES

Geosat was once a top-secret military satellite. Today, scientists use *Geosat* to study the ocean floor indirectly. Underwater features, such as mountains and trenches, affect the height of the ocean surface. Scientists can use *Geosat* to measure the height of the water on the ocean surface. They can use these measurements to make detailed maps of the ocean floor.

What Are the Features of the Ocean Floor?

Because most of the ocean floor is covered by kilometers of water, it has not been thoroughly explored. However, scientists do know that many of the features of the ocean floor are caused by plate tectonics.

Scientists divide the ocean floor into two major regions: the continental margin and the deep-ocean basin. The *continental margin* is the edge of a continent that is covered by ocean water. The *deep-ocean basin* begins at the edge of the continental margin and extends under the deepest parts of the ocean. ☑

☑ **READING CHECK**

5. Identify What are the two major regions of the ocean floor?

_____

_____

SECTION 2 The Ocean Floor *continued*

CONTINENTAL SHELF, SLOPE, AND RISE

There are three main parts of the continental margin: the continental shelf, the continental slope, and the continental rise. The **continental shelf** is the part of the margin that begins at the shoreline and slopes gently toward the open ocean. It continues until the ocean floor begins to slope more steeply. The depth of the continental shelf can reach 200 m. ☑

The **continental slope** is the steepest part of the continental margin. It begins at the edge of the continental shelf and continues down to the flattest part of the ocean floor. The depth of the continental slope ranges from about 200 m to about 4,000 m.

The **continental rise** is the base of the continental slope. It is made of large piles of sediment. The continental rise covers the boundary between the continental margin and the deep-ocean basin.

READING CHECK

6. Identify What are the three parts of the continental margin?

_____

_____

_____

The Continental Margin

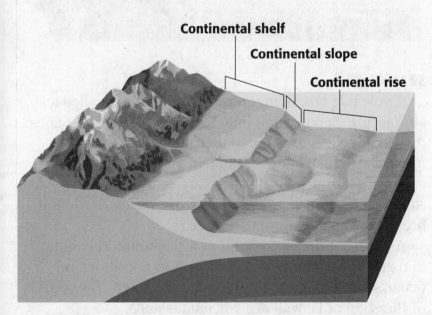

Continental shelf

Continental slope

Continental rise

TAKE A LOOK

7. Arrange Write the names of the three parts of the continental margin in order from shallowest to deepest.

_____

_____

_____

The deep-ocean basin includes the abyssal plain, mid-ocean ridges, rift valleys, seamounts, and ocean trenches. The **abyssal plain** is the flat, almost level part of the ocean basin. The abyssal plain is covered by layers of sediment. Some of the sediment comes from land. Some of it is made of the remains of dead sea creatures. These remains settle to the ocean floor when the creatures die.

SECTION 2 The Ocean Floor *continued*

PLATE TECTONICS AND OCEAN FLOOR FEATURES

Remember that tectonic plates move over Earth's surface. Where the plates touch, they can slide past each other, collide with each other, or move away from each other. The movement of tectonic plates creates features, such as mountains, on land. Tectonic movements can also form features on the ocean floor. These features include seamounts, mid-ocean ridges, rift valleys, and ocean trenches.

STANDARDS CHECK

ES 1b Lithospheric plates on the scales of continents and oceans constantly move at rates of centimeters per year in <u>response</u> to movements in the mantle. <u>Major</u> geological events, such as earthquakes, volcanic eruptions, and mountain building result from these plate motions.

Word Help: <u>response</u>
an action brought on by another action; a reaction

Word Help: <u>major</u>
of great importance or large scale

8. Infer What would the ocean floor look like if plate tectonics did not happen?

_____

_____

_____

_____

The Deep-Ocean Basin

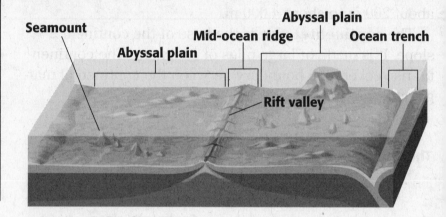

Seamount — Abyssal plain — Mid-ocean ridge — Abyssal plain — Ocean trench — Rift valley

SEAMOUNTS

A volcanic mountain on the ocean floor is called a **seamount**. Some seamounts form when magma pushes its way between tectonic plates and erupts to form a mountain. Other seamounts form far away from the edges of tectonic plates. These areas are called *hot spots*. At these locations, magma rises from within Earth and breaks through a tectonic plate. ☑

As lava continues to erupt at a seamount, the mountain gets taller. If it gets tall enough to rise above the ocean's surface, it is called a volcanic island. The islands of the state of Hawaii are volcanic islands.

READING CHECK

9. Define What is a seamount?

_____

_____

MID-OCEAN RIDGES

A long mountain chain that forms on the floor of the ocean is a **mid-ocean ridge**. Mid-ocean ridges form where tectonic plates move apart. This motion produces a crack in the ocean floor called a **rift valley**. Magma rises through the rift and cools to form new rock. The ridge is made of this new rock. ☑

Most mid-ocean ridges are far below the ocean surface. However, Iceland is one place on Earth where the mid-ocean ridge has risen above the ocean's surface.

✓ **READING CHECK**

10. Identify Where do mid-ocean ridges form?

_____

_____

OCEAN TRENCHES

Long, narrow valleys in the deep-ocean basins are called **ocean trenches**. Trenches form where tectonic plates are moving together. As the plates move toward each other, one plate sinks under the other plate in a process called *subduction*. ☑

Subduction causes pressure to increase on the plate that is sinking. Water and other fluids are squeezed out of the rock and sediments on this plate. These fluids cause mantle rock to melt and form magma. The magma rises to the surface and erupts to form a chain of volcanoes.

Ocean trenches are some of the deepest places on Earth. For example, the Marianas Trench in the Pacific Ocean is nearly 11,000 m deep.

✓ **READING CHECK**

11. Define Write your own definition for *ocean trench*.

_____

_____

_____

_____

Feature	What it is	Where it is found
Seamount		
	flat part of the ocean basin, covered in sediment	in the deep-ocean basins
		where plates move apart
Rift valley	crack in the crust at a mid-ocean ridge	
Ocean trench	long, narrow valley on the ocean floor	

TAKE A LOOK
12. Describe Fill in the table to compare the different features of the ocean floor.

Name _____ Class _____ Date _____

Section 2 Review

SECTION VOCABULARY

abyssal plain a large, flat, almost level area of the deep-ocean basin

continental rise the gently sloping section of the continental margin located between the continental slope and the abyssal plain

continental shelf the gently sloping section of the continental margin located between the shoreline and the continental slope

continental slope the steeply inclined section of the continental margin located between the continental rise and the continental shelf

mid-ocean ridge a long, undersea mountain chain that forms along the floor of the major oceans

ocean trench a long, narrow, and steep depression on the ocean floor that forms when one tectonic plate subducts beneath another plate; trenches run parallel to volcanic island chains or to the coastlines of continents; also called a trench or a deep-ocean trench

rift valley a long, narrow valley that forms as tectonic plates separate

seamount a submerged mountain on the ocean floor that is at least 1,000 m high and that has a volcanic origin

1. List Name four ways that scientists study the ocean floor.

_____

_____

_____

_____

2. Describe How do plate movements form ocean trenches?

_____

_____

_____

_____

3. Identify Where does the sediment on the abyssal plain come from? Give two sources.

_____

_____

_____

4. Explain How do mid-ocean ridges form?

_____

_____

_____

_____

_____

CHAPTER 13 | Exploring the Oceans

SECTION 3 **Life in the Ocean**

BEFORE YOU READ

After you read this section, you should be able to answer these questions:

• What are plankton, nekton, and benthos?
• What are two kinds of marine environments?

What Are Three Groups of Marine Life?

Studying the organisms that live in the ocean can be difficult for scientists because the ocean is so large. There are probably many kinds of *marine*, or ocean-dwelling, organisms that scientists have not discovered yet. There are many other marine organisms that scientists know little about.

In order to make studying marine organisms easier, scientists classify them into three main groups. The groups are based on where the organisms live and how they move. The three groups of marine life are plankton, nekton, and benthos.

Plankton are organisms that float or drift near the ocean's surface. Most plankton are very tiny. Plankton that can make their own food like plants are called *phytoplankton*. Plankton that cannot make their own food are called *zooplankton*.

Organisms that actively swim in the open ocean are called **nekton**. Kinds of nekton include mammals, such as whales and dolphins, and many kinds of fish.

Benthos are organisms that live on or in the ocean floor. Some types of benthos are crabs, coral, and clams.

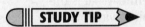

STUDY TIP

Describe As you read this section, draw a diagram of the ocean. Show the benthic and pelagic environments and label the zones that make up these environments.

Say It

Learn New Words Use a dictionary to look up words that start with the prefixes *phyto-* and *zoo-*. Discuss the words and their definitions in a small group. Together, try to figure out what the two prefixes mean.

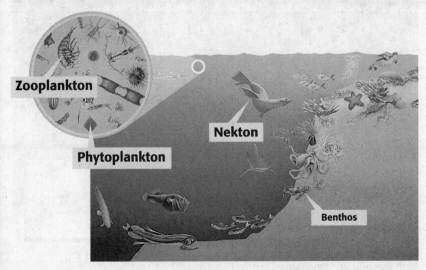

Zooplankton

Nekton

Phytoplankton

Benthos

TAKE A LOOK
1. Identify What are three groups of marine life?

_____

_____

Where Do Benthos Live?

The bottom area of the ocean is called the **benthic environment**. This environment includes the ocean floor and the area near it. It also includes the benthos, the organisms that live on or in the ocean floor. Scientists divide the benthic environment into five zones based on the organisms that live in each zone. These zones are the intertidal zone, the sublittoral zone, the bathyal zone, the abyssal zone, and the hadal zone. ☑

THE INTERTIDAL ZONE

The *intertidal zone* is the shallowest benthic zone. It is located on the beach between the low-tide and high-tide limits. At high tide, the intertidal zone is covered with ocean water. At low tide, the zone is exposed to the air and sun. Therefore, organisms in the intertidal zone are adapted to living both underwater and out of the water. Organisms in the intertidal zone must also protect themselves from crashing waves. ☑

The organisms in the intertidal zone have different adaptations. Some organisms, such as starfish, attach themselves to rocks to avoid being washed away. Other organisms, such as clams, have hard shells to protect them from strong waves and sunlight. Some animals can burrow into the sand or between rocks to avoid drying out. Some plants have strong, rootlike structures called *holdfasts* that help to keep the plants from washing away.

READING CHECK

2. Define What is the benthic environment?

_____

_____

_____

READING CHECK

3. Identify Where is the intertidal zone found?

_____

_____

TAKE A LOOK

4. Identify How do sea anemones and starfish prevent themselves from being washed away?

_____

_____

Organisms such as sea anemones and starfish attach themselves to rocks and reefs. These organisms must be able to survive both wet and dry conditions.

Intertidal zone

SECTION 3 Life in the Ocean *continued*

THE SUBLITTORAL ZONE

The *sublittoral zone* begins where the intertidal zone ends—at the low-tide limit. The sublittoral zone ends at the edge of the continental shelf, about 200 m below sea level. The environment in this zone is more constant than in the intertidal zone. Organisms here are always covered by water. The water temperature, water pressure, and sunlight stay fairly constant. The kind of sediment on the ocean floor affects where organisms live in this zone.

Math Focus

5. Convert How deep does the sublittoral zone extend in feet?

1 ft = 0.31 m

Corals can live in both the sublittoral and the intertidal zone. However, they are most common in the sublittoral zone.

TAKE A LOOK

6. Compare Give one difference between the sublittoral zone and the intertidal zone.

THE BATHYAL ZONE

The *bathyal zone* extends from the edge of the continental shelf to the abyssal plain. It is between 200 m and 4,000 m below sea level. Because sunlight does not reach most of this zone, very few plants live here. Animals that live in this zone include sponges, sea stars, and octopuses.

Critical Thinking

7. Infer Where do you think the animals living in the bathyal zone get their food if there are few plants in this zone?

Octopuses are animals that are commonly found in the bathyal zone.

SECTION 3 Life in the Ocean *continued*

THE ABYSSAL ZONE

The *abyssal zone* is located on the abyssal plain. It can be as deep as 6,000 m below sea level. Most of the ocean floor lies in this zone, so it is the largest zone in the ocean. No plants and few other organisms live in the abyssal zone. The organisms that do live in this zone include crabs, sponges, and worms. Many of these animals live near hot-water vents in the ocean floor called *black smokers.* ☑

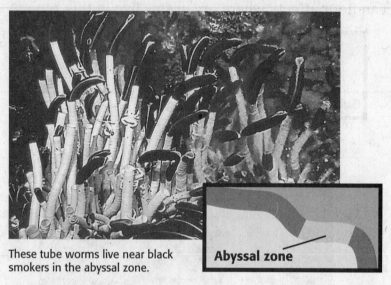

These tube worms live near black smokers in the abyssal zone.

Abyssal zone

THE HADAL ZONE

The *hadal zone* is the deepest zone in the benthic environment. This zone is found in ocean trenches. It can be as deep as 11,000 m below sea level. It is completely dark, very cold, and has high water pressure. It is very difficult to study, so scientists do not know a lot about the hadal zone. Some organisms known to live in this zone are sponges, worms, and clams.

Hadal zone

These clams are some of the few organisms known to live in the hadal zone.

✓ READING CHECK

8. Identify On what part of the ocean floor is the abyssal zone found?

_____

_____

TAKE A LOOK
9. Define What is a black smoker?

_____

_____

TAKE A LOOK
10. Describe Where is the hadal zone located?

_____

_____

Where Do Plankton and Nekton Live?

The **pelagic environment** is the open ocean water that lies above the benthic environment. It is beyond the intertidal zone and above the abyssal zone. Plankton and nekton live in the pelagic environment. There are two main zones in the pelagic environment—the neritic zone and the oceanic zone. ☑

THE NERITIC ZONE

The *neritic zone* is the warm, shallow water that covers the continental shelf. Many marine organisms live here because this zone receives more sunlight than other ocean zones. The sunlight allows phytoplankton to grow. The phytoplankton provide food for many marine animals.

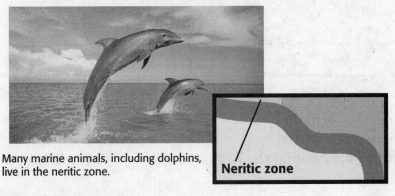

Many marine animals, including dolphins, live in the neritic zone.

Neritic zone

THE OCEANIC ZONE

The *oceanic zone* is the water that covers the bathyal zone, abyssal zone, and hadal zone. The oceanic zone is colder and has higher water pressure than the neritic zone. Since the oceanic zone is so large, the animals living there may be very spread out. Some animals live in both the neritic zone and the oceanic zone. However, some strange animals live at the deepest part of the oceanic zone.

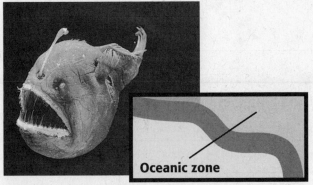

Oceanic zone

This angler fish lives in the deep parts of the oceanic zone. It uses the wormlike growth on its head as a lure to attract prey.

READING CHECK

11. Define What is the pelagic environment?

_____

_____

TAKE A LOOK

12. Describe Where is the neritic zone?

_____

_____

TAKE A LOOK

13. Identify In what zone does the angler fish live?

_____

Section 3 Review

SECTION VOCABULARY

benthic environment the region near the bottom of a pond, lake, or ocean	**pelagic environment** in the ocean, the zone near the surface or at middle depths, beyond the sublittoral zone and above the abyssal zone
benthos organisms that live at the bottom of oceans or bodies of fresh water	**plankton** the mass of mostly microscopic organisms that float or drift freely in freshwater and marine environments
nekton all organisms that swim actively in open water, independent of currents	

1. Compare How are plankton different from nekton?

_____

_____

_____

2. Define What are the two ocean environments? In your own words, describe where each is located.

_____

_____

_____

_____

3. Explain Why does the neritic zone contain the largest amount of marine life in the ocean?

_____

_____

4. Explain Why don't plants grow in the abyssal zone or the hadal zone?

_____

_____

5. Identify What unique environmental conditions do organisms in the intertidal zone have to be adapted to?

_____

_____

6. List What are the two zones in the pelagic environment?

_____

_____

CHAPTER 13 Exploring the Oceans

SECTION
4

Resources from the Ocean

BEFORE YOU READ

After you read this section, you should be able to answer these questions:

• What are the living resources from the ocean?

• What are the nonliving resources from the ocean?

What Are the Living Resources of the Ocean?

People have been harvesting plants and animals from the ocean for thousands of years. Today, harvesting food from the ocean is a multi-billion-dollar industry. As the population of humans on Earth has grown, the demand for these resources has increased. However, the availability of these resources has not increased as much.

Harvesting fish from the ocean can harm the environment. Usually, fish can reproduce faster than people can catch them. However, new technology, such as drift nets, has allowed people to catch more fish in less time. This may allow people to take fish faster than they can reproduce. This could cause the population of fish in the oceans to decrease. Also, other animals, such as dolphins and turtles, can be caught in fishing nets.

Recently, laws have been passed that control fishing more strictly. These laws are supposed to help reduce the damage fishing can cause to the environment. As a result of these laws, people have begun to raise fish and other types of seafood, such as shellfish and seaweed, in farms near the shore. By growing seafood in these farms, people can conserve and protect wild organisms.

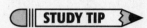 **STUDY TIP**

Summarize As you read, make a chart describing the resources that people use from the ocean.

Critical Thinking

1. Explain When can fish be considered a renewable resource? When would fish not be considered a renewable resource?

_____

_____

_____

_____

_____

_____

New technology, such as these drift nets, allows people to catch more fish in less time. However, other animals, such as dolphins and turtles, can sometimes get caught in the nets.

SECTION 4 Resources from the Ocean *continued*

What Are the Nonliving Resources of the Ocean?

Fish and other seafood are important resources that people take from the oceans. However, people also take many nonliving resources from the oceans. These resources include energy resources and material resources.

FRESH WATER

Fresh water is often considered a renewable resource. However, in parts of the world where the climate is dry, fresh water is limited. In these parts of the world, ocean water is desalinated to provide fresh drinking water. **Desalination** is the process of removing salt from sea water. ☑

Most desalination plants heat ocean water to cause the water to evaporate. The water vapor, which is not salty, is collected and condensed into liquid fresh water. Another method of desalination involves passing the ocean water through a membrane to leave the salts behind. However, no matter what process is used, desalination is expensive and can be slow.

TIDAL ENERGY

The ocean is constantly moving as tides come in and go out. People can use the motion of the water to generate electricity. Energy that is generated from the movement of tides is called *tidal energy*. Tidal energy is clean, inexpensive, and renewable. However, it can only be used in certain parts of the world.

READING CHECK

2. Define What is desalination?

_____

_____

Critical Thinking

3. Infer Why can't tidal energy be used everywhere?

_____

_____

_____

TAKE A LOOK

4. Explain Why is tidal energy considered to be renewable?

_____

_____

High tide

Gate closes

Gate closed

❶ As the tide rises, water enters a bay behind a dam. At high tide, the gate closes to keep the water in the bay.

❷ The gate stays closed as the tide goes out.

Low tide

Gate opens

❸ At low tide, the gate opens and the water rushes out of the bay. As it moves through the dam, the water turns wheels called *turbines* that generate electricity.

OIL AND NATURAL GAS

Oil and natural gas are considered the most valuable resources in the ocean. Oil and natural gas form from the remains of tiny plants and animals. These remains take millions of years to turn into oil and natural gas. Therefore, oil and gas are nonrenewable resources.

Many deposits of oil and natural gas are found in rock near the continental margins. In order to obtain these resources, engineers must drill wells through the rock. About one-fourth of the world's oil is now obtained from wells in rock beneath the oceans.

Oil is refined by manufacturers to make gasoline. Gasoline powers vehicles and generators that make electricity. Oil is also used to make plastic and other products. ☑

MINERALS

Many different kinds of minerals can be found on the ocean floor. These minerals are commonly in the form of nodules. *Nodules* are potato-shaped lumps of minerals that form from chemicals dissolved in ocean water.

Nodules can be made of many different kinds of minerals. Most nodules contain the element manganese. Manganese can be used to make certain kinds of steel. Some nodules contain the valuable metals iron, copper, nickel, or cobalt. Some contain phosphorus, which can be used in fertilizer.

Nodules can be very large. They may contain a large amount of valuable minerals. However, they form in the very deep parts of the ocean. For this reason, they are difficult to mine. ☑

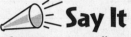

Say It

Discuss In a small group, talk about different ways that you use resources from the ocean every day.

READING CHECK

5. List Give two ways that people use oil or natural gas.

_____

_____

READING CHECK

6. Explain Why are nodules hard to mine?

_____

_____

Minerals can be found on the ocean floor in the form of nodules. These nodules are difficult to mine because they are found in very deep water.

Section 4 Review

SECTION VOCABULARY

desalination a process of removing salt from ocean water	

1. List Name two living resources from the ocean.

_____

_____

2. Define Write your own definition for desalination.

_____

_____

3. Describe How can the tides be used to generate electricity?

_____

_____

_____

_____

4. Identify Give five minerals that may be found in nodules.

_____

_____

_____

_____

5. List Give four nonliving ocean resources.

_____

_____

_____

_____

6. Infer Why are people starting to farm the oceans instead of harvesting wild organisms?

_____

_____

CHAPTER 13 Exploring the Oceans

SECTION
5 **Ocean Pollution**

BEFORE YOU READ

After you read this section, you should be able to answer these questions:

- What are the different types of pollution in the ocean?
- How can we preserve ocean resources?

What Pollutes the Ocean?

Many human activities produce pollution that can harm the oceans. Some of this pollution comes from a specific source. Pollution that can be traced to one source is called **point-source pollution**. However, some pollution comes from many sources. Pollution that cannot be traced to a single source is called **nonpoint-source pollution**. ☑

TRASH DUMPING

People dump trash in many places, including the ocean. In the 1980s, scientists became alarmed by the kinds of trash that were washing up on beaches. Bandages, vials of blood, and other medical wastes were found among the trash.

The Environmental Protection Agency (EPA) found that hospitals in the United States were dumping medical wastes into the oceans. Much of this waste is now buried in sanitary landfills. However, other kinds of trash are still dumped into the oceans.

Barges like this one carry garbage out to sea. The garbage is dumped into the open ocean.

> **STUDY TIP**
>
> **Summarize** As you read, underline the main ideas in each paragraph. When you finish reading, write a short summary of the section using the ideas you underlined.

> ☑ **READING CHECK**
>
> **1. Define** Write your own definition for nonpoint-source pollution.
>
> _____
>
> _____

TAKE A LOOK
2. Identify How does trash get into the oceans?

_____

_____

SECTION 5 Ocean Pollution *continued*

EFFECTS OF TRASH DUMPING

Trash thrown into the ocean can affect the organisms that live there. It also affects the organisms, such as people, that depend on the ocean for food. For example, most plastic material that is thrown into the ocean does not break down for thousands of years. Animals can mistake plastic material for food and choke on it.

TAKE A LOOK

TAKE A LOOK

3. Describe How can trash harm the organisms that live in the oceans?

_____

_____

_____

_____

Marine animals, such as this bird, can choke on plastic trash that is thrown into the oceans.

SLUDGE DUMPING

Raw sewage is all of the liquid and solid wastes that are flushed down toilets and poured down drains. In most places, raw sewage is collected and sent to a treatment plant. The treatment removes solid waste and cleans the raw sewage. The solid waste that remains is called *sludge*.

In many places, people dump sludge into the ocean. Currents can stir up the sludge and move it closer to shore. The sludge can pollute beaches and kill ocean life. Many countries have banned sludge dumping. However, it still happens in many parts of the world.

Critical Thinking

4. Compare How is raw sewage different from sludge?

_____

_____

_____

_____

_____

Sludge that is dumped into the oceans can carry bacteria. It can make beaches dirty and harm marine organisms.

OIL SPILLS

Most of the world uses oil as an energy source. However, oil is only found in certain places around the world. Therefore, large tankers must transport billions of barrels of oil across the oceans. Sometimes, the tankers break open and the oil spills out of them.

Oil spills can cause many problems for the environment. Oil is poisonous to plants and animals. It is also very hard to clean up oil spills, so their effects can last for a long time. ☑

In 1990, the United States Congress passed the Oil Pollution Act. This law states that all oil tankers that travel in United States waters must have two hulls. If the outer hull of a ship is damaged, the inner hull can keep oil from spilling into the ocean.

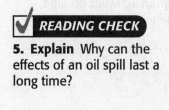

READING CHECK

5. Explain Why can the effects of an oil spill last a long time?

_____

_____

The Oil Pollution Act may help to prevent large oil spills. However, as the figure below shows, big spills only cause about 5% of the oil pollution in the ocean. Most of the oil in the ocean comes from nonpoint-source pollution on land.

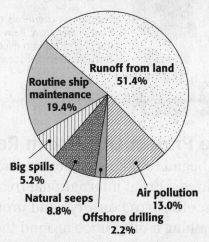

Runoff from land 51.4%

Routine ship maintenance 19.4%

Big spills 5.2%

Natural seeps 8.8%

Offshore drilling 2.2%

Air pollution 13.0%

Math Focus

6. Read a Graph What are the three largest sources of oil pollution in the oceans?

_____

_____

_____

SECTION 5 Ocean Pollution *continued*

NONPOINT-SOURCE POLLUTION

Nonpoint-source pollution is pollution that comes from many sources instead of a single place. Most ocean pollution is nonpoint-source pollution. Things that people do on land can pollute rivers. The rivers can carry the pollution into the oceans. ☑

Nonpoint-source pollution is hard to control because it enters the water in many different ways. However, there are things that people can do to help reduce nonpoint-source pollution. For example, we can throw away chemicals, such as used motor oil, properly instead of pouring them into sewers.

READING CHECK

7. Explain How can human activities on land cause ocean pollution?

_____

_____

_____

Oil and gasoline can leak out of cars and trucks and onto the streets. Rain can wash the oil and gasoline into rivers, which carry them to the oceans.

Boats and other watercraft can leak oil, gasoline, and other chemicals into the water.

TAKE A LOOK

8. Identify Give three examples of nonpoint-source pollution.

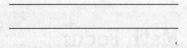

_____

_____

_____

People use chemicals to help their lawns grow. Rain can wash the chemicals into rivers, which carry them to the oceans.

How Can We Protect Our Ocean Resources?

People have begun to take steps to save and protect our ocean resources. From international treaties to volunteer cleanups, efforts to conserve and protect ocean resources are making a difference around the world.

SECTION 5 Ocean Pollution *continued*

NATIONS TAKE NOTICE

In the 1970s and 1980s, ocean pollution was very bad. Many countries realized that they would need to work together to reduce ocean pollution. In 1989, 64 countries signed a treaty that bans the dumping of many harmful materials into the ocean.

Many other treaties and laws have also been passed to help protect the oceans. For example, Congress passed the Clean Water Act in 1972. This law gave the Environmental Protection Agency more control over the trash that is dumped into the ocean.

Another law, the U.S. Marine Protection, Research, and Sanctuaries Act, was also passed in 1972. This law forbids people from dumping harmful materials into the oceans. These laws have helped to reduce the pollution entering the oceans. However, waste dumping and oil spills still happen.

CITIZENS TAKE CHARGE

Citizens of many different countries have demanded that their governments do more to prevent ocean pollution. They have also begun to take the matter into their own hands. For example, people began to organize beach cleanups. Millions of tons of trash have been gathered from beaches. Also, people are helping to spread the word about the problems with dumping wastes into the oceans. ☑

Critical Thinking

9. Infer Why do countries need to work together to reduce ocean pollution?

_____

_____

_____

_____

_____

_____

✓ **READING CHECK**

10. Explain How can individual people help to reduce ocean pollution?

_____

_____

_____

During a beach cleanup, many people work together to remove trash from a beach. This helps make the beach safer for people, other animals, and plants.

Section 5 Review

SECTION VOCABULARY

nonpoint-source pollution pollution that comes from many sources rather than from a single, specific site	point-source pollution pollution that comes from a specific site

1. Compare How is point-source pollution different from nonpoint-source pollution?

_____

_____

2. Describe Fill in the table below to describe different sources of ocean pollution.

Type of Pollution	Description
Trash dumping	
Oil spills	
Chemical leaks	

3. Infer Most of the trash in the United States is buried instead of being dumped into the oceans. However, there is still trash in the oceans. Why is this?

_____

_____

_____

4. Identify What type of pollution is most ocean pollution?

_____

5. Describe Why is nonpoint-source pollution hard to control?

_____

_____

_____

6. List Give two United States laws that protect ocean resources.

_____

_____

CHAPTER 14 | The Movement of Ocean Water

SECTION
1 **Currents**

National Science Education Standards
ES 1j

BEFORE YOU READ

After you read this section, you should be able to answer these questions:

• What factors affect ocean currents?

• Why are ocean currents important?

What Are Surface Currents?

Imagine that you are stranded on an island. You write a note and put it into a bottle. You throw the bottle into the ocean to communicate with the outside world. Can you predict where the bottle will end up? If you understand ocean currents, you can! The oceans contain many streamlike movements of water called **ocean currents**. There are two main kinds of ocean currents: surface currents and deep currents.

Surface currents are horizontal, streamlike movements of water that are found at or near the surface of the ocean. Surface currents can be up to several hundred meters deep. They can be as long as several thousand kilometers. Three factors affect surface currents: global winds, the Coriolis effect, and continental deflections.

STUDY TIP

Summarize As you read, make a diagram showing the types of ocean currents and the factors that affect them.

READING CHECK

1. Define Write your own definition for *surface current*.

_____
_____
_____

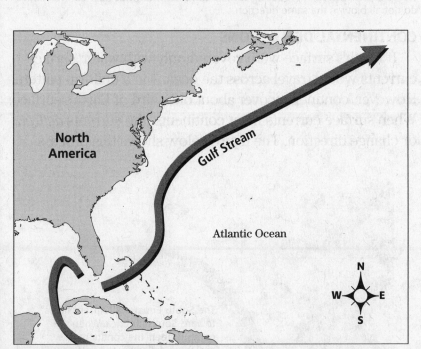

North America

Gulf Stream

Atlantic Ocean

N
W — E
S

The Gulf Stream is one of the largest surface currents in the world. Every year, it transports at least 25 times as much water as all of the rivers on Earth combined!

TAKE A LOOK

2. Read a Map In what direction does the Gulf Stream flow?

_____

GLOBAL WINDS

Have you ever blown across a bowl of hot soup? You may have noticed that your breath pushes the soup across the surface of the bowl. In much the same way, winds that blow across the surface of Earth's oceans push water across Earth's surface. This process causes surface currents in the ocean. ☑

Many winds blow across Earth's surface, but they do not all blow in the same direction. Near the equator, the winds blow mostly east to west. Between 30° and 60° latitude, the winds blow mostly west to east.

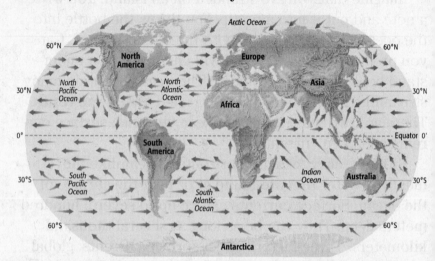

Winds are important in producing surface currents. The winds near Earth's surface do not all blow in the same direction.

CONTINENTAL DEFLECTIONS

If Earth's surface were covered only with water, surface currents would travel across the oceans in a uniform pattern. However, continents cover about one-third of Earth's surface. When surface currents meet continents, the currents *deflect*, or change direction. The figure below shows this process.

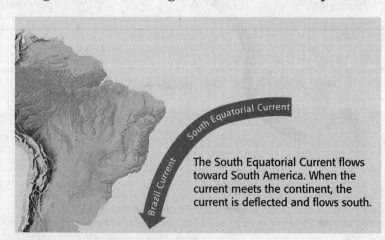

The South Equatorial Current flows toward South America. When the current meets the continent, the current is deflected and flows south.

READING CHECK

3. Explain How do winds cause surface currents?

_____
_____
_____
_____

TAKE A LOOK

4. Predict Consequences What would probably happen to the South Equatorial Current if South America were not there?

_____
_____
_____

THE CORIOLIS EFFECT

Earth's rotation also affects the paths of surface currents. If Earth did not rotate, surface currents would flow in straight lines. However, because Earth does rotate, the currents travel along curved paths. This deflection of moving objects from a straight path because of Earth's rotation is called the **Coriolis effect**. ☑

As Earth rotates, places near the equator travel faster than places closer to the poles. This difference in speed causes the Coriolis effect. Wind or water moving from the poles to the equator is deflected to the west. Wind or water moving from the equator to the poles is deflected east. The figure below shows examples of these paths.

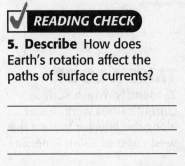
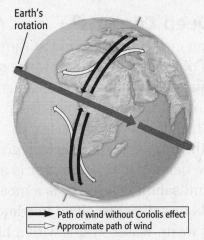

Earth's rotation

| ➡ Path of wind without Coriolis effect |
| ⇨ Approximate path of wind |

The Coriolis effect causes wind and water to move along curved paths.

TAKE A LOOK

6. Apply Concepts A surface current starts at the equator near the west coast of Africa and begins moving north. In which direction will the current end up moving?

_____

The Coriolis effect is most noticeable for things that travel very fast or travel over long distances. Over short distances or with slow-moving objects, the rotation of the Earth does not make much of a difference.

HOW SURFACE CURRENTS DISTRIBUTE HEAT

Surface currents help to move heat from one part of Earth's surface to another. Water near the equator absorbs heat energy from the sun. Then, warm-water currents carry the heat from the equator to other parts of the ocean. The heat from the warm-water currents moves into colder water or into the atmosphere. The figure on the next page shows Earth's main surface currents.

SECTION 1 Currents *continued*

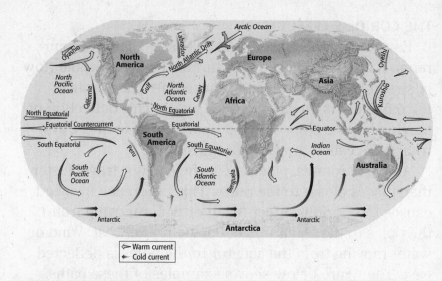

This map shows Earth's major surface currents. Surface currents help to distribute heat across Earth's surface.

TAKE A LOOK
7. Identify Which surface current carries warm water along the equator toward the west coast of South America?

_____
_____

What Are Deep Currents?

Not all ocean currents are found at the surface. Movements of ocean water far below the surface are called **deep currents**. Unlike surface currents, deep currents are not controlled by wind. Deep current movements are controlled by water density. ☑

Density is the amount of matter in a given space or volume. The density of ocean water is affected by temperature and salinity. *Salinity* is a measure of the amount of salts or solids dissolved in a liquid. Cold water is denser than warm water. Water with a high salinity is denser than water with a low salinity.

Deep currents form when the density of ocean water increases and it sinks toward the bottom of the ocean. There are three main ways that the density of ocean water can increase. The figure below shows one way. The figures on the next page show two other ways.

✓ **READING CHECK**

8. Compare How are deep currents different from surface currents? Give two ways.

_____
_____
_____
_____

Decreasing Temperature
Near the poles, heat moves from ocean water into the colder air. The water becomes colder. The particles in the water slow down and move closer together. The volume of the water decreases, which makes the water denser.

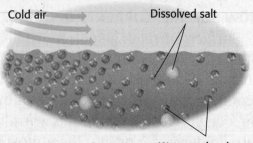

Cold air   Dissolved salt

Water molecules

SECTION 1 Currents *continued*

Increasing Salinity Through Freezing When ocean water freezes, the salt in the ocean water does not become part of the ice. The salt remains in the water that has not frozen. This process increases the salinity of the water, and the water becomes denser.

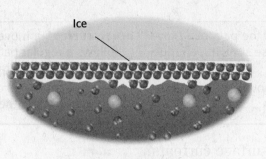
Ice

TAKE A LOOK
9. Explain How does freezing cause ocean water to become denser?

_____
_____
_____
_____
_____

Increasing Salinity Through Evaporation When ocean water evaporates, the salt in the water remains in the liquid. This process increases the salinity of the water, and the water becomes denser.

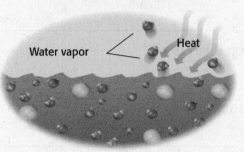
Water vapor   Heat

There are several main deep currents in the ocean. The deepest and densest water in the ocean is the *Antarctic Bottom Water*, which forms near Antarctica. *North Atlantic Deep Water* is less dense and forms in the North Atlantic Ocean. Water that is less dense stays above denser water. Therefore, North Atlantic Deep Water stays on top of Antarctic Bottom Water when the two meet.

Deep currents and surface currents are closely linked. The warm water in surface currents sinks as it cools and becomes the cold water in deep currents. The figure below shows how this happens.

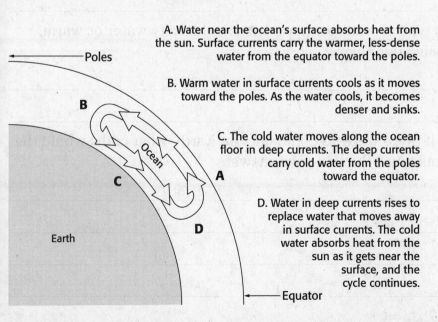

A. Water near the ocean's surface absorbs heat from the sun. Surface currents carry the warmer, less-dense water from the equator toward the poles.

B. Warm water in surface currents cools as it moves toward the poles. As the water cools, it becomes denser and sinks.

C. The cold water moves along the ocean floor in deep currents. The deep currents carry cold water from the poles toward the equator.

D. Water in deep currents rises to replace water that moves away in surface currents. The cold water absorbs heat from the sun as it gets near the surface, and the cycle continues.

Critical Thinking
10. Infer Why do most deep currents form near the poles?

_____
_____
_____
_____

Section 1 Review

SECTION VOCABULARY

Coriolis effect the curving of the path of a moving object from an otherwise straight path due to the Earth's rotation	**ocean current** a movement of ocean water that follows a regular pattern
deep current a streamlike movement of ocean water far below the surface	**surface current** a horizontal movement of ocean water that is caused by wind and that occurs at or near the ocean's surface

1. Identify What causes surface currents?

_____

2. Identify What causes deep currents?

_____

3. Describe What three factors control the path of a surface current?

_____

_____

_____

4. List Give three ways that the density of ocean water can increase.

_____

_____

_____

5. Explain What causes the Coriolis effect?

_____

_____

6. Apply Concepts Which type of water is more dense: cold, salty water or warm, less salty water? Explain your answer.

_____

_____

6. Predict Consequences If there were no continents on Earth, what paths would the ocean's surface currents take? Explain your answer.

_____

_____

_____

_____

CHAPTER 14 The Movement of Ocean Water

SECTION
2 **Currents and Climate**

BEFORE YOU READ

After you read this section, you should be able to answer these questions:

• How do surface currents affect climate?

• How do changes in surface currents affect climate?

How Do Surface Currents Affect Climate?

Surface currents can have a large impact on climate. The temperature of the water at the surface of the ocean affects the air above it. Warm water can heat air and produce warmer air temperatures. Cold water can absorb heat and produce cooler air temperatures.

WARM-WATER CURRENTS AND CLIMATE

Surface currents can make coastal areas warmer than inland areas at the same latitude. For example, Great Britain and Newfoundland, Canada, are located at about the same latitude. However, the Gulf Stream flows close to Great Britain. The warm water of the Gulf Stream warms the air around Great Britain. As a result, Great Britain has a milder climate than Newfoundland.

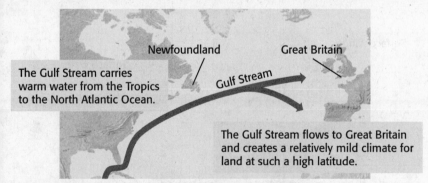

Newfoundland   Great Britain

The Gulf Stream carries warm water from the Tropics to the North Atlantic Ocean.

Gulf Stream

The Gulf Stream flows to Great Britain and creates a relatively mild climate for land at such a high latitude.

COLD-WATER CURRENTS AND CLIMATE

Cold-water currents also affect coastal areas. Coastal areas near cold currents tend to have cooler climates than inland areas at the same latitude. For example, the California Current is a cold-water current that flows near the West Coast of the United States. As a result, the climate along the West Coast is usually cooler than the climate of areas further inland. The figure on the top of the next page shows the location of the California Current.

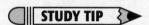

STUDY TIP

Summarize in Pairs Read this section quietly to yourself. Then, talk about the section with a partner. Together, try to answer any questions that you have.

STANDARDS CHECK

ES 1j Global patterns of atmospheric movement influence local weather. Oceans have a <u>major</u> effect on climate, because water in the oceans holds a large amount of heat.

Word Help: <u>major</u>
of great importance or large scale

1. Explain Why is Great Britain's climate milder than Newfoundland's?

_____

_____

_____

_____

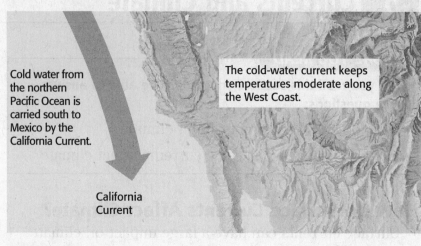

Cold water from the northern Pacific Ocean is carried south to Mexico by the California Current.

The cold-water current keeps temperatures moderate along the West Coast.

California Current

TAKE A LOOK
2. Identify In which direction does the California Current flow?

_____

UPWELLING

Ocean **upwelling** happens when cold, nutrient-rich water from the deep ocean replaces warm surface water. Upwelling is caused by local winds. These winds blow toward the equator along the northwest coast of South America and west coast of North America. The winds cause the local surface currents to move away from the shore. Cold water then replaces the warm surface water. ☑

Upwelling is important for ocean life. Nutrients support the growth of *plankton*, which are the base of the food chain in the ocean. Climate disturbances, such as El Niño, can interrupt the process of upwelling. This causes the diversity of organisms near the ocean's surface to decrease.

✓ **READING CHECK**
3. Define What is upwelling?

_____

_____

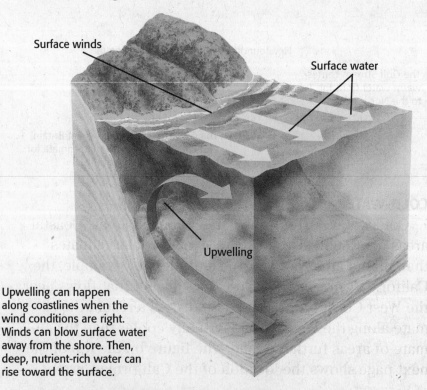

Surface winds

Surface water

Upwelling

TAKE A LOOK
4. Explain How do winds cause upwelling?

_____

_____

_____

_____

Upwelling can happen along coastlines when the wind conditions are right. Winds can blow surface water away from the shore. Then, deep, nutrient-rich water can rise toward the surface.

SECTION 2 Currents and Climate *continued*

How Do El Niño and La Niña Affect Climate?

Every 2 to 12 years, the South Pacific trade winds move less warm water to the western Pacific than usual. As a result, surface-water temperatures along the west coast of South America rise. Over time, this warming spreads westward. This periodic change in the location of warm and cool surface waters is called **El Niño**. El Niño events can last a year or longer.

Sometimes, El Niño is followed by **La Niña**. La Niña happens when surface-water temperatures in the eastern Pacific become unusually cool. La Niña also affects weather patterns.

Name	When does it happen?
El Niño	
La Niña	

EFFECTS OF EL NIÑO

El Niño can have a major effect on weather patterns. Flash floods and mudslides may happen in areas of the world that usually receive little rain, such as Peru. Other areas of the world, such as Indonesia and Australia, may receive less rain than usual.

El Niño changes the way the ocean and atmosphere interact. Changes in the weather during El Niño show how the atmosphere, ocean, and weather patterns are related. Scientists can predict the changes in the weather on land that might be caused by El Niño by studying the atmosphere and the ocean.

To study El Niño, scientists collect data with buoys anchored to the ocean floor along the equator. The buoys record data about water temperature, air temperature, currents, and winds. The data sometimes show that the South Pacific trade winds are weaker than usual. The data may also show that the surface-water temperatures in the oceans have increased. Either of these changes can tell scientists that El Niño is likely to happen.

Say It

Discuss You may have heard news reports about the effects of El Niño. In a small group, talk about some of the effects of El Niño that you heard about on the news.

TAKE A LOOK
5. Describe Fill in the blank spaces in the table.

Critical Thinking

6. Apply Concepts El Niño happens when there is warmer water near the west coast of South America. Why do scientists collect information about air temperatures in order to help them predict El Niño and La Niña?

_____

_____

_____

Section 2 Review

SECTION VOCABULARY

El Niño a change in the surface water temperature in the Pacific Ocean that produces a warm current	**La Niña** a change in the eastern Pacific Ocean in which the surface water temperature becomes unusually cool
	upwelling the movement of deep, cold, and nutrient-rich water to the surface

1. Explain Why do surface-water temperatures on the west coast of South America rise during El Niño?

_____

_____

_____

2. Apply Concepts City A and City B are the same height above sea level. Based on the figure below, make a prediction about the average temperature in City A compared to City B. Explain your answer.

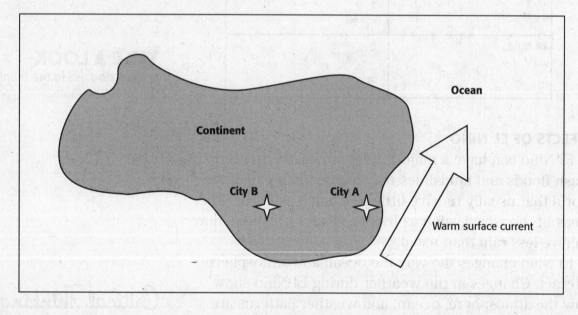

_____

_____

_____

3. Explain Why is upwelling important for marine life?

_____

_____

_____

_____

SECTION
3 # Waves

National Science
Education Standards
ES 1b

BEFORE YOU READ

**After you read this section, you should be able to answer
these questions:**

• How do waves form?

• What are the parts of a wave?

• How do waves move?

How Do Ocean Waves Form?

A *wave* is any disturbance that carries energy through
matter or empty space. Waves in the ocean carry energy
through water.

Ocean waves form when energy is transferred from a
source to the ocean water. The source of energy for most
ocean waves is the wind. Most ocean waves form as wind
blows across the water's surface. However, the energy
for some waves comes from earthquakes or meteorite
impacts. ☑

Ocean waves can travel at different speeds. They can
be very small or extremely large. The size and speed of a
wave depend on the amount of energy the wave carries.

PARTS OF A WAVE

Waves are made up of two main parts: crests and
troughs. A *crest* is the highest point of the wave. A *trough*
is the lowest point of the wave. The distance between
one crest and the next, or between one trough and the
next, is the *wavelength*. The distance in height between
the crest and the trough is called the *wave height*.

STUDY TIP

Summarize As you read,
underline the main ideas in
each paragraph. When you
finish reading, write a short
summary of the section using
the ideas you underlined.

✓ READING CHECK

1. Identify Give two sources
of energy that can cause
ocean waves.

_____

_____

_____

Parts of a Wave

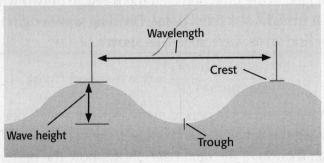

Wavelength

Crest

Wave height

Trough

TAKE A LOOK
2. Define Write your own
definition for *wavelength*.

_____

_____

_____

SECTION 3 Waves *continued*

Critical Thinking

3. Predict Consequences
People who own boats often leave the boats anchored a short distance away from the shore. The boats stay in about the same place over many days. What would happen to these boats if waves caused water to move horizontally?

_____

_____

_____

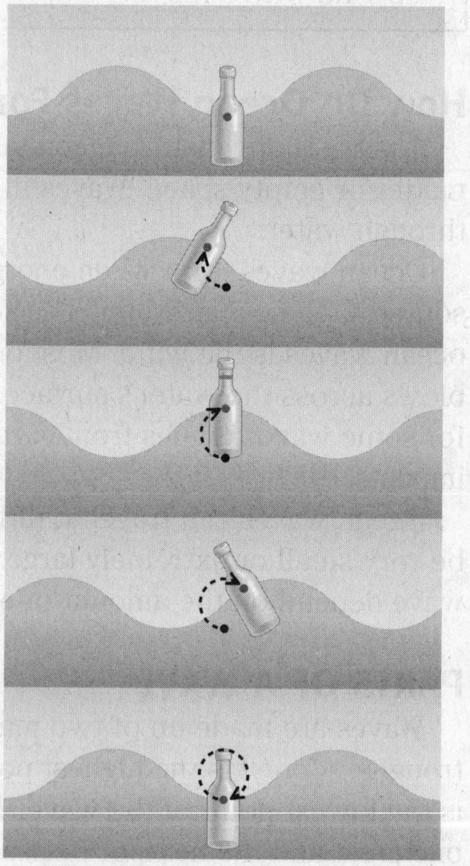

TAKE A LOOK
4. Describe What is the shape of the path that the bottle takes as the wave passes by it?

_____

_____

Math Focus

5. Calculate A water wave has a speed of 5 m/s. If its wavelength is 50 m, what is its wave period? Show your work.

_____

_____

_____

MOVEMENT OF WAVES

If you have ever watched ocean waves, you may have noticed that water seems to move across the ocean's surface. However, this movement is only an illusion. The energy in the wave causes the water to rise and fall in circular movements. The water does not move horizontally very much. The figure below shows how waves can move energy without moving water horizontally.

The bottle shows the circular motion of matter when a wave moves in the ocean. The energy in the wave makes matter near the surface move in circular motions. The matter does not move horizontally.

WAVE SPEED

Waves travel at different speeds. To calculate wave speed, scientists must know the wavelength and the wave period. *Wave period* is the time between the passage of two wave crests or troughs at a fixed point. Dividing wavelength by wave period gives wave speed, as shown below.

$$\frac{\text{wavelength (m)}}{\text{wave period (s)}} = \text{wave speed (m/s)}$$

Increasing the wave period decreases the wave speed. Decreasing the wave period increases the wave speed. The figure on the top of the next page shows how the period of a wave can be measured.

SECTION 3 Waves *continued*

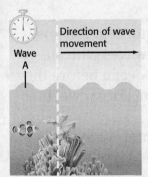 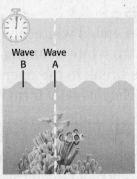 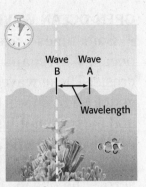

Direction of wave movement

Wave A

Wave B Wave A

Wave B Wave A

Wavelength

1. The waves are moving from left to right. The reef is a fixed point because it is not moving. The dotted line marks the center of the reef.

2. The timer begins running as the crest of Wave A passes the center of the reef.

3. The timer stops when the crest of Wave B passes the center of the reef. The time that the timer recorded, 5 s, is the wave period.

TAKE A LOOK
6. Identify What part of Wave B is passing the reef when the timer is stopped?

_____

_____

DEEP-WATER WAVES AND SHALLOW-WATER WAVES

You may have seen ocean waves get taller as they move toward the shore. This happens because the depth of the water affects the size and shape of the waves. *Deep-water waves* are waves that move in water deeper than one-half their wavelength. When waves reach water shallower than one-half their wavelength, they begin to interact with the ocean floor. This produces *shallow-water waves*.

As waves begin to touch the ocean floor, they transfer energy from the water to the ocean floor. As a result, the water at the bottom of the waves slows down. However, the water at the top of the wave continues to travel at the original speed. Eventually, the wave crest crashes onto the shore as a *breaker*. The area where breakers start to form is called the *breaker zone*. The area between the breaker zone and the shore is called the *surf*.

Critical Thinking

7. Apply Concepts An ocean wave has a wavelength of 60 m. It is traveling through water that is 40 m deep. Is it a shallow-water wave or a deep-water wave?

_____

_____

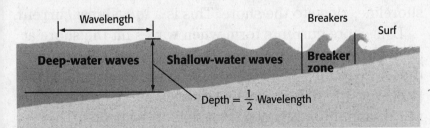

Wavelength

Breakers

Surf

Deep-water waves   **Shallow-water waves**   **Breaker zone**

Depth = ½ Wavelength

Deep-water waves become shallow-water waves when they reach depths of less than half of their wavelength.

TAKE A LOOK
8. Describe What are breakers?

_____

_____

OPEN-OCEAN WAVES

Sometimes waves called whitecaps form in the open ocean. **Whitecaps** are white, foaming waves with steep crests. These waves break in the open ocean before they get close to shore. They usually form in stormy weather, and most do not last very long.

Winds that are far from shore can form waves called swells. **Swells** are rolling waves that move steadily across the ocean. They have longer wavelengths than whitecaps. Swells can travel for thousands of kilometers.

What Are Some Effects of Waves?

After waves crash on the beach, the water glides back to the ocean. It flows underneath the incoming waves. This kind of water movement produces a current called **undertow**. Undertow carries sand and pieces of rock away from the shore.

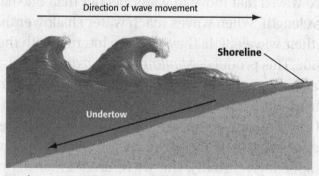

Direction of wave movement

Shoreline

Undertow

Head-on waves create an undertow.

LONGSHORE CURRENTS

Sometimes, water moves in a current parallel to the shoreline, close to the shore. This is a **longshore current**.

Longshore currents form when waves hit the shore at an angle instead of head-on. The waves wash sand onto the shore at the same angle that the waves are moving. However, when the waves wash back into the ocean, they move sand directly down the slope of the beach. This causes the sand to move in a zigzag pattern. The figure on the next page shows how longshore currents form.

Longshore currents transport most of the sediment on beaches. This movement of sand erodes and builds up the coastline. Longshore currents can also carry and spread trash and pollution along the shore.

Critical Thinking

9. Compare How are whitecaps different from swells? Give two ways.

_____

_____

_____

_____

TAKE A LOOK

10. Identify Does an undertow current move toward the shore or away from the shore?

_____

_____

SECTION 3 Waves *continued*

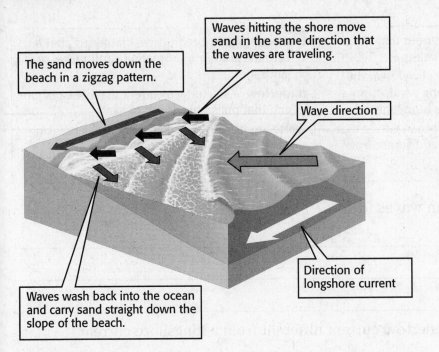

The sand moves down the beach in a zigzag pattern.

Waves hitting the shore move sand in the same direction that the waves are traveling.

Wave direction

Waves wash back into the ocean and carry sand straight down the slope of the beach.

Direction of longshore current

11. Infer Why don't longshore currents form in places where waves hit the shore head-on?

_____

_____

_____

_____

TSUNAMIS

Tsunamis are waves that form when a large volume of ocean water suddenly moves. Most tsunamis are caused by movement from underwater earthquakes. However, a volcanic eruption, a landslide, an explosion, or a meteorite impact can also cause a tsunami. Most tsunamis occur in the Pacific Ocean because of the many earthquakes there.

STORM SURGES

Severe storms, such as hurricanes, can blow ocean water into a large "pile" near the shore This causes sea level to rise in the area near the storm. The local rise in sea level near the shore is called a **storm surge**. As the storm moves onto shore, so does the giant mass of water beneath it. This huge amount of water can cause serious flooding. A storm surge can be the most destructive part of a hurricane.

Storm surges contain a lot of energy and can reach about 8 m in height. That is as tall as a two-story building! Storm surges often disappear as quickly as they form. This makes them difficult to study.

Critical Thinking

12. Apply Concepts Do most tsunamis probably form near plate boundaries or far from them? Explain your answer.

_____

_____

_____

_____

Name ______________________ Class ______________ Date ______________

Section 3 Review

NSES ES 1b

SECTION VOCABULARY

longshore current a water current that travels near and parallel to the shoreline

storm surge a local rise in sea level near the shore that is caused by strong winds from a storm, such as those from a hurricane

swell one of a group of long ocean waves that have steadily traveled a great distance from their point of generation

tsunami a giant ocean wave that forms after a volcanic eruption, submarine earthquake, or landslide

undertow a subsurface current that is near shore and that pulls objects out to sea

whitecap the bubbles in the crest of a breaking wave

1. Describe How do ocean waves form?

____

____

____

2. Compare How is an undertow current different from a longshore current?

____

____

____

3. Calculate A wave has a wave period of 20 s and a wavelength of 100 m. What is its speed? How would the speed change if the wave period increased?

____

____

4. List Name five events that can cause a tsunami.

____

____

____

____

____

5. Summarize What is a storm surge? Why are storm surges difficult to study?

____

____

____

Name ______________________ Class ______________ Date ______________

CHAPTER 14 The Movement of Ocean Water

SECTION 4

Tides

BEFORE YOU READ

After you read this section, you should be able to answer these questions:

- What causes tides?
- How do tides vary?

National Science Education Standards

ES 3c

What Are Tides?

Remember that wind can move ocean water and produce waves. Other forces can also move ocean water in regular patterns, such as tides. **Tides** are daily changes in the level of the ocean water. Both the sun and the moon influence the level of tides.

STUDY TIP

Compare As you read, make a chart describing the causes and features of high tides, low tides, spring tides, and neap tides.

WHY TIDES HAPPEN

The moon's gravity pulls on every particle on Earth. However, the moon's gravity doesn't pull on every particle with the same strength. The moon's gravitational pull on Earth decreases with distance from the moon. Therefore, the pull on some parts of Earth is stronger than on others.

The part of Earth that faces the moon is pulled toward the moon with the greatest force. Therefore, the water on the side of Earth that faces the moon bulges toward the moon. The water on Earth's opposite side is pulled toward the moon the least. Therefore, it bulges away from the moon. The figure below shows these bulges.

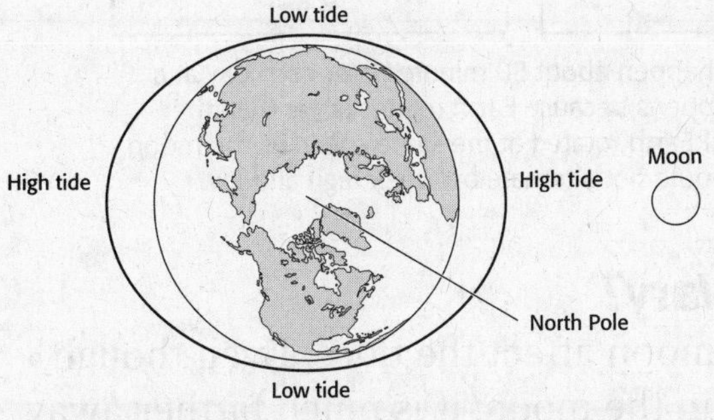

Water bulges toward the moon on the side of Earth that faces the moon. Water bulges away from the moon on Earth's far side. As a result, these two sides of Earth experience high tide. In this image, the sizes and locations of Earth, the oceans, and the moon are not drawn to scale.

STANDARDS CHECK

ES 3c Gravity is the force that keeps planets in orbit around the sun and governs the rest of the motion in the solar system. Gravity alone holds us to the earth's surface and explains the phenomena of the tides.

Word Help: phenomenon any fact or event that can be sensed or described scientifically (plural, *phenomena*)

1. Identify What causes the tides?

____

____

____

The difference in the moon's pull is more noticeable in liquids than in solids because liquids can move more easily. Therefore, the effects of the moon's pull on the oceans are more noticeable than on the land.

SECTION 4 Tides *continued*

HIGH TIDES AND LOW TIDES

The bulges that form in the oceans because of the moon's pull are called *high tides*. In high-tide areas, the water level is higher than average sea level. In areas between high tides, *low tides* form. In low-tide areas, the water level is lower than average sea level. This happens because the water is pulled toward high-tide areas.

Remember that Earth rotates on its axis. As a result, high tides happen in different places on Earth at different times of day. However, because Earth's rotation is predictable, the tides are also predictable. Many places on Earth experience two high tides and two low tides every day. ☑

TIMING THE TIDES

The moon revolves around Earth more slowly than Earth rotates. A place that is facing the moon takes 24 h and 50 min to rotate to face the moon again. Therefore, high and low tides at that place happen about 50 minutes later each day.

✓ **READING CHECK**

2. Describe Why are the tides predictable?

_____

_____

_____

_____

Math Focus

3. Calculate An area experiences high tide at 9:30 A.M. on Monday. At about what time will it experience high tide on Thursday?

_____

_____

_____

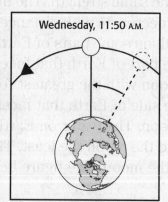

High and low tides happen about 50 minutes later each day at a given place. This happens because Earth rotates faster than the moon orbits Earth. If Earth rotated at the same speed as the moon orbits Earth, tides would not alternate between high and low.

How Do Tides Vary?

The sun and the moon affect the tides. Even though the sun is bigger than the moon, it is much farther away from Earth than the moon is. Therefore, the sun's effect on tides is less than the moon's. The combined forces of the sun and the moon on Earth produce different tidal ranges. A **tidal range** is the difference between levels of ocean water at high tide and low tide. ☑

✓ **READING CHECK**

4. Explain Why does the sun affect the tides less than the moon does?

_____

_____

_____

SECTION 4 Tides *continued*

SPRING TIDES

Tides that have the largest daily tidal range are **spring tides**. Spring tides happen when the sun, Earth, and the moon are aligned, as shown in the figures below. Spring tides happen during the new-moon and full-moon phases, or every 14 days. During these times, the pull of the sun and moon produces one pair of very large tidal bulges.

Spring tides happen when the sun, the moon, and Earth are aligned. This can happen in two ways. One way is when the moon is between Earth and the sun, as shown in the left-hand figure. The other way is when Earth is between the moon and the sun, as shown in the right-hand figure.

TAKE A LOOK

5. Describe Draw an oval around Earth in each picture to show where the tides are highest and where they are lowest during spring tides.

NEAP TIDES

Tides that have the smallest daily tidal range are called **neap tides**. Neap tides happen when the sun, Earth, and the moon form a 90° angle, as shown in the figures below. They happen halfway between the spring tides, during the first-quarter and third-quarter phases of the moon. During these times, the pull of the sun and moon produces smaller tidal bulges.

Neap tides happen when the sun, the moon, and Earth form a 90° angle.

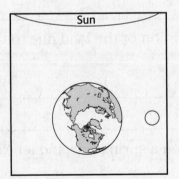 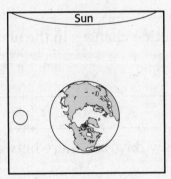

Critical Thinking

6. Apply Ideas If you have a calendar that shows only the phases of the moon, can you predict when spring tides and neap tides will happen? Explain your answer.

_____

_____

_____

_____

_____

Section 4 Review

NSES ES 3c

SECTION VOCABULARY

neap tide a tide of minimum range that occurs during the first and third quarters of the moon	**tidal range** the difference in levels of ocean water at high tide and low tide
spring tide a tide of increased range that occurs two times a month, at the new and full moons	**tide** the periodic rise and fall of the water level in the oceans and other large bodies of water

1. Explain How are high tides different from low tides?

_____

_____

_____

2. Describe Fill in the blank spaces in the table below.

Tide	Tidal range: small or large?	When it happens
Neap tide		
Spring tide		

3. Explain Why do high tides happen in different places at different times of day?

_____

_____

4. Identify What produces tidal ranges?

_____

_____

5. Explain Why don't we notice changes in the elevation of the land due to the moon's pull?

_____

_____

6. Apply Concepts How many days are there between a spring tide and a neap tide? Explain your answer.

_____

_____

_____

CHAPTER 15 The Atmosphere

SECTION 1

Characteristics of the Atmosphere

National Science
Education Standards
ES 1h

BEFORE YOU READ

After you read this section, you should be able to answer these questions:

• What is Earth's atmosphere made of?

• How do air pressure and temperature change as you move away from Earth's surface?

• What are the layers of the atmosphere?

What Is Earth's Atmosphere Made Of?

An **atmosphere** is a layer of gases that surrounds a planet or moon. On Earth, the atmosphere is often called just "the air." When you take a breath of air, you are breathing in atmosphere.

The air you breathe is made of many different things. Almost 80% of it is nitrogen gas. The rest is mostly oxygen, the gas we need to live. There is also water in the atmosphere. Some of it is invisible, in the form of a gas called *water vapor*. ☑

Water is also found in the atmosphere as water droplets and ice crystals, like those that make up clouds. The atmosphere also contains tiny *particles*, or solid pieces. These particles are things like dust and dirt from continents, salt from oceans, and ash from volcanoes.

STUDY TIP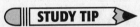

Define When you come across a word you don't know, circle it. When you figure out what it means, write the word and its definition in your notebook.

☑ **READING CHECK**

1. List Which two gases make up most of Earth's atmosphere?

_____

_____

Gases in Earth's Atmosphere

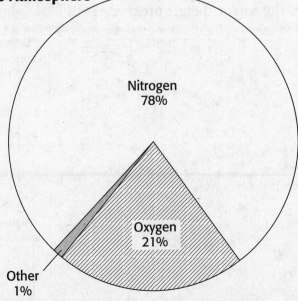

Nitrogen
78%

Oxygen
21%

Other
1%

Math Focus

2. Analyze Data About what fraction of the Earth's atmosphere is NOT made of nitrogen? Give your answer as a reduced fraction.

_____

Where Do the Gases in the Atmosphere Come From?

The gases in Earth's atmosphere come from many different sources. The table below shows some of those sources.

Gas	Where the gas comes from
Oxygen	Plants give off oxygen as they grow.
Nitrogen	Nitrogen is given off when dead plants and animals decay.
Water vapor	Liquid water evaporates and becomes water vapor. Plants give off water vapor as they grow. Water vapor comes out of the Earth during volcanic eruptions.
Carbon dioxide	Carbon dioxide comes out of the Earth during volcanic eruptions. When animals breathe, they give off carbon dioxide. Carbon dioxide is given off when we burn things that were once plant or animal material.

TAKE A LOOK

3. Identify Name two gases that volcanoes contribute to the atmosphere.

_____

_____

Why Does Air Pressure Change with Height?

Air pressure is how much the air above you weighs. It is a measure of how hard air molecules push on a surface. We don't normally notice air pressure, because our bodies are used to it. ☑

As you move up from the ground and out toward space, there are fewer gas molecules pressing down from above. Therefore, the air pressure drops. The higher you go, the lower the air pressure gets.

☑ READING CHECK

4. Define Write your own definition for air pressure.

_____

_____

TAKE A LOOK

5. Compare How is the air pressure around the tree different from the air pressure around the plane?

_____

_____

_____

Lower pressure

Higher pressure

Why Does Air Temperature Change with Height?

Like air pressure, air temperature changes as you move higher in the atmosphere. Air pressure always gets lower as you move higher, but air temperature can get higher or lower. The air can get hotter or colder. ☑

There are different layers of the atmosphere. Each layer is made of a different combination of gases. Air temperature depends on the gases in the atmosphere. Some gases absorb energy from the sun better than others. When a gas absorbs energy from the sun, the air temperature goes up.

What Are the Layers of the Atmosphere?

There are four main layers of the atmosphere: troposphere, stratosphere, mesosphere, and thermosphere. You cannot actually see these different layers. The divisions between the layers are based on how each layer's temperature changes with height.

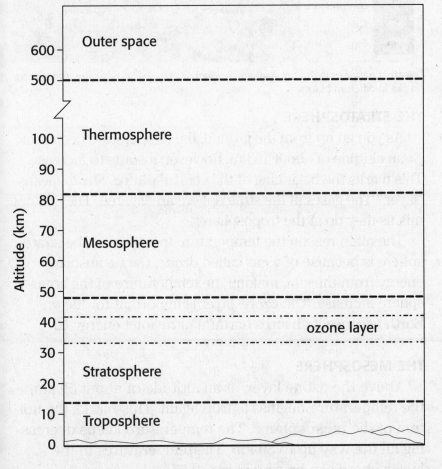

✓ **READING CHECK**

6. Compare How are the changes in air temperature with height different from changes in air pressure with height?

_____

_____

_____

📣 **Say It**

Make Up a Memory Trick In groups of two or three, make up a sentence to help you remember the order of the layers of the atmosphere. The words in the sentence should start with T, S, M, and T. For example, "Tacos Sound Mighty Tasty." A sentence like this is called a *mnemonic*.

TAKE A LOOK

7. Identify At what altitude does the mesosphere end and the thermosphere begin?

_____

SECTION 1 Characteristics of the Atmosphere *continued*

Critical Thinking

8. Explain Why is the troposphere important to people?

_____

_____

_____

_____

TAKE A LOOK

9. Analyze What does the map tell you about the air temperature in the troposphere?

_____

_____

_____

THE TROPOSPHERE

The troposphere is the layer of the atmosphere that we live in. It is where most of the water vapor, carbon dioxide, pollution, and living things on Earth exist. Weather conditions such as wind and rain all take place in the troposphere.

The troposphere is also the densest layer of the atmosphere. This is because the troposphere is at the bottom with all the other layers pushing down from above. Almost 90% of the gases in the atmosphere are in the troposphere. As you move higher into the troposphere (say, to the top of a mountain), both air temperature and air pressure decrease.

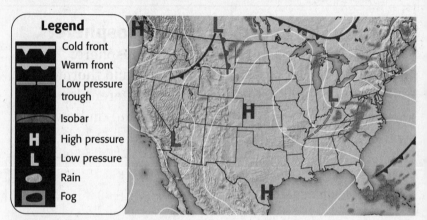

Weather happens in the troposphere. A weather map shows what the troposphere is like in different places.

THE STRATOSPHERE

As you go up from the ground, the temperature decreases. At an altitude of about 15 km, however, it starts to increase. This marks the beginning of the stratosphere. *Strato* means "layer." The gases in the stratosphere are layered. They do not mix as they do in the troposphere.

The main reason the temperature increases in the stratosphere is because of a gas called *ozone*. Ozone absorbs energy from the sun, making the temperature of the atmosphere increase. The ozone layer is important for life on Earth because it absorbs harmful ultraviolet energy. ☑

READING CHECK

10. Explain Why is ozone in the stratosphere important for living things?

_____

_____

_____

THE MESOSPHERE

Above the ozone layer, at an altitude of about 50 km, the temperature begins to drop again. This marks the bottom of the mesosphere. The temperature keeps decreasing all the way up to 80 km. The temperatures in the mesosphere can be as low as −93°C.

THE THERMOSPHERE

The **thermosphere** is the uppermost layer of the atmosphere. In the thermosphere, temperatures begin to rise again. The thermosphere gets its name from its extremely high temperatures, which can be above 1,000°C. *Therm* means "heat." The temperatures in the thermosphere are so high because it contains a lot of oxygen and nitrogen, which absorb energy from the sun. ☑

THE IONOSPHERE—ANOTHER LAYER

The troposphere, stratosphere, mesosphere, and thermosphere are the four main layers of the atmosphere. However, scientists also sometimes study a region called the ionosphere. The *ionosphere* contains the uppermost part of the mesosphere and the lower part of the thermosphere. It is made of nitrogen and oxygen *ions*, or electrically charged particles.

The ionosphere is where auroras occur. *Auroras* are curtains and ribbons of shimmering colored lights. They form when charged particles from the sun collide with the ions in the ionosphere. The ionosphere is important to us because it can reflect radio waves. An AM radio wave can travel all the way around the Earth by bouncing off the ionosphere.

✓ **READING CHECK**

11. Explain Why is the thermosphere called the thermosphere?

_____

_____

_____

Layer	How temperature and pressure change as you move higher	Important features
Troposhere	temperature decreases pressure decreases	
Stratosphere		gases are arranged in layers contains the ozone layer
		has the lowest temperatures
Thermosphere	temperature increases pressure decreases	

TAKE A LOOK

12. Identify Use the information from the text to fill in the table.

Section 1 Review

SECTION VOCABULARY

air pressure the measure of the force with which air molecules push on a surface	**stratosphere** the layer of the atmosphere that is above the troposphere and in which temperature increases as altitude increases
atmosphere a mixture of gases that surrounds a planet or moon	**thermosphere** the uppermost layer of the atmosphere, in which temperature increases as altitude increases
mesosphere the layer of the atmosphere between the stratosphere and the thermosphere and in which temperature decreases as altitude increases	**troposphere** the lowest layer of the atmosphere, in which temperature decreases at a constant rate as altitude increases

1. **Define** Write your own definition for atmosphere.

_____

_____

_____

2. **Explain** Why does air temperature change as you move up from the Earth's surface?

_____

_____

_____

3. **Make a Graph** The graph below shows how the temperature changes as you move up through the atmosphere. On the graph, draw a curve showing how the pressure changes.

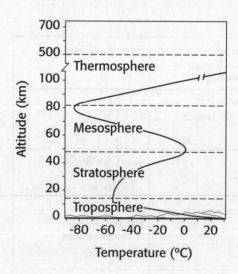

4. **Identify Relationships** How does the sun affect air temperatures?

_____

_____

CHAPTER 15 The Atmosphere

SECTION
2 # Atmospheric Heating

After you read this section, you should be able to answer these questions:

• How does energy travel from the sun to Earth?
• What are the differences between radiation, conduction, and convection?
• Why is Earth's atmosphere so warm?

How Does Energy Travel from the Sun to Earth?

Most of the heat energy on Earth's surface comes from the sun. Energy travels from the sun to Earth by **radiation**, which means that it travels through space as waves. As solar energy (energy from the sun) is absorbed by air, water, and land, it turns into heat energy. This energy causes winds, the water cycle, ocean currents, and changes in the weather.

What Happens to Radiation from the Sun?

Not all of the radiation from the sun reaches Earth's surface. Much of it gets absorbed by the atmosphere. Some of it is scattered and reflected by clouds and gases.

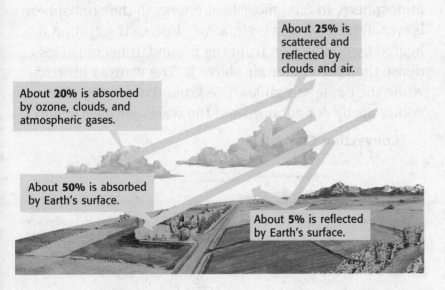

About **25%** is scattered and reflected by clouds and air.

About **20%** is absorbed by ozone, clouds, and atmospheric gases.

About **50%** is absorbed by Earth's surface.

About **5%** is reflected by Earth's surface.

STUDY TIP

Outline In your notebook, write an outline of this chapter. Use the questions in bold to make your outline. As you read, fill in information about each question.

TAKE A LOOK
1. Identify How much of the sunlight that gets to Earth is absorbed by Earth's surface?

_____

_____

2. Summarize What happens to the sunlight that is not absorbed by Earth's surface?

_____

_____

_____

How Is Heat Transferred by Contact?

Once sunlight is absorbed by Earth's surface, it is *converted*, or changed, into heat energy. Then, the heat can be transferred to other objects and moved to other places. When a warm object touches a cold object, heat moves from the warm object to the cold one. This movement of heat is called **thermal conduction**.

When you touch the sidewalk on a hot, sunny day, heat energy is conducted from the sidewalk to you. The same thing happens to air molecules in the atmosphere. When they touch the warm ground, the air molecules heat up. ☑

How Is Heat Energy Transferred by Motion?

If you have ever watched a pot of water boil, you have seen convection. During **convection**, warm material, such as air or water, carries heat from one place to another.

When you turn on the stove under a pot of water, the water closest to the pot heats up. As the water heats up, its density decreases. The warm water near the pot is not as dense as the cool water near the air. Therefore, the cool water sinks while the warm water rises.

As it rises, the warm water begins to cool. When it cools, its density increases. It becomes denser than the layer below, so it sinks back to the bottom of the pot. This forms a circular movement called a *convection current*.

Convection currents also move heat through the atmosphere. In fact, most heat energy in the atmosphere is transferred by convection. Air close to the ground is heated by conduction from the ground. It becomes less dense than the cooler air above it. The warmer air rises while the cooler air sinks. The ground warms up the cooler air by conduction, and the warm air rises again.

✓ READING CHECK

3. List Name two ways that air gets heated.

_____

_____

Critical Thinking

4. Apply Concepts Before the water in the pot can heat up, the pot itself must heat up. Does the pot heat up by conduction, convection, or radiation? Explain your answer.

_____

_____

_____

_____

TAKE A LOOK

5. Describe What happens to warm air as it moves through the atmosphere?

_____

_____

Convection Current

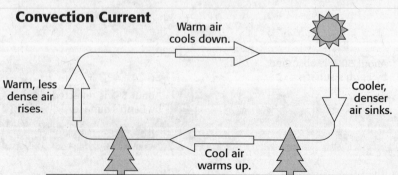

Warm air cools down.

Warm, less dense air rises.

Cooler, denser air sinks.

Cool air warms up.

How Does the Earth Stay Warm?

A gardener who needs to keep plants warm uses a glass building called a greenhouse. Light travels through the glass into the building, and the air and plants inside absorb the energy. The energy is converted to heat, which cannot travel back through the glass as easily as light came in. Much of the heat energy stays trapped within the greenhouse, keeping the air inside warmer than the air outside.

Earth's atmosphere acts like the glass walls of a greenhouse. Sunlight travels through the atmosphere easily, but heat does not. Gases in the atmosphere, such as water vapor and carbon dioxide, absorb heat energy coming from Earth. Then, they radiate it back to Earth's surface. This is known as the **greenhouse effect**. ☑

The Greenhouse Effect

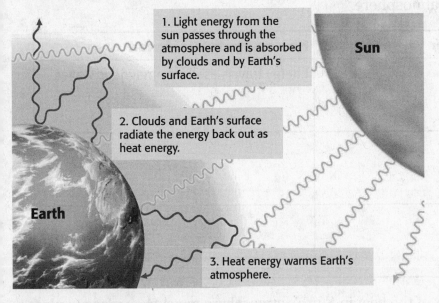

1. Light energy from the sun passes through the atmosphere and is absorbed by clouds and by Earth's surface.

Sun

2. Clouds and Earth's surface radiate the energy back out as heat energy.

Earth

3. Heat energy warms Earth's atmosphere.

READING CHECK

6. List Name two gases in Earth's atmosphere that absorb heat.

_____

_____

TAKE A LOOK
7. Identify On the drawing, label the light coming from the sun with an **L**. Label the heat energy that is trapped by Earth's atmosphere with an **H**.

What Is Global Warming?

Many scientists are worried that Earth has been getting warmer over the past hundred years. This increase in temperatures all over the world is called **global warming**.

Scientists think that human activities may be causing global warming. When we burn fossil fuels, we release greenhouse gases, such as carbon dioxide, into the atmosphere. Because greenhouse gases trap heat in the atmosphere, adding more of them can make Earth even warmer. Global warming can have a strong effect on weather and climate.

Say It

Predict How might global warming affect your community? What can you do to slow global warming? In groups of two or three, discuss how global warming might affect your lives.

Section 2 Review

SECTION VOCABULARY

convection the transfer of thermal energy by the circulation or movement of a liquid or gas	**radiation** the transfer of energy as electro-magnetic waves
global warming a gradual increase in average global temperature	**thermal conduction** the transfer of energy as heat through a material
greenhouse effect the warming of the surface and lower atmosphere of Earth that occurs when water vapor, carbon dioxide, and other gases absorb and reradiate thermal energy	

1. Apply Concepts A person is camping outside. The person toasts a marshmallow by holding it above the flames of the fire. Does the marshmallow cook because of convection, conduction, or radiation? Explain your answer.

_____

_____

2. Compare Fill in the table below to name and describe the three ways energy is transferred in Earth's atmosphere.

Type of energy transfer	How energy is transferred
	Energy travels as electromagnetic waves.
Conduction	

3. Explain How does most of the heat in Earth's atmosphere move from place to place?

_____

_____

4. Identify Relationships Explain how global warming and the greenhouse effect are related.

_____

_____

_____

CHAPTER 15 The Atmosphere

SECTION 3
Global Winds and Local Winds

National Science
Education Standards
ES 1j

> **BEFORE YOU READ**
>
> After you read this section, you should be able to answer these questions:
>
> • What causes wind?
>
> • What is the Coriolis effect?
>
> • What are the major global wind systems on Earth?

What Causes Wind?

Wind is moving air caused by differences in air pressure. Air moves from areas of high pressure to areas of low pressure. The greater the pressure difference, the faster the air moves, and the stronger the wind blows. ☑

You can see how air moves if you blow up a balloon and then let it go. The air inside the balloon is at a higher pressure than the air around the balloon. If you open the end of the balloon, air will rush out.

What Causes Differences in Air Pressure?

Most differences in air pressure are caused by differences in air temperature. Temperature differences happen because some parts of Earth get more energy from the sun than others. For example, the sun shines more directly on the equator than on the poles. As a result, the air is warmer near the equator. ☑

The warm air near the equator is not as dense as the cool air near the poles. Because it is less dense, the air at the equator rises, forming areas of low pressure. The cold air near the poles sinks, forming areas of high pressure. The air moves in large circular patterns called *convection cells*. The drawing on the next page shows these convection cells.

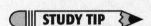
STUDY TIP

Underline Each heading in this section is a question. Underline the answer to each question when you find it in the text.

READING CHECK

1. Define What is wind?

_____

_____

TAKE A LOOK
2. Identify On the drawing, label the high-pressure area with an **H** and the low-pressure area with an **L**.

READING CHECK

3. Explain Why isn't all the air on Earth at the same temperature?

_____

_____

_____

SECTION 3 Global Winds and Local Winds *continued*

Convection Cells

TAKE A LOOK
4. Describe Is air rising or sinking in areas of high pressure?

_____

_____

✓ **READING CHECK**

5. Identify What are the three main global wind belts?

_____

_____

_____

What Are the Major Global Wind Systems?

Global winds are large-scale wind systems. There are three pairs of major global wind systems, or wind belts: trade winds, westerlies, and polar easterlies. ☑

Trade winds are wind belts that blow from 30° latitude almost to the equator. They curve to the west as they blow toward the equator. **Westerlies** are wind belts that are found between 30° and 60° latitude. The westerlies blow toward the poles from west to east. Most of the United States is located in the belt of westerly winds. These winds can carry moist air over the United States, producing rain and snow.

Polar easterlies are wind belts that extend from the poles to 60° latitude. They form as cold, sinking air moves away from the poles. In the Northern Hemisphere, polar easterlies can carry cold arctic air over the United States. This can produce snow and freezing weather.

Wind belt	Location (latitude)	Toward the equator or toward the poles?
Trade winds	0° to 30°	toward the equator
Westerlies		
	60° to 90°	

TAKE A LOOK
6. Describe Fill in the blanks in the table.

The figure on the next page shows the locations of these different wind belts. Notice that the winds do not move in straight lines. The paths of the wind belts are controlled by convection cells and by the Earth's rotation.

The trade winds meet and rise near the equator in a region known as the doldrums. The wind in the doldrums is very weak.

The region between the trade winds and the westerlies is known as the horse latitudes. Here, cool air sinks, creating a region of high pressure. The winds here are very weak.

→ Cool air
→ Warm air
→ Wind direction

There are three pairs of major global wind belts on Earth: the polar easterlies, the westerlies, and the trade winds.

STANDARDS CHECK

ES 1j Global patterns of atmospheric movement influence local weather. Oceans have a <u>major</u> effect on climate, because water in the oceans holds a large amount of heat.

Word Help: <u>major</u>
of great importance or large scale

7. Explain Use the map to explain why surface winds are generally very weak near the equator.

_____

_____

_____

_____

_____

Why Do Global Winds Curve?

Remember that pressure differences can cause air to move and form winds. If Earth did not rotate, these winds would blow in straight lines. However, because Earth does rotate, the winds follow curved paths. This *deflection*, or curving, of moving objects from a straight path because of Earth's rotation is called the **Coriolis effect**. ☑

As Earth rotates, places near the equator travel faster than places closer to the poles. This difference in speed causes the Coriolis effect. Wind moving from the poles to the equator is deflected to the west. Wind moving from the equator to the poles is deflected east.

READING CHECK

8. Describe How does Earth's rotation affect the paths of global winds?

_____

_____

_____

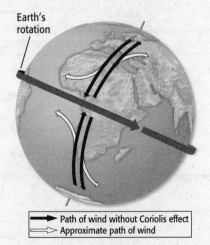

→ Path of wind without Coriolis effect
⇨ Approximate path of wind

The Coriolis effect causes wind and water to move along curved paths.

TAKE A LOOK

9. Apply Ideas If air is moving south from California, which way will it tend to curve?

_____

SECTION 3 Global Winds and Local Winds *continued*

What Are Jet Streams?

The polar easterlies, prevailing westerlies, and trade winds are all winds that we feel on the ground. However, wind systems can also form at high altitude. **Jet streams** are narrow belts of very high-speed winds in the upper troposphere and lower stratosphere. They blow from west to east all the way around the Earth. ☑

Jet streams can reach speeds of 400 km/h. Pilots flying east over the United States or the Atlantic Ocean try to catch a jet stream. This wind pushes airplanes along, helping them fly faster and use less fuel. Pilots flying west try to avoid the jet streams.

The global wind systems are always found in about the same place every day. Unlike these global wind systems, jet streams can be in different places on different days. Because jet streams can affect the movements of storms, meteorologists try to track the jet streams. They can sometimes predict the path of a storm if they know where the jet streams are.

☑ **READING CHECK**

10. Identify In what two layers of the atmosphere are the jet streams found?

_____

_____

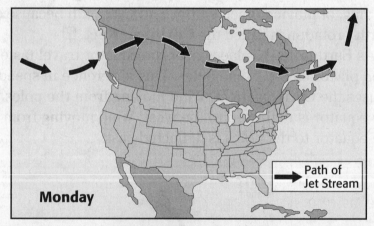
Monday
Path of Jet Stream

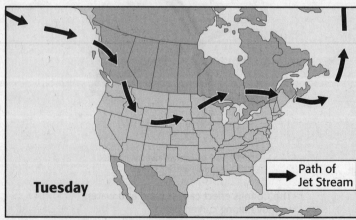
Tuesday
Path of Jet Stream

TAKE A LOOK

11. Infer Why would a pilot flying across North America take a different route on Tuesday than on Monday?

_____

_____

_____

_____

Jet streams form between hot and cold air masses. Unlike the other wind systems, jet streams are found in slightly different places every day.

What Are Local Winds?

Most of the United States is in the belt of prevailing westerly winds, which move from west to east. However, you've probably noticed that the wind in your neighborhood does not always blow from the west to the east. This is because global winds are not the only winds that blow. Local winds are also important. *Local winds* are winds that generally move over short distances and can blow from any direction.

Like the other wind systems, local winds are caused by differences in temperature. Many of these temperature differences are caused by geographic features, such as mountains and bodies of water. The figure below shows how water and mountains can affect local winds.

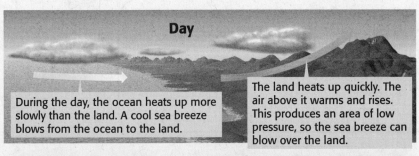

Day

During the day, the ocean heats up more slowly than the land. A cool sea breeze blows from the ocean to the land.

The land heats up quickly. The air above it warms and rises. This produces an area of low pressure, so the sea breeze can blow over the land.

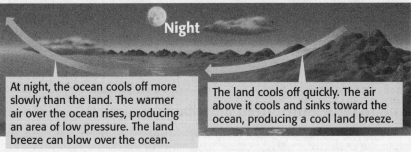

Night

At night, the ocean cools off more slowly than the land. The warmer air over the ocean rises, producing an area of low pressure. The land breeze can blow over the ocean.

The land cools off quickly. The air above it cools and sinks toward the ocean, producing a cool land breeze.

MOUNTAIN BREEZES AND VALLEY BREEZES

Mountain and valley breezes are other examples of local winds caused by geography. During the day, the sun warms the air on mountain slopes. The warm air rises up the mountain slopes, producing a warm valley breeze. At night, the air on the slopes cools. The cool air moves down the slopes, producing a cool mountain breeze.

Section 3 Review

SECTION VOCABULARY

Coriolis effect the curving of the path of a moving object from an otherwise straight path due to the Earth's rotation	**trade winds** prevailing winds that blow from east to west from 30° latitude to the equator in both hemispheres
jet stream a narrow band of strong winds that blow in the upper troposphere	**westerlies** prevailing winds that blow from west to east between 30° and 60° latitude in both hemispheres
polar easterlies prevailing winds that blow from east to west between 60° and 90° latitude in both hemispheres	**wind** the movement of air caused by differences in air pressure

1. Identify The drawing below shows a convection cell. Put arrows on the cell to show which way the air is moving. Label high pressure areas with an **H** and low pressure areas with an **L**. Label cold air with a **C** and warm air with a **W**.

2. Identify Which global wind system blows toward the poles between 30° and 60° latitude?

_____

3. Explain Why does wind tend to blow down from mountains at night?

_____

_____

4. Apply Concepts Would there be winds if Earth's surface were the same temperature everywhere? Explain your answer.

_____

_____

_____

_____

CHAPTER 15 The Atmosphere

SECTION 4 **Air Pollution**

BEFORE YOU READ

After you read this section, you should be able to answer these questions:

• What is air pollution?

• What causes air pollution?

• How does air pollution affect the environment?

• How can people reduce air pollution?

What Is Air Pollution?

Air pollution is the addition of harmful substances to the atmosphere. An *air pollutant* is anything in the air that can damage the environment or make people or other organisms sick. Some air pollution comes from natural sources. Other forms of air pollution are caused by things people do.

There are two kinds of air pollutants: primary pollutants and secondary pollutants. Primary pollutants are pollutants that are put directly into the air. Dust, sea salt, volcanic ash, and pollen are primary pollutants that come from natural sources. Chemicals from paint and other materials and vehicle exhaust are primary pollutants that come from human activities.

Secondary pollutants form when primary pollutants react with each other or with other substances in the air. Ozone is an example of a secondary pollutant. It forms on sunny days when chemicals from burning gasoline react with each other and with the air. Ozone damages human lungs and can harm other living things as well. ☑

STUDY TIP

Describe As you read, make a table describing the sources of air pollution discussed in this section.

READING CHECK

1. Explain Why is ozone called a secondary pollutant?

_____

_____

_____

_____

TAKE A LOOK

2. Describe Fill in the blanks in the table.

Pollutant	Primary pollutant or secondary pollutant?	Natural or caused by people?
Car exhaust	primary	human-caused
Dust		
Ozone		
Paint chemicals		
Pollen		
Sea salt		
Volcanic ash		

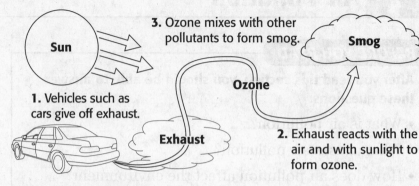

1. Vehicles such as cars give off exhaust.

Exhaust

2. Exhaust reacts with the air and with sunlight to form ozone.

Ozone

3. Ozone mixes with other pollutants to form smog.

Smog

Sun

What Is Smog?

On a hot, still, sunny day, yellowish brown air can cover a city. This is called *smog*. Smog forms when ozone mixes with other pollutants. During summer in cities such as Los Angeles, a layer of warm air can trap smog near the ground. In the winter, a storm can clear the air.

 **Say It**

This is what Los Angeles looks like on a clear day.

This is what Los Angeles looks like when smog is trapped near the ground.

How Do Humans Cause Air Pollution?

Many of our daily activities cause air pollution. The main source of human-caused air pollution in the United States is motor vehicles. Cars, motorcycles, trucks, buses, trains, and planes all give off exhaust. *Exhaust* is a gas that contains pollutants that create ozone and smog. ☑

Factories and power plants that burn coal, oil, and gas also give off pollutants. Businesses that use chemicals, such as dry cleaners and auto body shops, can add to air pollution.

 **READING CHECK**

SECTION 4 Air Pollution *continued*

What Causes Air Pollution Indoors?

Sometimes the air inside a building can be more polluted than the air outside. There is no wind to blow pollutants away and no rain to wash them out of the air indoors. Therefore, they can build up inside. It is important to air out buildings by opening the windows or using fans that bring fresh air in from outside. ☑

Sources of Indoor Air Pollution

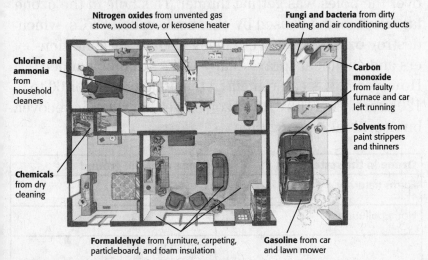

Nitrogen oxides from unvented gas stove, wood stove, or kerosene heater

Fungi and bacteria from dirty heating and air conditioning ducts

Chlorine and ammonia from household cleaners

Carbon monoxide from faulty furnace and car left running

Solvents from paint strippers and thinners

Chemicals from dry cleaning

Formaldehyde from furniture, carpeting, particleboard, and foam insulation

Gasoline from car and lawn mower

5. Explain Why can air pollution indoors be worse than air pollution outdoors?

_____

_____

_____

TAKE A LOOK

6. Identify Name two sources of indoor air pollution shown here that may be in your own home.

_____

_____

What Is Acid Precipitation?

Acid precipitation is rain, sleet, or snow that contains acids from air pollution. When we burn fossil fuels, such as coal, pollutants such as sulfur dioxide are released into the air. These pollutants combine with water in the atmosphere to form acids.

Acid precipitation can kill or damage plants, damage soil, and poison water. When acid rain flows into lakes, it can kill fish and other aquatic life.

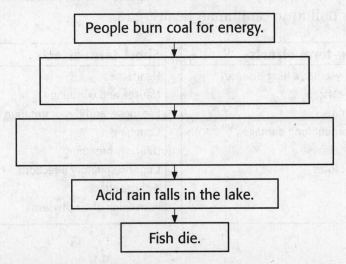

People burn coal for energy.

Acid rain falls in the lake.

Fish die.

TAKE A LOOK

7. Sequence Complete the graphic organizer to show how burning coal can cause fish to die.

SECTION 4 Air Pollution *continued*

What Is the Ozone Hole?

Close to the ground, ozone is a pollutant formed by human activities. However, high in the stratosphere, ozone is an important gas that forms naturally. The ozone layer absorbs harmful ultraviolet (UV) radiation from the sun. Ultraviolet radiation can harm living things. For example, it can cause skin cancer in humans. ☑

In the 1980s, scientists noticed that the ozone layer over the poles was getting thinner. This hole in the ozone layer was being caused by chemicals called CFCs, which destroy ozone. CFCs were being used in air conditioners and chemical sprays. Many CFCs are now banned. However, CFCs can remain in the atmosphere for 60 to 120 years. Therefore, the ozone layer may slowly recover, but it will take a long time.

✔ **READING CHECK**

8. Explain How is the ozone layer helpful to humans?

_____

_____

_____

_____

TAKE A LOOK
9. Compare Fill in the chart to show the differences between ozone in the atmosphere and ozone near the ground.

Ozone in the statosphere	Ozone near the ground
Forms naturally	
Not a pollutant	
	harmful to living things

How Does Air Pollution Affect Human Health?

Air pollution can cause many health problems. Some are short-term problems. They happen quickly and go away when the air pollution clears up or the person moves to a cleaner location. Others are long-term health problems. They develop over long periods of time and are not cured easily. The table below lists some of the effects of air pollution on human health. ☑

✔ **READING CHECK**

10. Compare What is the difference between short-term effects and long-term effects of air pollution?

_____

_____

_____

_____

Long-term effects	Short-term effects
Emphysema (a lung disease)	Headache
Lung cancer	Nausea and vomiting
Asthma	Eye, nose, and throat irritation
Permanent lung damage	Coughing
Heart disease	Difficulty breathing
Skin cancer	Upper respiratory infections
	Asthma attacks
	Worsening of emphysema

SECTION 4 Air Pollution *continued*

What Can We Do About Air Pollution?

Air pollution in the United States is not as bad now as it was 30 years ago. People today are much more aware of how they can cause or reduce air pollution. Air pollution can be reduced by new laws, by technology, and by people changing their lifestyles.

The United States government and the governments of other countries have passed laws to control air pollution. These laws limit the amount of pollution that sources such as cars and factories are allowed to release. For example, factories and power plants now have scrubbers on smokestacks. A *scrubber* is a tool that helps remove pollutants from smoke before it leaves the smokestack.

Many cars are more efficient now than they used to be, so they produce less pollution. Individuals can do a lot on their own to reduce air pollution, as well. For example, we can walk or bike instead of driving.

Critical Thinking

11. Analyze Processes
Electric cars don't give off any exhaust. They don't cause pollution in the cities where they are driven. However, driving them can cause pollution in other places. How?
(Hint: Where does most electricity come from?)

_____

_____

_____

_____

_____

_____

_____

In Copenhagen, Denmark, companies lend bicycles for anyone to use for free. The program helps reduce automobile traffic and air pollution.

Section 4 Review

SECTION VOCABULARY

acid precipitation rain, sleet, or snow that contains a high concentration of acids	**air pollution** the contamination of the atmosphere by the introduction of pollutants from human and natural sources

1. Identify Relationships How are fossil fuels related to air pollution and acid precipitation?

_____

_____

_____

2. Compare Complete the table below to compare different pollutants.

Pollutant	Source	Negative effects	Solutions
CFCs			banning CFCs
Ozone			
Sulfur dioxide	burning of fossil fuels		

3. Infer Name three things, other than humans, that can be harmed by air pollution.

_____

_____

_____

4. Explain Why is the hole in the ozone layer dangerous?

_____

_____

_____

CHAPTER 16 | Understanding Weather
SECTION
1 | **Water in the Air**

National Science
Education Standards
ES 1f, 1i

BEFORE YOU READ

After you read this section, you should be able to answer these questions:

• What is weather?

• How does water in the air affect the weather?

What Is Weather?

Knowing about the weather is important in our daily lives. Your plans to go outside can change if it rains. Being prepared for extreme weather conditions, such as hurricanes and tornadoes, can even save your life.

Weather is the condition of the atmosphere at a certain time and place. Weather depends a lot on the amount of water in the air. Therefore, to understand weather, you need to understand the water cycle. ☑

THE WATER CYCLE

The movement of water between the atmosphere, the land, and the oceans is called the *water cycle*. The sun is the main source of energy for the water cycle. The sun's energy heats Earth's surface. This causes liquid water to *evaporate*, or change into water vapor (a gas). When the water vapor cools, it may change back into a liquid and form clouds. This is called **condensation**. The liquid water may fall as rain, snow, sleet, or hail on the land.

STUDY TIP

Outline Before you read, make an outline of this section using the questions in bold. As you read, fill in the main ideas of the chapter in your outline.

✓ READING CHECK

1. Define Write your own definition for *weather*.

_____

_____

_____

_____

STANDARDS CHECK

ES 1i Clouds, formed by the condensation of water vapor, <u>affect</u> weather and climate.

Word Help: <u>affect</u>
to change; to act upon

2. Identify By what process do clouds form?

_____

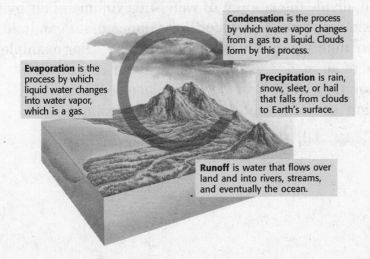

Condensation is the process by which water vapor changes from a gas to a liquid. Clouds form by this process.

Evaporation is the process by which liquid water changes into water vapor, which is a gas.

Precipitation is rain, snow, sleet, or hail that falls from clouds to Earth's surface.

Runoff is water that flows over land and into rivers, streams, and eventually the ocean.

301

What Is Humidity?

Water vapor makes up only a small fraction of the mass of the atmosphere. However, this small amount of water vapor has an important effect on weather and climate.

When the sun's energy heats up Earth's surface, water in oceans and water bodies evaporates. The amount of water vapor in the air is called **humidity**. Warmer air can hold more water vapor than cooler air can. ☑

READING CHECK

3. Identify How does air temperature affect how much water vapor the air can hold?

_____

_____

Math Focus

4. Read a Graph How much water vapor can air at 30°C hold?

_____

_____

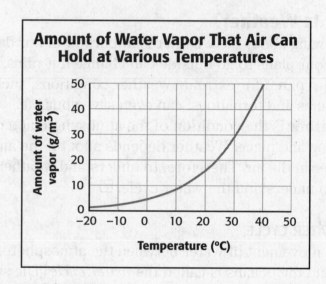

Amount of Water Vapor That Air Can Hold at Various Temperatures

Amount of water vapor (g/m³) vs. Temperature (°C)

Math Focus

5. Calculate What is the relative humidity of 25°C air that contains 10 g/m³ of water vapor? Show your work.

_____

_____

_____

RELATIVE HUMIDITY

Scientists often describe the amount of water in the air using relative humidity. **Relative humidity** is the ratio of the amount of water vapor in the air to the greatest amount the air can hold.

There are two steps to calculating relative humidity. First, divide the amount of water in a volume of air by the maximum amount of water that volume of air can hold. Then, multiply by 100 to get a percentage. For example, 1 m³ of air at 25°C can hold up to about 23 g of water vapor. If air at 25°C in a certain place contains only 18 g/m³ of water vapor, then the relative humidity is:

$$\frac{18 \text{ g/m}^3}{23 \text{ g/m}^3} \times 100 = 78\% \text{ relative humidity}$$

SECTION 1 | Water in the Air *continued*

FACTORS AFFECTING RELATIVE HUMIDITY

Temperature and humidity can affect relative humidity. As humidity increases, relative humidity increases if the temperature stays the same. Relative humidity decreases as temperature rises and increases as temperature drops if the humidity stays the same.

MEASURING RELATIVE HUMIDITY

Scientists measure relative humidity using special tools. One of these tools is called a *psychrometer*. A psychrometer contains two thermometers. The bulb of one thermometer is covered with a wet cloth. This is called a *wet-bulb thermometer*. The other thermometer bulb is dry. This thermometer is a *dry-bulb thermometer*.

You are probably most familiar with dry-bulb thermometers. Wet-bulb thermometers work differently than dry-bulb thermometers. As air passes through the cloth on a wet-bulb thermometer, some of the water in the cloth evaporates. As the water evaporates, the cloth cools. The wet-bulb thermometer shows the temperature of the cloth.

If humidity is low, the water evaporates more quickly. Therefore, the temperature reading on the wet-bulb thermometer is much lower than the reading on the dry-bulb thermometer. If the humidity is high, less water evaporates. Therefore, the temperature changes very little.

Critical Thinking

6. Compare How is relative humidity different from humidity?

_____
_____
_____
_____
_____

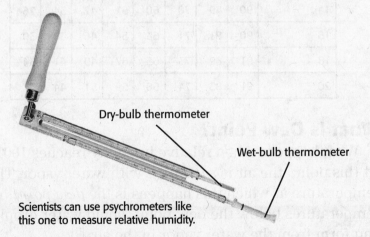

Dry-bulb thermometer

Wet-bulb thermometer

Scientists can use psychrometers like this one to measure relative humidity.

TAKE A LOOK
7. Identify What are two parts of a psychrometer?

_____
_____

The difference in temperature readings between the dry-bulb and wet-bulb thermometers is a measure of the relative humidity. The larger the difference between the readings, the lower the relative humidity.

SECTION 1 Water in the Air *continued*

USING A RELATIVE-HUMIDITY TABLE

Scientists use tables like the one below to determine relative humidity. Use the table to work through the following example.

The dry-bulb thermometer on a psychrometer reads 10°C. The wet-bulb thermometer reads 7°C. Therefore, the difference between the thermometer readings is 3°C. In the first column of the table, find the row head for 10°C, the dry-bulb reading. Then, find the column head for 3°C, the difference between the readings. Find the place where the row and column meet. The number in the table at this point is 66, so the relative humidity is 66%.

Relative Humidity (%)								
Dry-bulb reading (°C)	Difference between wet-bulb reading and dry-bulb reading (°C)							
	1	2	3	4	5	6	7	8
0	81	64	46	29	13			
2	84	68	52	37	22	7		
4	85	71	57	43	29	16		
6	86	73	60	48	35	24	11	
8	87	75	63	51	40	29	19	8
10	88	77	66	55	44	34	24	15
12	59	78	68	58	48	39	29	21
14	90	79	70	60	51	42	34	26
16	90	81	71	63	54	46	38	30
18	91	82	73	65	57	49	41	34
20	91	83	74	66	59	51	44	37

TAKE A LOOK
8. Apply Concepts The dry-bulb reading on a psychrometer is 8°C. The wet-bulb reading is 7°C. What is the relative humidity?

_____

_____

What Is Dew Point?

What happens when relative humidity reaches 100%? At this point, the air is *saturated* with water vapor. The temperature at which this happens is the *dew point*. At temperatures below the dew point, liquid water droplets can form from the water vapor in the air. ☑

Condensation happens when air is saturated with water vapor. Air can become saturated if water evaporates and enters the air as water vapor. Air can also become saturated when it cools below its dew point.

✓ **READING CHECK**

9. Explain What happens when the temperature of air is below its dew point?

_____

_____

SECTION 1 Water in the Air *continued*

AN EVERYDAY EXAMPLE

You have probably seen air become saturated because of a temperature decrease. For example, when you add ice cubes to a glass of juice, the temperatures of the juice and the glass decrease. The glass absorbs heat from the air, so the temperature of the air near the glass decreases. When the air's temperature drops below its dew point, water vapor condenses on the glass. The condensed water forms droplets on the glass.

The glass absorbs heat from the air. The air cools to below its dew point. Water vapor condenses onto the side of the glass.

How Do Clouds Form?

A **cloud** is a group of millions of tiny water droplets or ice crystals. Clouds form as air rises and cools. When air cools below the dew point, water droplets or ice crystals form. Water droplets form when water condenses above 0°C. Ice crystals form when water condenses below 0°C.

DIFFERENT KINDS OF CLOUDS

Scientists classify clouds by shape and altitude. The three main cloud shapes are stratus clouds, cumulus clouds, and cirrus clouds. The three altitude groups are low clouds, middle clouds, and high clouds. The figure on the next page shows these different cloud types. ☑

Critical Thinking

10. Apply Concepts People who wear glasses may notice that their glasses fog up when they come indoors on a cold day. Why does this happen?

_____
_____
_____
_____
_____
_____

TAKE A LOOK

11. Describe Where did the liquid water on the outside of the glass come from?

_____
_____
_____

✓ **READING CHECK**

12. Explain How are clouds classified?

_____
_____
_____

SECTION 1 Water in the Air *continued*

Say It

Observe and Describe Look at the clouds every day for a week. Each day, write down the weather and what the clouds looked like. At the end of the week, share your observations with a small group. How was the weather related to the kinds of clouds you saw each day?

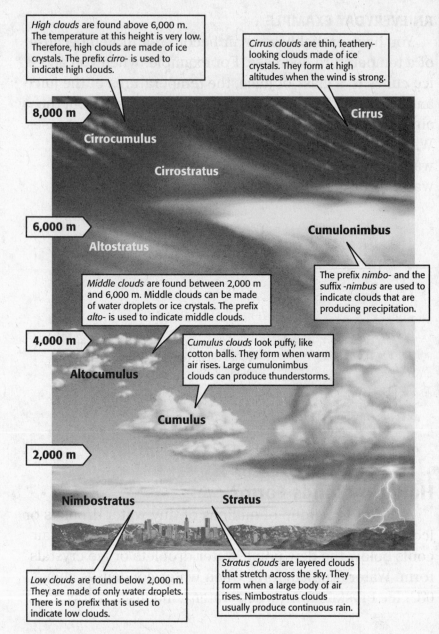

High clouds are found above 6,000 m. The temperature at this height is very low. Therefore, high clouds are made of ice crystals. The prefix *cirro-* is used to indicate high clouds.

Cirrus clouds are thin, feathery-looking clouds made of ice crystals. They form at high altitudes when the wind is strong.

8,000 m

Cirrus

Cirrocumulus

Cirrostratus

6,000 m

Cumulonimbus

Altostratus

Middle clouds are found between 2,000 m and 6,000 m. Middle clouds can be made of water droplets or ice crystals. The prefix *alto-* is used to indicate middle clouds.

The prefix *nimbo-* and the suffix *-nimbus* are used to indicate clouds that are producing precipitation.

4,000 m

Cumulus clouds look puffy, like cotton balls. They form when warm air rises. Large cumulonimbus clouds can produce thunderstorms.

Altocumulus

Cumulus

2,000 m

TAKE A LOOK

13. Compare How is a nimbostratus cloud different from a stratus cloud?

Nimbostratus

Stratus

_____

_____

_____

Low clouds are found below 2,000 m. They are made of only water droplets. There is no prefix that is used to indicate low clouds.

Stratus clouds are layered clouds that stretch across the sky. They form when a large body of air rises. Nimbostratus clouds usually produce continuous rain.

What Is Precipitation?

Water in the air can return to Earth's surface through precipitation. **Precipitation** is solid or liquid water that falls to Earth's surface from clouds. There are four main kinds of precipitation: rain, snow, sleet, and hail. Rain and snow are the most common kinds of precipitation. Sleet and hail are less common. ☑

READING CHECK

14. Define What is precipitation?

_____

_____

SECTION 1 Water in the Air *continued*

RAIN

Water droplets in clouds are very tiny. Each droplet is smaller than the period at the end of this sentence. These tiny droplets can combine with each other. As the droplets combine, they become larger. When a droplet reaches a certain size, it can fall to Earth's surface as *rain*. ☑

SLEET

Sleet forms when rain falls through a layer of very cold air. If the air is cold enough, the rain freezes in the air and becomes falling ice. Sleet can make roads very slippery. When it lands on objects, sleet can coat the objects in ice.

SNOW

Snow forms when temperatures are so low that water vapor turns directly into a solid. That is, the water vapor in the cloud turns into an ice crystal without becoming a liquid first. Snow can fall as single ice crystals. In many cases, the crystals join together to form larger snowflakes. ☑

HAIL

Balls or lumps of ice that fall from clouds are called *hail*. Hail forms in cumulonimbus clouds. Hail can become very large. Hail grows larger in a cycle, as shown in the chart below.

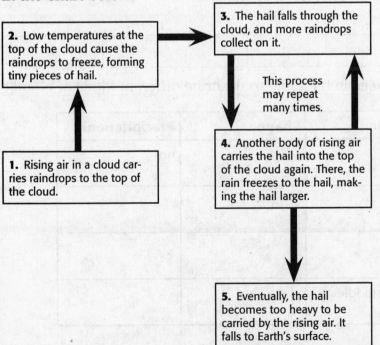

2. Low temperatures at the top of the cloud cause the raindrops to freeze, forming tiny pieces of hail.

3. The hail falls through the cloud, and more raindrops collect on it.

This process may repeat many times.

1. Rising air in a cloud carries raindrops to the top of the cloud.

4. Another body of rising air carries the hail into the top of the cloud again. There, the rain freezes to the hail, making the hail larger.

5. Eventually, the hail becomes too heavy to be carried by the rising air. It falls to Earth's surface.

✔ **READING CHECK**

15. Explain What happens to water droplets in clouds when they combine?

_____

_____

✔ **READING CHECK**

16. Identify What is a snowflake?

_____

_____

TAKE A LOOK

17. Identify When does hail fall to the ground?

_____

_____

Name ______________________ Class ______________ Date ______________

Section 1 Review

NSES ES 1f, 1i

SECTION VOCABULARY

cloud a collection of small water droplets or ice crystals suspended in the air, which forms when the air is cooled and condensation occurs

condensation the change of state from a gas to a liquid

humidity the amount of water vapor in the air

precipitation any form of water that falls to Earth's surface from the clouds

relative humidity the ratio of the amount of water vapor in the air to the amount of water vapor needed to reach saturation at a given temperature

weather the short-term state of the atmosphere, including temperature, humidity, precipitation, wind, and visibility

1. Identify Relationships How is dew point related to condensation?

____

____

2. Identify What is the main source of energy for the water cycle?

____

3. Explain How do clouds form?

____

____

____

____

4. Compare What is the difference between sleet and snow?

____

____

5. Apply Concepts Fill in the spaces in the table to describe different kinds of clouds.

Name	Altitude	Shape	Precipitation?
Cirrostratus	high		no
Altocumulus		puffy	
Nimbostratus			
Cumulonimbus	low to middle		

CHAPTER 16 | Understanding Weather

SECTION 2 | Air Masses and Fronts

National Science
Education Standards
ES 1j

BEFORE YOU READ

After you read this section, you should be able to answer these questions:

• How is an air mass different from a front?

• How do fronts affect weather?

What Are Air Masses?

Have you ever been caught outside when it suddenly started to rain? What causes such an abrupt change in the weather? Changes in weather are caused by the movement of bodies of air called air masses. An **air mass** is a very large volume of air that has a certain temperature and moisture content.

There are many types of air masses. Scientists classify air masses by the water content and temperature of the air. These features depend on where the air mass forms. The area over which an air mass forms is called a *source region*. One source region is the Gulf of Mexico. Air masses that form over this source region are wet and warm. ☑

Each type of air mass forms over a certain source region. On maps, meteorologists use two-letter symbols to represent different air masses. The first letter indicates the water content of the air mass. The second letter indicates its temperature. The figure below shows the main air masses that affect North America.

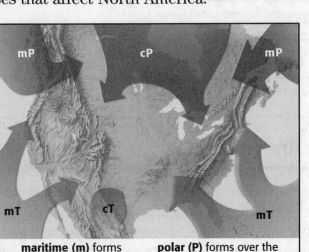

| maritime (m) forms over water; wet | polar (P) forms over the polar regions; cold |
| continental (c) forms over land; dry | tropical (T) forms over the Tropics; warm |

mP   cP   mP

mT   cT   mT

STUDY TIP

Summarize As you read, make a chart comparing the four kinds of fronts. In your chart, describe how each kind of front forms and what kind of weather it can cause.

READING CHECK

1. Identify How do scientists classify air masses?

_____

_____

TAKE A LOOK

2. Apply Concepts Describe the temperature and moisture content of a cT air mass.

_____

_____

SECTION 2 Air Masses and Fronts *continued*

COLD AIR MASSES

Most of the cold winter weather in the United States comes from three polar air masses. Continental polar (cP) air masses form over northern Canada. They bring extremely cold winter weather. In the summer, cP air masses can bring cool, dry weather. ☑

Maritime polar (mP) air masses form over the North Pacific Ocean. They are cool and very wet. They bring rain and snow to the Pacific Coast in winter. They bring fog in the summer.

Maritime polar air masses also form over the North Atlantic Ocean. They bring cool, cloudy weather and precipitation to New England.

WARM AIR MASSES

Four warm air masses influence the weather in the United States. Maritime tropical (mT) air masses form over warm areas in the Pacific Ocean, the Gulf of Mexico, and the Atlantic Ocean. They move across the East Coast and into the Midwest. In summer they bring heat, humidity, hurricanes, and thunderstorms to these areas.

Continental tropical air masses (cT) form over deserts and move northward. They bring clear, dry, hot weather in the summer.

Air mass	How it affects weather
cP from northern Canada	
mP from the North Pacific Ocean	
mT from the Gulf of Mexico	
cT from the deserts	

What Are Fronts?

The place where two or more air masses meet is called a **front**. When air masses meet, the less dense air mass rises over the denser air mass. Warm air is less dense than cold air. Therefore, a warm air mass will generally rise above a cold air mass. There are four main kinds of fronts: cold fronts, warm fronts, occluded fronts, and stationary fronts. ☑

READING CHECK

3. Identify What is the source region for cP air masses?

_____

Critical Thinking

4. Infer Why don't warm air masses form over the North Atlantic or Pacific oceans?

_____
_____
_____

TAKE A LOOK
5. Identify Fill in the blank spaces in the table.

READING CHECK

6. Define What is a front?

_____
_____

COLD FRONTS

A *cold front* forms when a cold air mass moves under a warm air mass. The cold air pushes the warm air mass up. The cold air mass replaces the warm air mass. Cold fronts can move quickly and bring heavy precipitation. When a cold front has passed, the weather is usually cooler. This is because a cold, dry air mass moves in behind the cold front.

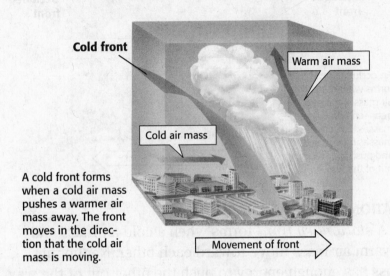

A cold front forms when a cold air mass pushes a warmer air mass away. The front moves in the direction that the cold air mass is moving.

TAKE A LOOK
7. Describe What happens to the warm air mass at a cold front?

_____

_____

WARM FRONTS

A *warm front* forms when a warm air mass moves in over a cold air mass that is leaving an area. The warm air replaces the cold air as the cold air moves away. Warm fronts can bring light rain. They are followed by clear, warm weather. ☑

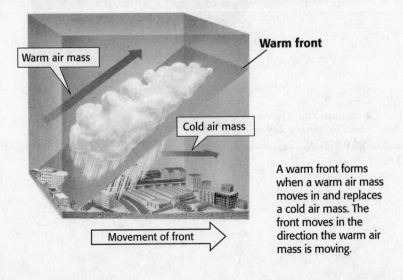

A warm front forms when a warm air mass moves in and replaces a cold air mass. The front moves in the direction the warm air mass is moving.

✓ READING CHECK

8. Define What is a warm front?

_____

_____

_____

OCCLUDED FRONTS

An *occluded front* forms when a warm air mass is caught between two cold air masses. Occluded fronts bring cool temperatures and large amounts of rain and snow.

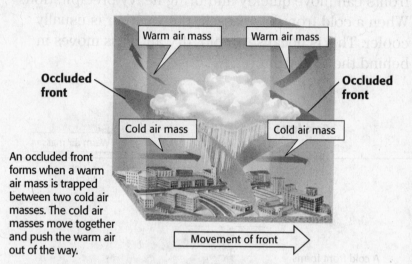

An occluded front forms when a warm air mass is trapped between two cold air masses. The cold air masses move together and push the warm air out of the way.

TAKE A LOOK
9. Describe What happens to the warm air mass in an occluded front?

_____

_____

_____

STATIONARY FRONT

A *stationary front* forms when a cold air mass and a warm air mass move toward each other. Neither air mass has enough energy to push the other out of the way. Therefore, the two air masses remain in the same place. Stationary fronts cause many days of cloudy, wet weather.

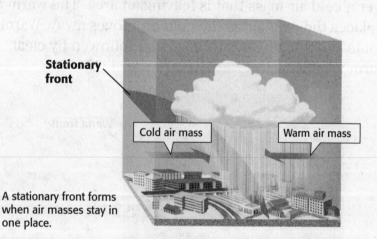

A stationary front forms when air masses stay in one place.

TAKE A LOOK
10. Infer What do you think is the reason that stationary fronts bring many days of the same weather?

_____

_____

_____

_____

How Does Air Pressure Affect Weather?

Remember that air produces pressure. However, air pressure is not always the same everywhere. Areas with different pressures can cause changes in the weather. These areas may have lower or higher air pressure than their surroundings.

A **cyclone** is an area of the atmosphere that has lower pressure than the surrounding air. The air in the cyclone rises. As the air rises, it cools. Clouds can form and may cause rainy or stormy weather.

An **anticyclone** is an area of the atmosphere that has higher pressure than the surrounding air. Air in anticyclones sinks and gets warmer. Its relative humidity decreases. This warm, sinking air can bring dry, clear weather.

Cyclones and anticyclones can affect each other. Air moving out from the center of an anticyclone moves toward areas of low pressure. This movement can form a cyclone. The figure below shows how cyclones and anti-cyclones can affect each other.

Critical Thinking

11. Compare Give two differences between cyclones and anticyclones.

_____
_____
_____
_____
_____
_____

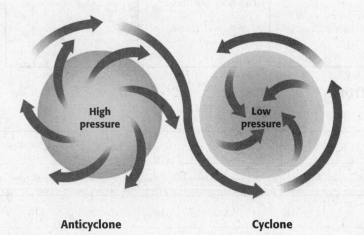

High pressure

Low pressure

Anticyclone

Cyclone

TAKE A LOOK

12. Identify In which direction does air move: from a cyclone to an anticyclone, or from an anticyclone to a cyclone?

_____

Name ______________________ Class ______________ Date ______________

Section 2 Review

NSES **NSES ES 1j**

SECTION VOCABULARY

air mass a large body of air throughout which temperature and moisture content are similar

anticyclone the rotation of air around a high pressure center in the direction opposite to Earth's rotation.

cyclone an area in the atmosphere that has lower pressure than the surrounding areas and has winds that spiral toward the center

front the boundary between air masses of different densities and usually different temperatures

1. **Identify Relationships** How are fronts and air masses related?

____

____

2. **Compare** Fill in the table to describe cyclones and anticyclones.

Name	Compared to surrounding air pressure, the pressure in the middle is...	What does the air inside it do?	What kind of weather does it cause?
cyclone	...lower than surrounding pressure.		
anticyclone		sinks and warms	

3. **List** What are four kinds of fronts?

____

____

____

____

4. **Identify** What are the source regions for the mT air masses that affect weather in the United States?

____

____

____

5. **Describe** What kind of air mass causes hot, clear, dry summer weather in the United States?

____

CHAPTER 16 | Understanding Weather
SECTION
3 | **Severe Weather**

<table>
</table>

BEFORE YOU READ

After you read this section, you should be able to answer these questions:

• What are some types of severe weather?

• How can you stay safe during severe weather?

National Science Education Standards
ES 1i, 1j

What Causes Thunderstorms?

A **thunderstorm** is an intense storm with strong winds, heavy rain, lightning, and thunder. Many thunderstorms happen along cold fronts. However, thunderstorms can also happen in other areas. Two conditions are necessary for a thunderstorm to form: warm, moist air near Earth's surface and an unstable area of the atmosphere.

The atmosphere is unstable when a body of cold air is found above a body of warm air. The warm air rises and cools as it mixes with the cool air. When the warm air reaches its dew point, the water vapor condenses and forms cumulus clouds. If the warm air keeps rising, the clouds may become dark cumulonimbus clouds.

STUDY TIP

Describe After you read this section, make a flow chart showing how a tornado forms.

Critical Thinking

1. Infer Why does air near the surface have to be moist in order for a thunderstorm to form?

_____

_____

LIGHTNING

As a cloud grows bigger, parts of it begin to develop electrical charges. The upper parts of the cloud tend to become positively charged. The lower parts tend to become negatively charged. When the charges get big enough, electricity flows from one area to the other. Electricity may also flow between the clouds and the ground. These electrical currents are **lightning**. ☑

READING CHECK

2. Describe How does lightning form?

_____

_____

_____

Different parts of thunderclouds and the ground can have different electrical charges. When electricity flows between these areas, lightning forms.

SECTION 3 Severe Weather *continued*

THUNDER

You have probably seen large lightning bolts that travel between the clouds and the ground. When lightning moves through the air, the air gets very hot. The hot air expands rapidly. As it expands, it makes the air vibrate. The vibrations release energy in the form of sound waves. The result is **thunder**. ☑

SEVERE THUNDERSTORMS

Severe thunderstorms can cause a lot of damage. They can produce strong winds, hail, flash floods, or tornadoes. Hail can damage crops, cars, and windows. Flash flooding from heavy rain can cause serious property damage. Flash flooding is the leading cause of weather-related deaths. Lightning can start fires and cause injuries and deaths.

How Do Tornadoes Form?

Fewer than 1% of thunderstorms produce tornadoes. A **tornado** can form when a rapidly spinning column of air, called a *funnel cloud*, touches the ground. The air in the center of a tornado has low pressure. When the area of low pressure touches the ground, material from the ground can be sucked up into the tornado. ☑

A tornado begins as a funnel cloud that pokes through the bottom of a cumulonimbus cloud. The funnel cloud becomes a tornado when the funnel cloud touches the ground. The pictures below show how a tornado forms.

❶ Wind moving in opposite directions causes a layer of air in the middle of a cloud to begin to spin.

❷ Strong vertical winds cause the spinning column of air to turn into a vertical position.

❸ The spinning column of air moves to the bottom of the cloud and forms a funnel cloud.

❹ The funnel cloud becomes a tornado when it touches down on the ground.

READING CHECK

3. Define What is thunder?

_____

_____

_____

_____

READING CHECK

4. Explain Why can material be sucked up into a tornado?

_____

_____

TAKE A LOOK

5. Describe When does a funnel cloud become a tornado?

_____

_____

SECTION 3 Severe Weather *continued*

TORNADO FACTS

About 75% of the world's tornadoes happen in the United States. Most happen in the spring and early summer. During these times, cold, dry air from Canada meets warm, moist air from the Tropics. This causes the thunderstorms that produce tornadoes.

Most tornadoes last for only a few minutes. However, their strong, spinning winds can cause a lot of damage. An average tornado has wind speeds between 120 km/h and 180 km/h, but some can be much higher. Winds from tornadoes can tear up trees and destroy buildings. They can even be strong enough to lift cars and trailers up into the air. The area damaged by a tornado is usually about 8 km long and 10 to 60 m wide.

How Do Hurricanes Form?

A **hurricane** is a large, rotating tropical weather system. Hurricanes have wind speeds of over 120 km/h. They can be 160 km to 1,500 km in diameter and can travel for thousands of miles. They are the most powerful storms on Earth. Hurricanes are also called typhoons and cyclones.

Most hurricanes form between 5°N and 20°N latitude or between 5°S and 20°S latitude. They form over the warm, tropical oceans found at these latitudes. At higher latitudes, the water is too cold for hurricanes to form. ☑

Math Focus

6. Convert What is the average wind speed in a tornado in miles per hour?

1 km = 0.62 mi.

_____

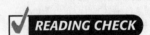

READING CHECK

7. Explain Why don't hurricanes form at high latitudes?

_____

_____

Hurricanes can be so large that they are visible from space. This photograph of a hurricane was taken by a satellite.

SECTION 3 Severe Weather *continued*

HOW HURRICANES FORM

A hurricane begins as a group of thunderstorms traveling over tropical ocean waters. Winds traveling in two different directions meet and cause the storm to spin. Because of the Coriolis effect, hurricanes rotate counterclockwise in the Northern Hemisphere and clockwise in the Southern Hemisphere. ☑

Hurricanes are powered by solar energy. The sun's energy causes ocean water to evaporate. As the water vapor rises in the air, it cools and condenses. A group of thunderstorms form and produce a large, spinning storm. A hurricane forms as the storm gets stronger.

✔ **READING CHECK**

8. Explain What causes hurricanes to rotate in different directions in the Northern and Southern Hemispheres?

_____

_____

TAKE A LOOK
9. Define What is the eye of a hurricane?

_____

_____

_____

At the center of the hurricane is the eye. The *eye* is a core of warm, relatively calm air with low pressure and light winds. There are updrafts and downdrafts in the eye. An *updraft* is a current of rising air. A *downdraft* is a current of sinking air.

Around the eye is a group of cumulonimbus clouds called the *eye wall*. These clouds produce heavy rain and strong winds. The winds can be up to 300 km/h. The eye wall is the strongest part of the hurricane.

Outside the eye wall are spiraling bands of clouds called *rain bands*. These bands also produce heavy rain and strong wind. They circle the center of the hurricane.

The hurricane will continue to grow as long as it is over warm ocean water. When the hurricane moves over colder waters or over land, the storm loses energy. This is why hurricanes are not common in the middle of continents. The storms lose their energy quickly when they move over land.

SECTION 3 Severe Weather *continued*

DAMAGE CAUSED BY HURRICANES

Hurricanes can cause serious damage when they move near or onto land. The strong winds from hurricanes can knock down trees and telephone poles. They can damage or destroy buildings and homes.

Many people think that the winds are the most damaging part of a hurricane. However, most of the damage from hurricanes is actually caused by flooding from heavy rains and storm surges. A *storm surge* is a rise in sea level that happens during a storm. A storm surge from a hurricane can be up to 8 m high. The storm-surge flooding from Hurricane Katrina in 2005 caused more damage than the high-speed winds from the storm. ☑

How Can You Stay Safe During Severe Weather?

Severe weather can be very dangerous. During severe weather, it is important for you to listen to a local TV or radio station. Severe-weather announcements will tell you where a storm is and if it is getting worse. Weather forecasters use watches and warnings to let people know about some kinds of severe weather. A *watch* means that severe weather may happen. A *warning* means that severe weather is happening somewhere nearby.

The table below gives ways to stay safe during different kinds of severe weather. ☑

Severe weather	How to stay safe
Thunderstorms	If you are outside, stay away from tall objects that can attract lightning. If you are in an open area, crouch down. Stay away from water. If you are inside, stay away from windows.
Tornadoes	During a tornado warning, find shelter quickly in a basement or cellar. If you cannot get to a basement, go to a windowless room in the center of the building (such as a closet or bathroom). If you are outside, lie down in an open field or a deep ditch.
Floods	Find a high place to wait out the flood. Always stay out of floodwaters.
Hurricanes	Protect the windows in your home by covering them with wood. Stay inside during the storm. If you are told to leave your home, do so quickly and calmly.

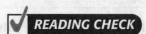

READING CHECK

10. Define What is a storm surge?

_____

_____

READING CHECK

11. Explain Why should you listen to weather reports during severe weather?

_____

_____

Section 3 Review

SECTION VOCABULARY

hurricane a severe storm that develops over tropical oceans and whose strong winds of more than 120 km/h spiral in toward the intensely low-pressure storm center	**thunder** the sound caused by the rapid expansion of air along an electrical strike
	thunderstorm a usually brief, heavy storm that consists of rain, strong winds, lightning, and thunder
lightning an electric discharge that takes place between two oppositely charged surfaces, such as between a cloud and the ground, between two clouds, or between two parts of the same cloud	**tornado** a destructive, rotating column of air that has very high wind speeds and that may be visible as a funnel-shaped cloud

1. Explain Why do thunder and lightning usually happen together?

_____

_____

_____

2. Identify How can severe thunderstorms cause damage?

_____

_____

3. Identify Where do most tornadoes happen?

_____

_____

4. Explain Why do most tornadoes happen in the spring and early summer?

_____

_____

_____

5. Analyze How does energy from the sun power hurricanes?

_____

_____

_____

_____

_____

6. Describe When do hurricanes lose energy?

_____

_____

7. Identify Give three ways to stay safe if you are caught outside in a thunderstorm.

_____

_____

_____

_____

CHAPTER 16 Understanding Weather

SECTION 4 Forecasting Weather

National Science
Education Standards
ES 1i, 1j

BEFORE YOU READ

After you read this section, you should be able to answer these questions:

• What instruments are used to forecast weather?

• How do you read a weather map?

What Is a Weather Forecast?

Weather affects how you dress and how you plan your day. Severe weather can put people in danger. Therefore, accurate weather forecasts are important. A *weather forecast* is a prediction of weather conditions over the next few days. Meteorologists make weather forecasts using information on atmospheric conditions. ☑

Meteorologists use special instruments to collect data. Some of these instruments are far above the ground. Others are tools you may be familiar with from everyday use.

WEATHER BALLOONS

Weather balloons carry electronic equipment. The equipment on a weather balloon can measure weather conditions as high as 30 km above Earth's surface. This equipment measures temperature, air pressure, and relative humidity. It transmits the information to meteorologists using radio signals. Meteorologists can track the path of the balloons to measure wind speed and direction.

STUDY TIP

Compare As you read this section, make a chart comparing the different tools that meteorologists use to collect weather data.

READING CHECK

1. Explain What do meteorologists use to forecast the weather?

_____

_____

TAKE A LOOK

2. Describe How do meteorologists obtain the information from weather balloons?

_____

_____

_____

Weather balloons carry equipment into the atmosphere. They use radio signals to transmit information on weather conditions to meteorologists on the ground.

SECTION 4 Forecasting Weather *continued*

THERMOMETERS AND BAROMETERS

Remember that air temperature and pressure can affect the weather. Therefore, meteorologists must be able to measure temperature and pressure accurately. They use **thermometers** to measure temperature, just like you do. They use tools called **barometers** to measure air pressure.

WINDSOCKS, WIND VANES, AND ANEMOMETERS

Meteorologists can use windsocks and wind vanes to measure wind direction. A *windsock* is a cone-shaped cloth bag that is open at both ends. The wind enters through the wide end and leaves through the narrow end. The wide end always points into the wind.

A *wind vane* is shaped like an arrow. It is attached to a pole. The wind pushes the tail of the arrow. The vane spins until the arrows points into the wind.

An **anemometer** measures wind speed. It has three or four cups connected to a pole with spokes. The wind pushes on the open sides of the cups. This makes them spin on the pole. The spinning of the pole produces an electric current, which is displayed on a dial. The faster the wind speed, the stronger the electric current, and the further the dial moves.

Critical Thinking

3. Infer Why is it important for meteorologists to be able to measure wind direction?

_____

_____

_____

TAKE A LOOK
4. Identify What is an anemometer?

_____

_____

Meteorologists use anemometers to measure wind speed.

RADAR AND SATELLITES

Scientists use *radar* to locate fronts and air masses. Radar can locate a weather system and show the direction it is moving. It can show how much precipitation is falling, and what kind of precipitation it is. Most television stations use radar to give information about weather systems. ☑

Weather satellites orbiting Earth produce images of weather systems. Satellites can also measure wind speeds, humidity, and temperatures from different altitudes. Meteorologists use weather satellites to track storms.

 **READING CHECK**

5. Describe Give two things that meteorologists can use radar to do.

_____

_____

What Are Weather Maps?

In the United States, two main groups of scientists collect weather data. One group is the National Weather Service (NWS). The other group is the National Oceanic and Atmospheric Administration (NOAA). These groups gather information from about 1,000 weather stations across the United States to produce weather maps. ☑

READING A WEATHER MAP

Some weather maps contain station models. A *station model* is a symbol that shows the weather at a certain location. Station models look like circles with numbers and symbols around them. The numbers and symbols stand for different measurements, as shown below.

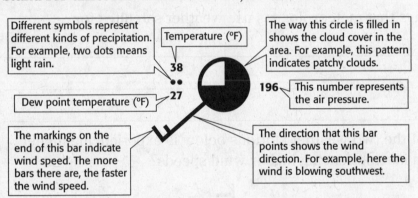

Different symbols represent different kinds of precipitation. For example, two dots means light rain.

Temperature (°F)

The way this circle is filled in shows the cloud cover in the area. For example, this pattern indicates patchy clouds.

This number represents the air pressure.

Dew point temperature (°F)

The markings on the end of this bar indicate wind speed. The more bars there are, the faster the wind speed.

The direction that this bar points shows the wind direction. For example, here the wind is blowing southwest.

Some weather maps, such as those you see on television, show lines called isobars. *Isobars* are lines that connect points of equal air pressure. They are similar to contour lines on a topographic map. Isobars that form closed circles represent areas of high (H) or low (L) pressure. Weather maps also show fronts.

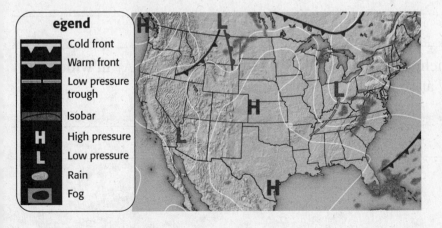

egend

Cold front
Warm front
Low pressure trough
Isobar
High pressure
Low pressure
Rain
Fog

☑ **READING CHECK**

6. Identify What are two groups that collect weather data in the United States?

_____

_____

TAKE A LOOK

7. Use a Model What is the dew point temperature for the station shown in the figure?

_____

8. Infer Will condensation happen in the air at the station in the figure? Explain your answer.

_____

_____

_____

TAKE A LOOK

9. Read a Map On the map, circle the areas of high pressure.

Name _____ Class _____ Date _____

Section 4 Review

NSES ES 1i, 1j

SECTION VOCABULARY

anemometer an instrument used to measure wind speed **barometer** an instrument that measures atmospheric pressure	**thermometer** an instrument that measures and indicates temperature

1. **Compare** How is an anemometer different from a windsock or a wind vane?

_____

2. **Identify** What three atmospheric conditions do weather balloons measure?

_____

_____

3. **Describe** Give three things that meteorologists use weather satellites for.

_____

_____

_____

4. **Apply Concepts** Which of the two weather stations below is experiencing higher air temperatures? Which is experiencing higher wind speeds?

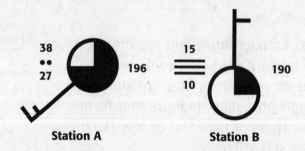

Station A   Station B

_____

5. **Apply Concepts** In which direction is the wind blowing at station A? In which direction is it blowing at station B?

_____

_____

SECTION 1 | What Is Climate?

National Science Education Standards
ES 1f, 1j, 3d

BEFORE YOU READ

After you read this section, you should be able to answer these questions:

- What is climate?
- What factors affect climate?
- How do climates differ around the world?

What Is Climate?

How is weather different from climate? **Weather** is the condition of the atmosphere at a certain time. The weather can change from day to day. In contrast, **climate** describes the average weather conditions in a region over a long period of time. The climate of an area includes the area's average temperature and amount of precipitation. Different parts of the world have different climates.

What Factors Affect Climate?

Climate is mainly determined by temperature and precipitation. Many factors affect temperature and precipitation, including latitude, wind patterns, landforms, and ocean currents. ☑

SOLAR ENERGY AND LATITUDE

Remember that the **latitude** of an area is its distance north or south of the equator. In general, the temperature of an area depends on its latitude. Latitudes closer to the poles tend to have colder climates. Latitude affects temperature because latitude determines how much direct solar energy an area gets, as shown in the figure below.

STUDY TIP

Ask Questions As you read this section, write down any questions that you have. When you finish reading, talk about your questions in a small group.

READING CHECK

1. List What are the two main things that determine climate?

_____

_____

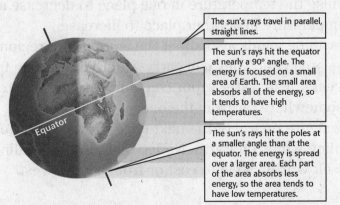

The sun's rays travel in parallel, straight lines.

The sun's rays hit the equator at nearly a 90° angle. The energy is focused on a small area of Earth. The small area absorbs all of the energy, so it tends to have high temperatures.

The sun's rays hit the poles at a smaller angle than at the equator. The energy is spread over a larger area. Each part of the area absorbs less energy, so the area tends to have low temperatures.

TAKE A LOOK

2. Explain Why do areas near the equator tend to have high temperatures?

_____

_____

SECTION 1 What Is Climate? *continued*

3. Explain Why don't areas near the equator have large seasonal changes in weather?

_____

_____

_____

_____

✓ READING CHECK

4. Identify What causes wind to form?

_____

_____

LATITUDE AND SEASONS

Most places in the United States have four seasons during the year. However, some places in the world do not have such large seasonal changes. For example, places near the equator have about the same temperatures and amounts of daylight all year.

Seasons happen because Earth is tilted on its axis by about 23.5°. This tilt affects how much solar energy an area gets as Earth orbits the sun. The figure below shows how Earth's tilt affects the seasons.

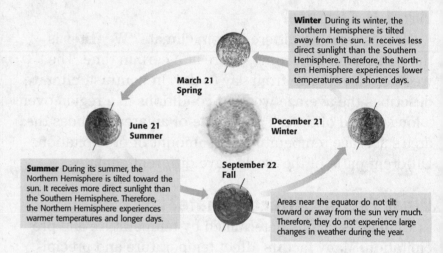

Winter During its winter, the Northern Hemisphere is tilted away from the sun. It receives less direct sunlight than the Southern Hemisphere. Therefore, the Northern Hemisphere experiences lower temperatures and shorter days.

March 21 Spring

June 21 Summer

December 21 Winter

Summer During its summer, the Northern Hemisphere is tilted toward the sun. It receives more direct sunlight than the Southern Hemisphere. Therefore, the Northern Hemisphere experiences warmer temperatures and longer days.

September 22 Fall

Areas near the equator do not tilt toward or away from the sun very much. Therefore, they do not experience large changes in weather during the year.

PREVAILING WINDS

Prevailing winds are winds that blow mainly in one direction. The wind patterns on Earth are caused by the uneven heating of Earth's surface. This uneven heating forms areas with different air pressures. *Wind* forms when air moves from areas of high pressure to areas of low pressure. ☑

Prevailing winds affect climate and weather because they move solar energy from one place to another. This can cause the temperature in one place to decrease and the temperature in another place to increase.

Prevailing winds also affect the amount of precipitation an area gets. They can carry water vapor away from the oceans. The water vapor can condense and fall to the land somewhere far from the ocean.

The figure on top of the next page shows the major prevailing winds on Earth. Notice that most prevailing winds blow from west to east or from east to west.

SECTION 1 What Is Climate? *continued*

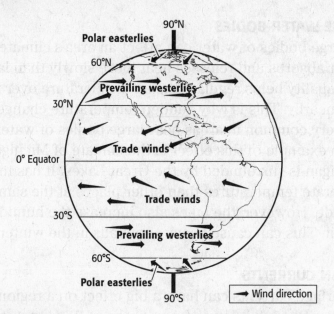

→ Wind direction

TOPOGRAPHY

The sizes and shapes of the land-surface features of a region form its *topography*. The topography of an area affects its climate because topography can affect temperature and precipitation. For example, elevation is a feature of topography that can have a large impact on temperature. **Elevation** is the height of an area above sea level. As elevation increases, temperature tends to decrease. ☑

Mountains can also affect precipitation. As air rises to move over a mountain, it cools. The cool air condenses, forming clouds. Precipitation may fall. This process causes the *rain-shadow effect*, which is illustrated in the figure below.

Air rises to flow over mountains. The air cools as it rises, and water vapor can condense to form clouds. The clouds can release the water as precipitation. Therefore, this side of the mountain tends to be wetter, with more vegetation.

The air on this side of the mountain contains much less water vapor. As the air sinks down the side of the mountain, it becomes warmer. The warm air absorbs moisture from the land. Therefore, this side of the mountain tends to be drier and more desert-like.

TAKE A LOOK
5. Read a Map In which direction do the Prevailing Westerlies blow?

_____

_____

✓ **READING CHECK**

6. Describe In general, how does elevation affect temperature?

_____

_____

_____

TAKE A LOOK
7. Explain Why do clouds form as air moves over a mountain?

_____

_____

_____

LARGE WATER BODIES

Large bodies of water can affect an area's climate. Water absorbs and releases heat more slowly than land. This quality helps regulate the air temperature over the land nearby. This is why sudden temperature changes are not very common in areas near large bodies of water. ☑

An example of this effect is the climate of Michigan. Michigan is surrounded by the Great Lakes. It has more-moderate temperatures than other places at the same latitude. However, the lakes also increase the humidity of the air. This can cause heavy snowfalls in the winter.

OCEAN CURRENTS

Surface currents can have a big effect on a region's climate. **Surface currents** are paths of flowing water found near the surface of the ocean. As surface currents move, they carry warm or cool water to different places. The temperature of the water affects the temperature of the air above it. For example, warm currents can heat the surrounding air.

An example of the effects of ocean currents on climate can be seen in Iceland. Iceland is an island near the Arctic Circle. The Gulf Stream, a warm surface current, flows past Iceland. The warm water in the Gulf Stream causes Iceland's climate to be fairly mild. In contrast, the island of Greenland is at a similar latitude but is not affected by the Gulf Stream. Greenland's climate is much colder than Iceland's.

☑ **READING CHECK**

8. Explain Why aren't sudden temperature changes common near large bodies of water?

_____

_____

_____

_____

Critical Thinking

9. Describe Processes Cool surface currents can cause the air above them to become cooler. Explain how this happens.

_____

_____

_____

_____

TAKE A LOOK

10. Identify What kind of surface current is found off the East Coast of the United States?

_____

_____

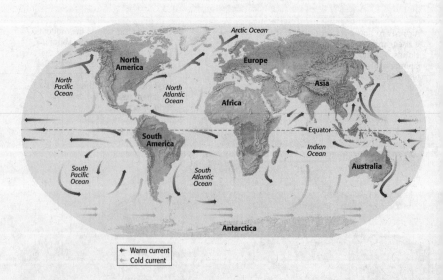

Warm current
Cold current

What Are the Different Climates Around the World?

Earth has three major climate zones: tropical, temperate, and polar. The figure below shows where these zones are found.

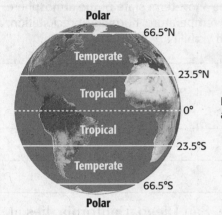

Polar
66.5°N
Temperate
23.5°N
Tropical
0°
Tropical
23.5°S
Temperate
66.5°S
Polar

Earth's three major climate zones are determined by latitude.

Each climate zone has a certain range of temperatures. The tropical zone, near the equator, has the highest temperatures. The polar zones, located at latitudes above 66.5°, have the lowest temperatures. ☑

BIOMES

Each climate zone contains several different kinds of climates. The different climates are the result of topography, winds, and ocean currents. The different climates affect the organisms that live in an area. A large area with a certain climate and types of organisms is called a **biome**. ☑

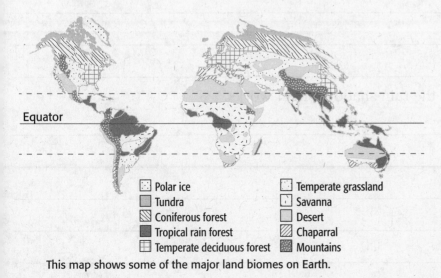

Equator

☐ Polar ice   ☐ Temperate grassland
☐ Tundra   ☐ Savanna
☐ Coniferous forest ☐ Desert
☐ Tropical rain forest ☐ Chaparral
☐ Temperate deciduous forest ☐ Mountains

This map shows some of the major land biomes on Earth.

TAKE A LOOK
11. Identify What determines Earth's major climate zones?

_____

READING CHECK

12. Describe Which climate zone has the highest temperatures?

_____

READING CHECK

13. Identify Relationships How are biomes and climate related?

_____

_____

_____

TAKE A LOOK
14. Explain Where are most tropical rain forest biomes located?

_____

Section 1 Review

NSES   ES 1f, 1j, 3d

SECTION VOCABULARY

biome a large region characterized by a specific type of climate and certain types of plant and animal communities	**prevailing winds** winds that blow mainly from one direction during a given period
climate the average weather conditions in an area over a long period of time	**surface current** a horizontal movement of ocean water that is caused by wind and that occurs at or near the ocean's surface
elevation the height of an object above sea level	**weather** the short-term state of the atmosphere, including temperature, humidity, precipitation, wind, and visibility
latitude the distance north or south from the equator; expressed in degrees	

1. Compare How is climate different from weather?

_____

_____

_____

2. Apply Concepts Nome, Alaska, lies at 64°N latitude. San Diego, California, lies at 32°N latitude. Which city receives more sunlight? Explain your answer.

_____

_____

3. Explain What causes some places on Earth to have seasons?

_____

_____

4. Identify What are four things that can affect climate?

_____

_____

_____

_____

5. Explain Describe how the rain-shadow effect works.

_____

_____

_____

_____

_____

Name _____ Class _____ Date _____

The Tropics

BEFORE YOU READ

After you read this section, you should be able to answer these questions:

- Where is the tropical zone?
- What are three biomes found in the tropical zone?

What Is the Tropical Zone?

Remember that latitudes near the equator receive more solar energy than other areas. The area between 23.5°N latitude and 23.5°S latitude receives the most solar energy. This region is called the **tropical zone**. It is also known as the *Tropics*. Because areas in the Tropics receive so much solar energy, they tend to have high temperatures.

There are three main biomes in the Tropics: tropical rain forest, tropical savanna, and tropical desert. All the tropical biomes have high temperatures. However, they receive different amounts of rain and have different types of soil. Therefore, different organisms live in each biome. The figure below shows where each of these biomes is found. ☑

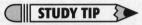

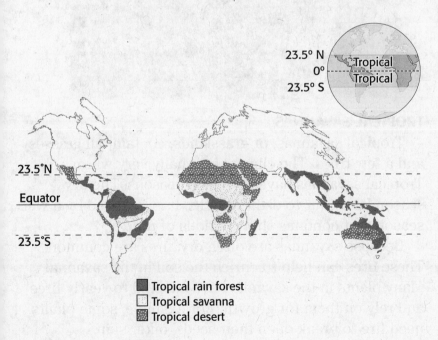

23.5° N
0°
23.5° S

Tropical
Tropical

23.5°N

Equator

23.5°S

■ Tropical rain forest
□ Tropical savanna
▨ Tropical desert

STUDY TIP

Compare After you read this section, make a chart comparing the three kinds of tropical biomes.

READING CHECK

1. Explain Why do the different tropical biomes have different organisms living in them, even though they all have high temperatures?

_____

_____

_____

TAKE A LOOK

2. Identify Where are the Tropics?

_____

_____

_____

SECTION 2 The Tropics *continued*

TROPICAL RAIN FORESTS

Tropical rain forests are warm and wet. They are located close to the equator, so they receive about the same amount of solar energy all year long. Therefore, there is little difference between the seasons. ☑

Tropical rain forests are homes to many different kinds of living things. Animals that live in tropical rain forests include monkeys, parrots, frogs, tigers, and leopards. Plants include mahogany trees, vines, ferns, and bamboo.

Many organisms live in tropical rain forests. When dead organisms decay, nutrients return to the soil. However, the nutrients are quickly used up by plants or washed away by rain. As a result, the soil is thin and poor in nutrients.

Tropical Rain Forest

• **Average Temperature Range 25°C to 28°C (77°F to 82°F)**
• **Average Yearly Precipitation 200 cm or more**
• **Soil Characteristics thin and nutrient-poor**

TROPICAL SAVANNAS

Tropical savannas, or grasslands, contain tall grasses and a few trees. The climate is usually very warm. Tropical savannas have two main seasons. The dry season lasts four to eight months. It is followed by a wet season that contains short periods of rain.

Because savannas are often dry, fires are common. These fires can help to enrich the soil in the savanna. Many plants in the savanna have adapted to yearly fires and rely on them for growth. For example, some plants need fire to break open their seeds' outer skin.

Animals that live in tropical savannas include giraffes, lions, crocodiles, and elephants. The figure on the top of the next page shows a tropical savanna.

✓ **READING CHECK**

3. Explain Why is there little difference between the seasons in a tropical rain forest?

_____

_____

_____

TAKE A LOOK

4. Explain Why is the soil in tropical rainforests thin and nutrient-poor?

_____

_____

_____

Critical Thinking

5. Predict Consequences What could happen to a tropical savanna if people stopped all fires from spreading? Explain your answer.

_____

_____

_____

_____

Tropical Savanna

• **Average Temperature Range**
27°C to 32°C (80°F to 90°F)

• **Average Yearly Precipitation**
100 cm

• **Soil Characteristics**
generally nutrient-poor

Math Focus

6. Convert About how many feet of rain does a tropical savanna get in a year?

1 in. = 2.54 cm

_____

TROPICAL DESERTS

A desert is an area that receives less than 25 cm of rainfall per year. Deserts are the driest places on Earth. Tropical desert plants, such as shrubs, are adapted to living in places with little water. Animals such as camels, lizards, snakes, and scorpions also have adaptations for living in the desert.

Most tropical deserts are very hot in the daytime. They can be up to 50°C (120°F) during the day. However, the temperatures at night may be much lower. Therefore, organisms that live in deserts are also adapted to changing temperatures. ☑

 **READING CHECK**

7. Explain Why do tropical desert organisms have to be adapted to changing temperatures?

_____

_____

Tropical Desert

• **Average Temperature Range**
16°C to 50°C (61°F to 120°F)

• **Average Yearly Precipitation**
0 cm to 25 cm

• **Soil Characteristics**
poor in organic matter

Section 2 Review

SECTION VOCABULARY

tropical zone the region that surrounds the equator and that extends from about 23° north latitude to 23° south latitude	

1. List What are the three biomes found in the Tropics?

_____

_____

_____

2. Identify What is one thing that all the biomes in the Tropics have in common?

_____

3. Compare Fill in the missing information about the features of each tropical biome.

Biome	Rainfall	Soil	Example of an animal found here	Example of a plant found here
Tropical rain forest		poor	parrot	
	100 cm per year		giraffe	
		poor		palm tree

4. Apply Concepts An area is located at 30°N latitude. It receives less than 25 cm per year of rain and has temperatures as high as 50°C during the day. Is the area a tropical desert? Explain your answer.

_____

_____

_____

5. Identify On which continent are most tropical savannas found?

_____

6. Identify Which tropical biome has the largest range of temperatures? Which tropical biome has the smallest range of temperatures?

_____

_____

CHAPTER 17 Climate

SECTION 3 Temperate and Polar Zones

After you read this section, you should be able to answer these questions:

- What biomes are found in the temperate zone?
- What biomes are found in the polar zone?
- What are two examples of microclimates?

What Is the Temperate Zone?

The climate zone between the tropical and the polar zones is the **temperate zone**. This zone extends from about 23.5° to about 66.5° north or south latitudes. Most of the continental United States is in the temperate zone. The temperate zone receives less solar energy than the Tropics. Therefore, temperatures in the temperate zone tend to be lower than those in the Tropics.

The four main biomes in the temperate zone are temperate forests, temperate grasslands, chaparrals, and temperate deserts. All of these biomes show seasonal changes in weather. However, some biomes have more extreme weather changes than others. For example, some areas in the United States have similar temperatures all year long. Other areas have very low temperatures in the winter and very high temperatures in the summer. ☑

STUDY TIP

Compare After you read this section, make a table comparing the four main temperate biomes.

READING CHECK

1. Identify What do the four main temperate biomes have in common?

_____

_____

_____

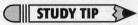

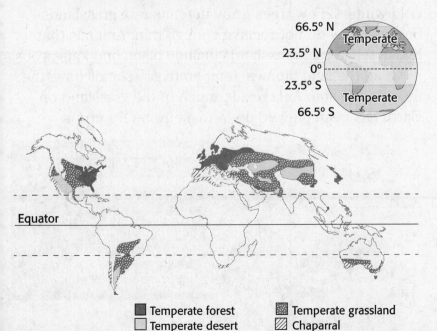

66.5° N
23.5° N
0°
23.5° S
66.5° S

Temperate

Temperate

Equator

■ Temperate forest
□ Temperate desert
▨ Temperate grassland
▧ Chaparral

TAKE A LOOK

2. Read a Map What kind of biome is found in northern and southern Africa?

_____

_____

TEMPERATE FORESTS

Temperate forests tend to have high amounts of rainfall and large seasonal temperature differences. The summers are warm, and the winters are cold. Animals that live in temperate forests include foxes, deer, and bears. Some trees in temperate forests lose their leaves each winter. These trees are called *deciduous* trees. Other trees, called *evergreens*, do not lose all of their leaves at once.

The soils in most temperate forests are very rich in nutrients. This is because the deciduous trees drop their leaves every winter. As the leaves decay, nutrients are added to the soil.

Critical Thinking

3. Infer A student visits a forest in Vermont in January. Most of the trees in the forest are covered with leaves. Are the trees probably deciduous trees or evergreens? Explain your answer.

_____

_____

_____

_____

_____

Temperate Forest

• Average Temperature Range 0°C to 28°C (32°F to 82°F)

• Average Yearly Precipitation 76 cm to 250 cm

• Soil Characteristics very fertile, organically rich

TEMPERATE GRASSLANDS

Temperate grasslands have warm summers and very cold winters. Few trees grow in temperate grasslands because they do not receive enough rain. Animals that live in temperate grasslands include bison and kangaroos.

Of all the land biomes, temperate grasslands have the most fertile soil. As a result, much of the grassland on Earth has been plowed up to make room for crops.

Temperate Grassland

• Average Temperature Range −6°C to 26°C (21°F to 78°F)

• Average Yearly Precipitation 38 cm to 76 cm

• Soil Characteristics most-fertile soils of all biomes

TAKE A LOOK

4. Identify What is the main kind of plant that grows in temperate grasslands?

_____

CHAPARRALS

Chaparral regions have cool, wet winters and hot, dry summers. Animals that live in the chaparral include mountain lions, coyotes, and quail.

Fires are common during the summers in chaparrals. Some chaparral plants are adapted to these fires. Chaparral plants also have adaptations that prevent water loss during dry conditions. For example, the main kinds of plants in the chaparral are evergreen shrubs. These shrubs have thick leaves with waxy coatings. The coatings help prevent the leaves from losing water. ☑

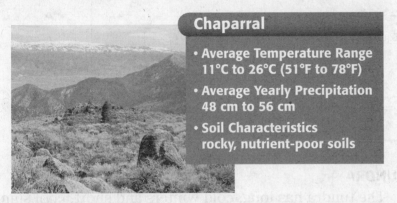

Chaparral

• **Average Temperature Range**
  11°C to 26°C (51°F to 78°F)

• **Average Yearly Precipitation**
  48 cm to 56 cm

• **Soil Characteristics**
  rocky, nutrient-poor soils

TEMPERATE DESERTS

Like tropical deserts, temperate deserts are hot in the daytime and receive little rainfall. However, temperate deserts tend to have much colder nights than tropical deserts. This is because temperate deserts tend to have low humidity and cloudless skies. These conditions allow solar energy to heat the surface a lot during the day. They also allow heat to move into the atmosphere at night. ☑

Plants that live in temperate deserts include cacti, shrubs, and thorny trees. Animals include lizards, snakes, bats, and toads.

Temperate Desert

• **Average Temperature Range**
  1°C to 50°C (34°F to 120°F)

• **Average Yearly Precipitation**
  0 cm to 25 cm

• **Soil Characteristics**
  poor in organic matter

What Is the Polar Zone?

The **polar zone** is located between 66.5° and 90° north and south latitudes, near the North and South Poles. This zone has the coldest temperatures of all climate zones. There are two biomes in the polar zone: tundra and taiga.

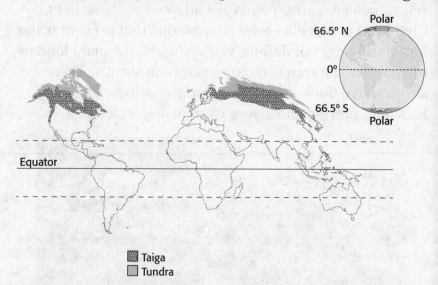

Taiga
Tundra

TAKE A LOOK

7. Identify On which continents is taiga found?

_____

_____

TUNDRA

The tundra has long, cold winters and short, cool summers. In the summer, only the top meter of soil thaws out. Below this depth is a permanently frozen layer called *permafrost*. It prevents water in the thawed soil from draining away. Therefore, the upper soil is muddy in the summer. Insects like mosquitoes thrive there. Birds migrate there in the summer to eat the insects. ☑

Other animals that live in the tundra include caribou, reindeer, and polar bears. Only small plants, such as mosses, live in the tundra.

✓ **READING CHECK**

8. Explain Why is the upper soil in the tundra muddy during the summer?

_____

_____

_____

Tundra

• **Average Temperature Range**
  −27°C to 5°C (−17°F to 41°F)

• **Average Yearly Precipitation**
  0 cm to 25 cm

• **Soil Characteristics**
  frozen

SECTION 3 Temperate and Polar Zones *continued*

TAIGA

Taiga biomes are found just south of tundra biomes in the Northern Hemisphere. The taiga has long, cold winters and short, warm summers. Animals that live in the taiga include moose, bears, and rabbits.

Evergreen trees called *conifers*, such as pine and spruce, are the main plants that grow in the taiga. The needle-like leaves from these trees contain acidic substances. When the needles die and decay on the ground, these substances make the soil acidic. Not very many plants can grow in acidic soils. Therefore, few plants grow on the forest floor of the taiga.

Say It

Share Experiences In a small group, talk about different biomes that you or your classmates have visited.

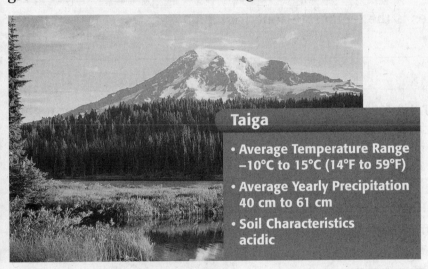

Taiga

- **Average Temperature Range** −10°C to 15°C (14°F to 59°F)
- **Average Yearly Precipitation** 40 cm to 61 cm
- **Soil Characteristics** acidic

Math Focus
9. Convert How much precipitation does the taiga get per year in inches?

1 in. = 2.54 cm

_____

What Are Microclimates?

Remember that latitude, topography, and water help determine the climate of an area. Local conditions can also affect the climate in smaller areas. A **microclimate** is the climate of a small area. Two examples of microclimates are alpine biomes and cities. ☑

Alpine biomes are cold microclimates found near the tops of mountains. In winter, the temperatures are below freezing. In summer, they range from 10°C to 15°C. It is the high elevations of alpine biomes that cause them to be so cold. Alpine biomes are even found on mountains in the Tropics.

Cities are also microclimates. Buildings and pavement are made of dark materials. They absorb solar energy and stay warm. City temperatures can be 1°C to 2°C warmer than temperatures in other areas.

✓ READING CHECK

10. Define What is a microclimate?

_____

_____

Section 3 Review

SECTION VOCABULARY

microclimate the climate of a small area **polar zone** the North or South Pole and the surrounding region	**temperate zone** the climate zone between the Tropics and the polar zone

1. List What are the four biomes of the temperate zone?

_____

_____

_____

2. Identify At what latitudes is the temperate zone found?

_____

_____

3. Explain Why are temperate deserts very hot during the day but very cold at night?

_____

_____

_____

4. Explain Why do cities often have higher temperatures than surrounding rural areas?

_____

_____

5. Explain Why are most taiga soils acidic?

_____

_____

6. Compare How are temperate deserts and the tundra similar?

_____

_____

7. Explain Why do few trees grow in temperate grasslands?

_____

_____

CHAPTER 17 | Climate

SECTION
4 **Changes in Climate**

National Science
Education Standards
ES 1k, 2a

BEFORE YOU READ

**After you read this section, you should be able to answer
these questions:**

• How has Earth's climate changed over time?

• What factors can cause climates to change?

How Was Earth's Climate Different in the Past?

The geologic record shows that Earth's climate in the
past was different from its climate today. During some
periods in the past, Earth was much warmer. During other
periods, Earth was much colder. In fact, much of Earth was
covered by sheets of ice during some times in the past.

An **ice age** happens when ice at high latitudes expands
toward lower latitudes. Scientists have found evidence of
many major ice ages in Earth's history. The most recent
one began about 2 million years ago. ☑

Many people think of an ice age as a time when the
temperature is always very cold. However, during an ice
age, there can be periods of colder or warmer weather.
A period of colder weather is called a *glacial period*. A
period of warmer weather is called an *interglacial period*.

During glacial periods, large sheets of ice grow. These
ice sheets form when ocean water freezes. Therefore,
sea level drops during glacial periods. The figure below
shows the coastlines of the continents during the last
glacial period. Notice that the continental coastlines
extended further into the ocean than they do today.

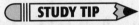

STUDY TIP

Learn New Words As you
read, underline any words
that you don't know. When
you figure out what they mean,
write the words and their
definitions in your notebook.

READING CHECK

1. Define Write your own
definition for *ice age*.

_____

_____

_____

_____

TAKE A LOOK

2. Explain Why is more land
exposed during glacial periods than at other times?

_____

_____

_____

_____

| | Extent of land mass at glacial maximum | | Extent of continental glaciation |
| | Current land mass | | Extent of sea ice |

SECTION 4 Changes in Climate *continued*

What Can Cause Climates to Change?

Scientists have several theories to explain ice ages and other forms of climate change. Factors that can cause climate change include Earth's orbit, plate tectonics, the sun's cycles, asteroid impacts, volcanoes, and human activities.

CHANGES IN EARTH'S ORBIT

A Serbian scientist, Milutin Milankovitch, found that changes in Earth's orbit and tilt can affect Earth's climate. He modeled the way Earth moves in space and found that Earth's movements change in a regular way. These changes happen over tens of thousands of years. For example, Earth's orbit around the sun is more circular at some times than others.

These variations in Earth's orbit and tilt affect how much sunlight Earth gets. Therefore, they can also affect climate. The figure below shows how these factors can change the amount of sunlight Earth gets.

Critical Thinking

3. Infer Could changes in climate over 100 years be caused by changes in Earth's orbit and tilt? Explain your answer.

_____

_____

_____

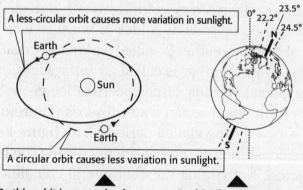

A less-circular orbit causes more variation in sunlight.

Earth

Sun

Earth

A circular orbit causes less variation in sunlight.

0°   23.5°
22.2° 24.5°
N

S

Earth's orbit is more circular at some times than at other times. The amount of solar energy that Earth gets from the sun varies more when Earth's orbit is less circular.

Earth's tilt on its axis can vary. When the tilt is greater, the poles get more solar energy.

TAKE A LOOK
4. Identify How does the shape of Earth's orbit change?

_____

_____

N

S

Earth's axis wobbles slightly. This affects how much sunlight Earth's surface gets at different times of the year.

SECTION 4 Changes in Climate *continued*

PLATE TECTONICS

Plate tectonics and continental drift also affect Earth's climate. When a continent is closer to the equator, its climate is warmer than when it is near the poles. Also, remember that continents can deflect ocean currents and winds. When continents move, the flow of air and water around the globe changes. These changes can strongly affect Earth's climate.

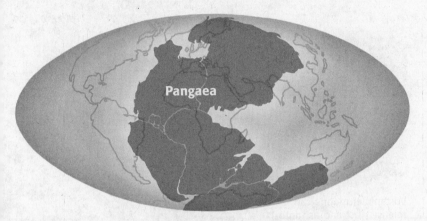

Pangaea

The locations of the continents can affect their climate. When India, Africa, South America, and Australia were part of Pangaea, they were covered with large ice sheets.

TAKE A LOOK
5. Identify How was the climate of India different when it was part of Pangaea?

_____

_____

THE SUN

Some changes in Earth's climate are caused by changes in the sun. Many people think that the sun is always the same, but this is not true. In fact, the amount of energy that the sun gives off can change over time. The sun follows a regular cycle in how much energy it gives off. Because the sun's energy drives most cycles on Earth, these changes can affect Earth's climate. ☑

✓ **READING CHECK**
6. Explain Why do changes in the sun's energy affect the climate on Earth?

_____

_____

_____

IMPACTS

Sometimes, objects from outer space, such as asteroids, crash into Earth. An *asteroid* is a small, rocky object that orbits the sun. If a large asteroid crashed into Earth, the climate of the whole planet could change.

When a large object hits Earth, particles of dust and rock fly into the atmosphere. This material can block some sunlight from reaching Earth's surface. This can cause temperatures on Earth to go down. In addition, plants may not be able to survive with less sunlight. Without plants, many animals would die off. Many scientists believe that an asteroid impact may have caused the dinosaurs to become extinct.

Critical Thinking
7. Identify Relationships Why may animals die off if there are fewer plants around?

_____

_____

VOLCANIC ERUPTIONS

Volcanic eruptions can affect Earth's climate for a short time. They send large amounts of dust and ash into the air. As with an asteroid impact, the dust and ash block sunlight from reaching Earth's surface. The figure below shows how volcanic dust can affect sunlight.

TAKE A LOOK
8. Compare How are the effects on climate of volcanic eruptions and asteroid impacts similar?

_____

_____

_____

_____

_____

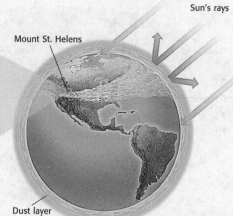

Volcanic eruptions can release dust and ash into the atmosphere. This plume of dust and ash was produced by the eruption of Mount St. Helens, in Washington, in 1980.

The dust and ash from the volcano can spread throughout the atmosphere. Sunlight reflects off this dust and ash. Less sunlight reaches Earth's surface.

What Is Global Warming?

A slow increase in global temperatures is called **global warming**. One thing that can cause global warming is an increase in the greenhouse effect. The **greenhouse effect** is Earth's natural heating process. During this process, gases in the atmosphere absorb energy in sunlight. This energy is released as heat, which helps to keep Earth warm. Without the greenhouse effect, Earth's surface would be covered in ice. ☑

One of the gases that absorbs sunlight in the atmosphere is carbon dioxide (CO_2). If there is more CO_2 in the atmosphere, the greenhouse effect can increase. This can cause global warming.

READING CHECK
9. Define What is global warming?

_____

_____

SECTION 4 Changes in Climate *continued*

WHERE CO_2 COMES FROM

Much of the CO_2 in the atmosphere comes from natural processes, such as volcanic eruptions and animals breathing. However, human activities can also increase the amount of CO_2 in the atmosphere. ☑

When people burn fossil fuels for energy, CO_2 is released into the atmosphere. When people burn trees to clear land for farming, CO_2 is released. In addition, plants use CO_2 for food. Therefore, when trees are destroyed, we lose a natural way of removing CO_2 from the atmosphere.

✓ **READING CHECK**

10. Identify What are two natural sources of carbon dioxide in the atmosphere?

_____

_____

PROBLEMS WITH GLOBAL WARMING

Many scientists think that if global warming continues, the ice at Earth's poles could melt. This could cause sea levels to rise. Many low-lying areas could flood. Global warming could also affect areas far from the oceans. For example, the Midwestern part of the United States could become warmer and drier. Northern areas, such as Canada, may become warmer. ☑

✓ **READING CHECK**

11. Explain Why may sea level rise if global warming continues?

_____

_____

_____

WHAT PEOPLE CAN DO

Many countries are working together to reduce the effects of global warming. Treaties and laws have helped to reduce pollution and CO_2 production. Most CO_2 is produced when people burn fossil fuels for energy. Therefore, reducing how much energy you use can reduce the amount of CO_2 produced. Here are some ways you can reduce your energy use:

• Turn off electrical devices, such as lights and computers, when you are not using them.

• Ride a bike, walk, or take public transportation instead of using a car to travel.

• Turn the heater to a lower temperature in the winter.

• Turn the air conditioner to a higher temperature in the summer.

Name ~~Peter Yu~~ Class ______ Date ______

Section 4 Review

NSES ES 1k, 2a

SECTION VOCABULARY

global warming a gradual increase in average global temperature **greenhouse effect** the warming of the surface and lower atmosphere of Earth that occurs when water vapor, carbon dioxide, and other gases absorb and reradiate thermal energy	**ice age** a long period of climatic cooling during which the continents are glaciated repeatedly

1. Identify Relationships How is global warming related to the greenhouse effect?

2. Describe What did Milutin Milankovitch's research show can affect Earth's climate?

3. Identify Give two ways that plate tectonics can affect an area's climate.

4. Predict Consequences How could global warming affect cities near the oceans? Explain your answer.

5. List Give three ways that human activities can affect the amount of CO_2 in the atmosphere.

CHAPTER 18 Studying Space

SECTION 1 Astronomy: The Original Science

National Science Education Standards
ES 3a, ES 3b, ES 3c

BEFORE YOU READ

After you read this section, you should be able to answer these questions:

- How do astronomers define a day, a month, and a year?
- What is the difference between the Ptolemaic and Copernican theories about the universe?
- What contributions did Brahe, Kepler, Newton, Galileo, and Hubble make to astronomy?

How Does Astronomy Affect Our Calendar?

Imagine that it is 5,000 years ago. You do not have a modern clock or calendar. How can you know what day it is? How can you know what month it is? One way is to study the movement of the moon, the planets, and the stars.

People in ancient cultures used the movements of the stars, planets, and moon to mark the passage of time. People observed that the objects in the solar system move in regular and predictable ways. Farmers used these cycles to figure out the best time of year to plant and harvest. Sailors used the stars to navigate their ships. ☑

The early observations of the night sky led to the first calendars. Our modern calendar is also based on the movements of the bodies in our solar system. In our modern calendar, a **year** is the amount of time it takes the Earth to orbit the sun once. A **month** is about the same amount of time that the moon takes to orbit the Earth once. A **day** is the time it takes for the Earth to rotate once on its axis.

Unit	Description
Day	
Month	
	the time it takes the Earth to orbit the sun once

Over time, the study of the night sky became the science of astronomy. **Astronomy** is the study of the universe. Scientists who study astronomy are called *astronomers*. Modern astronomy is based partly on the work of early astronomers.

STUDY TIP

Compare As you read, make a chart comparing the different scientists that are mentioned in this section. In your chart, describe each scientist's contributions to astronomy.

READING CHECK

1. Explain How did people in ancient cultures mark the passage of time?

_____

_____

_____

TAKE A LOOK
2. Identify Fill in the blank spaces in the table.

How Did Early Astronomers Affect Astronomy?

Almost everything that early astronomers knew came from what they could observe with their eyes. Therefore, most early astronomers thought the universe was made only of the moon, the planets, and the sun. They thought that all the stars were at the edge of the universe.

Early theories about the universe were incorrect in many ways. However, over time, more data became available to astronomers. As a result, theories about the universe began to change. ☑

PTOLEMY: AN EARTH-CENTERED UNIVERSE

Claudius Ptolemy was a Greek astronomer. In 140 CE, he wrote a book that brought together many ancient astronomical observations. He used these observations, together with careful calculations, to develop what is known as the *Ptolemaic theory*. According to this theory, the Earth is the center of the universe. The Ptolemaic theory also states that all other objects in the universe orbit the Earth.

Today, we know that the Ptolemaic theory is incorrect. However, Ptolemy's calculations predicted the motions of the planets better than any other theory at the time. The predictions fit the observations that other astronomers made. Therefore, the Ptolemaic theory was accepted as correct for more than 1,500 years.

COPERNICUS: A SUN-CENTERED UNIVERSE

In 1543, a Polish astronomer named Nicolaus Copernicus published a new theory. His theory stated that the sun is the center of the universe and that the planets revolve around the sun.

Scientists did not accept Copernicus's theory immediately. However, when it was accepted, it caused major changes in science and society. These changes were called the *Copernican revolution*.

✔ **READING CHECK**

3. Explain Why have astronomers changed their theories about the universe over time?

_____

_____

_____

Critical Thinking

4. Compare Today, scientists know that only part of Copernicus's theory is correct. Which part of Copernicus's theory is not correct?

_____

_____

TAKE A LOOK

5. Describe Fill in the blank spaces in the table.

Astronomer	Description of theory
Ptolemy	
	The sun is the center of the universe, and the planets orbit the sun.

TYCHO BRAHE: A WEALTH OF DATA

In the late 1500s, a Danish astronomer, Tycho Brahe, made the most detailed astronomical observations so far. Brahe thought the sun and moon revolved around the Earth, and the other planets revolved around the sun. Although his theory was incorrect, his precise observations helped future astronomers. ☑

JOHANNES KEPLER: LAWS OF PLANETARY MOTION

Johannes Kepler was Brahe's assistant. He continued to analyze Brahe's data after Brahe died. Kepler determined that the planets revolve around the sun in *elliptical*, or oval-shaped, orbits. He also developed three laws that describe planetary motion. These laws are still used today.

GALILEO: TURNING A TELESCOPE TO THE SKY

Galileo Galilei was one of the first people to use a telescope to observe objects in space. Before his time, astronomers observed space using only their eyes. Galileo made many important observations about the solar system. Some of these observations are listed below.

- There are craters and mountains on the surface of the Earth's moon.
- Jupiter has at least four moons.
- Dark spots sometimes appear on the surface of the sun.

These discoveries were important because they showed that the planets are physical bodies like the Earth. Until Galileo, people thought that the planets were stars that moved quickly through the sky.

ISAAC NEWTON: THE LAWS OF GRAVITY

In 1687, Sir Isaac Newton showed that all objects in the universe attract each other through a force called gravity. Heavy objects and objects that are close together have the strongest force of gravity. This explains why all the planets orbit the sun. The sun has more mass than any other object in the solar system.

EDWIN HUBBLE: BEYOND THE MILKY WAY

In 1924, Edwin Hubble used detailed observations to prove that other galaxies existed beyond the edge of our galaxy. His data confirmed that the universe is much larger than our own galaxy, the Milky Way.

✓ READING CHECK

6. Explain How did Tycho Brahe's work help astronomers?

_____

_____

STANDARDS CHECK

ES 3a The earth is the third planet from the sun in a system that includes the moon, the sun, eight other planets and their moons, and smaller objects, such as asteroids and comets. The sun, an average star, is the central and largest body in the solar system.

7. Identify What is the most massive object in the solar system?

_____

_____

Name ______________________ Class ______________ Date ______________

Section 1 Review

NSES ES 3a, ES 3b, ES 3c

SECTION VOCABULARY

astronomy the scientific study of the universe **day** the time required for Earth to rotate once on its axis	**month** a division of the year that is based on the orbit of the moon around the Earth **year** the time required for the Earth to orbit once around the sun

1. **Compare** What is the difference between a day, a month, and a year in terms of astronomy?

2. **Describe** What did people in ancient cultures observe about the motions of the planets, the moon, and the sun?

3. **Explain** Why was the Ptolemaic theory accepted for a long time?

4. **Infer** How did Tycho Brahe's work help Kepler develop his laws of planetary motion?

5. **Evaluate** What advantage did Galileo have over other, earlier astronomers?

6. **Identify** What did Edwin Hubble prove about the size of the universe?

CHAPTER 18	Studying Space

SECTION 2 | Telescopes

BEFORE YOU READ

After you read this section, you should be able to answer these questions:

- What are telescopes?
- How can telescopes help scientists study space?

How Can a Telescope Help Us Make Observations?

How much of the sky can you see when you gaze up at night? At most, you can see 3,000 stars. With a telescope, you can see millions of stars, as well as many other objects. A **telescope** is a tool that scientists use to study objects, such as stars, that are far away. A telescope collects light and other kinds of radiation from the sky and makes it brighter. In this way, telescopes make distant objects more visible.

An *optical telescope* is used to study visible light from objects in the universe. Simple optical telescopes, such as the one in the figure below, have two lenses. The *objective lens* collects light from distant objects. The objective lens focuses the light and forms an image at a focal point. A *focal point* is where rays of light that pass through a lens or reflect from a mirror come together. ☑

The second lens in a simple optical telescope is in the eyepiece. This lens *magnifies*, or makes bigger, the image that forms at the focal point.

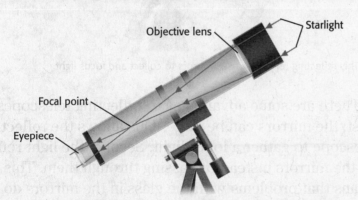

This simple refracting telescope has an objective lens that collects light and a lens in the eyepiece to magnify the image.

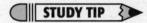

STUDY TIP

Learn New Words As you read, underline words you don't know. When you figure out what they mean, write the words and their definitions in your notebook.

READING CHECK

1. Define What is a focal point?

_____

_____

_____

TAKE A LOOK

2. Compare How is the objective lens different from the lens in the eyepiece?

_____

_____

_____

_____

REFRACTING TELESCOPES

Refracting telescopes are simple optical telescopes that use lenses to gather and focus light. The figure on the previous page shows a drawing of a refracting telescope. Refracting telescopes are the simplest telescopes, so they are usually easy to use. ☑

There are two disadvantages to refracting telescopes. First, lenses focus different colors of light at slightly different distances. This means that images cannot be focused well. Second, refracting telescopes cannot be very large. Large telescopes have large objective lenses. The bigger the objective lens, the more light the telescope can gather. However, if the lens is too large, it can bend under its own weight. This causes the image to look fuzzy.

REFLECTING TELESCOPES

Reflecting telescopes use curved mirrors to gather and focus light. Light enters the telescope and reflects off a large, curved mirror. The light then travels to a flat mirror near the eyepiece. The flat mirror focuses the image and reflects it to the eyepiece.

READING CHECK

3. Explain How do refracting telescopes gather and focus light?

_____

_____

TAKE A LOOK
4. Describe What does the flat mirror in a reflecting telescope do?

_____

_____

_____

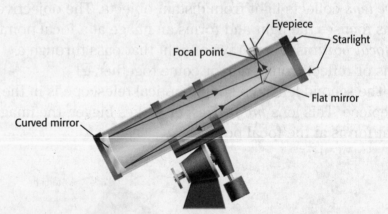

This reflecting telescope uses mirrors to collect and focus light.

There are some advantages to reflecting telescopes. First, the mirrors can be large. This allows the reflecting telescope to gather a lot of light. Second, the light reflects off the mirrors instead of passing through them. This means that problems with the glass in the mirrors do not affect the image. Third, mirrors can focus all colors of light to the same focal point. Therefore, the images can be better focused than with refracting telescopes.

SECTION 2 Telescopes *continued*

LARGE TELESCOPES AND CLEAR IMAGES

Some very large reflecting telescopes use several mirrors to collect and focus light. For example, the Keck Telescopes in Hawaii each use 36 mirrors to collect and focus light. However, even very large reflecting telescopes must be in a good location if they are to form clear images.

The light gathered by telescopes on the Earth is affected by the atmosphere. The motion of the air in the Earth's atmosphere causes starlight to shimmer and blur. Therefore, astronomers may place telescopes on mountain tops, where the air is thinner. There may also be less air and light pollution in these areas.

In order to avoid interference from the atmosphere, scientists have put telescopes in space. These telescopes can detect very faint objects because there is no air to blur the image.

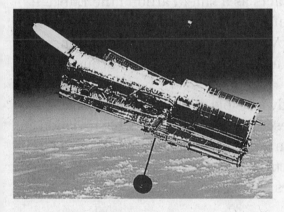

The mirrors in the Hubble Space Telescope are smaller than the mirrors in many telescopes on the Earth. However, the Hubble Telescope can produce images of very faint objects because the Earth's atmosphere does not blur the images.

What Is Light?

Optical telescopes make the visible light from objects in space easier for us to see. Visible light is a form of electromagnetic radiation. However, visible light is not the only form of electromagnetic radiation. Other examples of electromagnetic radiation are gamma rays, X rays, and radio waves. The **electromagnetic spectrum** is made up of all the kinds of electromagnetic radiation.

Electromagnetic radiation travels in waves. Each kind of radiation has a different wavelength. Gamma rays have the shortest wavelengths. Radio waves have the longest wavelengths. ☑

Most of the electromagnetic spectrum is invisible. For example, we cannot see gamma rays or radio waves. The figure on the top of the next page shows some of the different kinds of electromagnetic radiation.

Critical Thinking

5. Apply Concepts Scientists may place telescopes in deserts or other areas where the air is dry. This is because dry air often produces less blurry images than moist air. What is the most likely reason that dry air produces less blurry images than moist air?

_____

_____

_____

_____

READING CHECK

6. Identify Which kind of electromagnetic radiation has the shortest wavelength?

_____

SECTION 2 Telescopes *continued*

**Say It**

Discuss You may have heard or seen the terms "X ray," "infrared," and "ultraviolet" in other places. In a small group, talk about the ways that these words are used in other situations.

TAKE A LOOK

7. Identify Give two kinds of electromagnetic radiation that are invisible.

_____

_____

Shorter wavelengths ←——————————→ Longer wavelengths

Gamma ray · X ray · Ultraviolet · Visible (light we can see) · Infrared (heat) · Microwave · Radio

The electromagnetic spectrum is made up of all the kinds of electromagnetic radiation. Visible light is light that we can see. However, most electromagnetic radiation is invisible.

DETECTING ELECTROMAGNETIC RADIATION

The atmosphere acts as a shield around the Earth. It blocks most kinds of invisible radiation that come from objects in space. However, some types of radiation, such as radio waves and microwaves, can pass through the atmosphere.

Scientists can study invisible radiation using *nonoptical telescopes*. These telescopes can detect invisible radiation and focus it to produce an image. Astronomers study the entire electromagnetic spectrum because each type of radiation reveals different clues about an object. ☑

READING CHECK

8. Explain How do scientists study invisible radiation from objects in space?

_____

_____

RADIO TELESCOPES

Radio telescopes detect radio waves. Radio wavelengths are much longer than visible wavelengths. Therefore, radio telescopes have to be much larger than optical telescopes. However, the reflecting surfaces of radio telescopes do not have to be as smooth as those in optical telescopes. In addition, radio waves can be detected at night and during the day. Therefore, radio telescopes can be very useful, even though they are large.

Astronomers can use many radio telescopes together to get more detailed images. When radio telescopes are linked together, they work like a single giant telescope. The Very Large Array (VLA) consists of 27 radio telescopes spread over 30 km. Together, the VLA telescopes act as a single telescope that is 30 km across.

NONOPTICAL TELESCOPES IN SPACE

Most electromagnetic waves are blocked by the Earth's atmosphere. Therefore, scientists have placed some kinds of nonoptical telescopes in space. These telescopes produce images of objects in space using different kinds of electromagnetic radiation. For example, each figure below shows an image of our galaxy. The images look different because they were recorded from different types of electromagnetic radiation. ☑

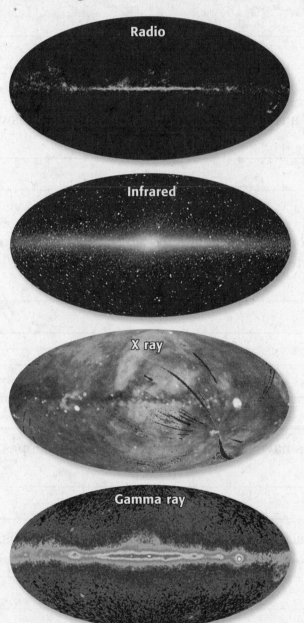

Each of these images shows our galaxy, the Milky Way. Different features of the galaxy are visible at different wavelengths of electromagnetic radiation.

☑ **READING CHECK**

9. Explain Why have scientists placed some nonoptical telescopes in space?

_____

_____

_____

_____

TAKE A LOOK

10. Compare On each image, circle a feature that is not found in any of the other images.

Section 2 Review

SECTION VOCABULARY

electromagnetic spectrum all of the frequencies or wavelengths of electromagnetic radiation **reflecting telescope** a telescope that uses a curved mirror to gather and focus light from distant objects	**refracting telescope** a telescope that uses a set of lenses to gather and focus light from distant objects **telescope** an instrument that collects electromagnetic radiation from the sky and concentrates it for better observation

1. Compare What is the main difference between a refracting telescope and a reflecting telescope?

_____

_____

_____

2. Describe What limits the size of a refracting telescope? Explain your answer.

_____

_____

_____

3. Identify List five types of electromagnetic radiation. Put them in order, from the longest wavelengths to the shortest.

_____

_____

_____

_____

4. Explain Why do radio telescopes have to be larger than optical telescopes?

_____

_____

5. Explain Why do astronomers place telescopes in space? Give two reasons.

_____

_____

_____

_____

CHAPTER 18 Studying Space

SECTION 3 Mapping the Stars

After you read this section, you should be able to answer these questions:

• What are constellations?

• How can we precisely locate stars in the night sky?

What Are Constellations?

People in ancient cultures grouped stars into patterns and named sections of the sky based on those patterns. **Constellations** are sections of the sky that contain recognizable star patterns.

Different civilizations had different names for the same constellations. For example, the Greeks saw a hunter (Orion) in the northern sky, but the Japanese saw a drum. Today, different cultures still see different shapes in the stars in the sky. However, astronomers have agreed on the names and locations of the constellations.

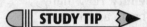

STUDY TIP

Summarize in Pairs Read this section quietly to yourself. With a partner, talk about what you learned. Together, try to figure out things that you don't understand.

The ancient Greeks saw a hunter in this set of stars. The Japanese saw the same set of stars as a drum.

Infer Why is it important for modern astronomers to agree on the names and locations of the constellations? In a small group, talk about what might happen if astronomers did not agree on these things. How might the science of astronomy be different?

CONSTELLATIONS: ORGANIZING THE SKY

Many people think of constellations as stick figures made by connecting bright stars with imaginary lines. However, to an astronomer, a constellation is an entire section of the sky. Just as Texas is a region of the United States, Ursa Major is a region of the sky. Each constellation shares a border with a neighboring constellation. Every star or galaxy in the sky is located within one of 88 constellations. ☑

✔ **READING CHECK**

1. Identify How many regions do astronomers break the sky into?

_____

_____

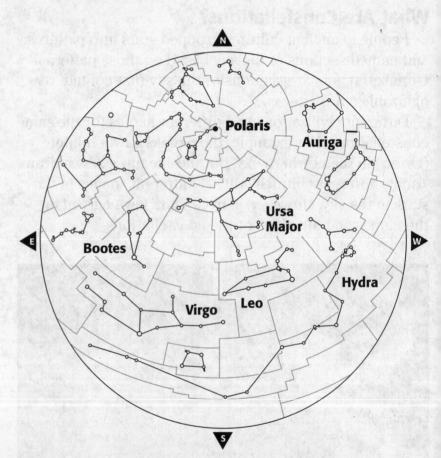

This sky map shows some of the constellations that are visible in the Northern Hemisphere at midnight in the spring.

TAKE A LOOK
2. Identify Which constellation takes up a large part of the southern and southwestern sky in the spring in the Northern Hemisphere?

_____

_____

✔ **READING CHECK**

3. Explain Why can't you use the same star map to show all of the stars that are visible to everyone on the Earth?

_____

_____

_____

SEASONAL CHANGES

The figure above shows what the midnight sky in the Northern Hemisphere looks like in the spring. However, as the Earth travels around the sun, different areas of the universe are visible. In addition, different constellations are visible from different points on the Earth. Therefore, this map is not accurate for the Northern Hemisphere during seasons other than spring, or for the Southern Hemisphere. ☑

THE PATH OF STARS ACROSS THE SKY

You know that the sun seems to move across the sky during the day. In the same way, the stars and planets rise and set during the night. This apparent motion is caused by Earth's rotation. As the Earth rotates, different parts of the universe become visible to people on the Earth. ☑

Near the poles, some stars can be seen at all times of year and all times of night. These stars are called *circumpolar stars. Circum* means "around" or "circling." Circumpolar stars seem to move through the sky in circles around the poles.

How Can You Describe the Location of a Star?

Have you ever tried to point out a star to someone? It can be very difficult to describe the exact location of an object in the sky. You can use a tool called an *astrolabe* to help you describe the location of such an object.

To use an astrolabe, you need to understand the differences between horizon, altitude, and zenith. The **horizon** is the line where the Earth and the sky seem to meet. An object's **altitude** is the angle between the object and the horizon. The **zenith** is an imaginary point in the sky that is directly above your head. The zenith always has an altitude of 90°. The figure below shows these three reference points.

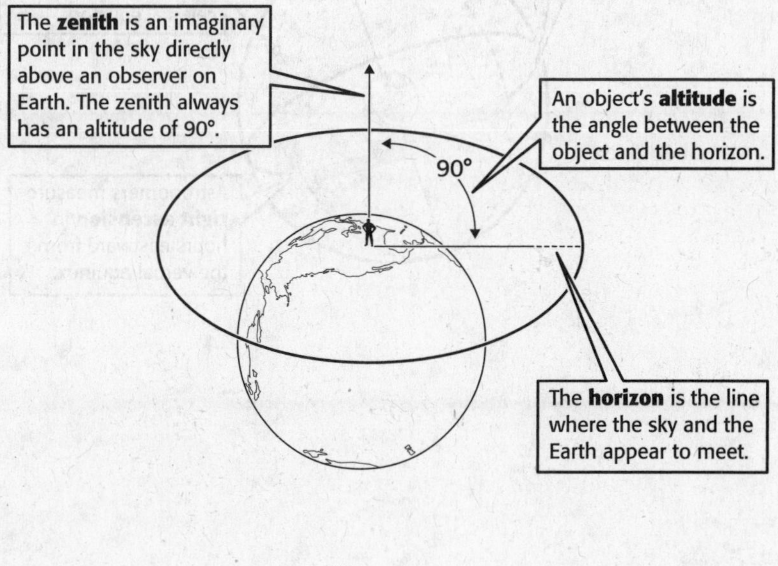

The **zenith** is an imaginary point in the sky directly above an observer on Earth. The zenith always has an altitude of 90°.

An object's **altitude** is the angle between the object and the horizon.

90°

The **horizon** is the line where the sky and the Earth appear to meet.

READING CHECK

4. Identify Why do the stars seem to move across the sky?

_____

_____

_____

Math Focus

5. Estimate Angles On the figure, draw a star to show the location of an object with an altitude of about 45°.

SECTION 3 Mapping the Stars *continued*

THE CELESTIAL SPHERE

To talk to each other about a star, astronomers must have a common method of describing the star's location. The method that astronomers use is based on the celestial sphere. The *celestial sphere* is an imaginary sphere that surrounds the Earth. Remember that we use latitude and longitude to describe the location of objects on the Earth's surface. In the same way, astronomers use declination and right ascension to plot positions in the sky.

Remember that latitude is a measure of the distance north or south of the equator. *Declination* is the distance of an object north or south of the celestial equator. The *celestial equator* is an imaginary circle formed by extending the Earth's equator into space, as shown in the figure below.

Remember that longitude is a measure of the distance east or west of the prime meridian. *Right ascension* is a measure of how far east an object is from the vernal equinox. The *vernal equinox* is the position of the sun on the first day of spring.

Critical Thinking

6. Compare How is latitude similar to declination? How are they different?

_____

_____

_____

_____

_____

_____

_____

The Celestial Sphere

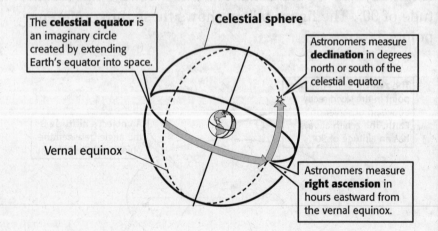

The **celestial equator** is an imaginary circle created by extending Earth's equator into space.

Celestial sphere

Astronomers measure **declination** in degrees north or south of the celestial equator.

Vernal equinox

Astronomers measure **right ascension** in hours eastward from the vernal equinox.

TAKE A LOOK
7. Define What is the celestial equator?

_____

_____

_____

SECTION 3 Mapping the Stars *continued*

How Big Is the Universe?

In the 1500s, Nicolaus Copernicus noticed that the planets appeared to move, but the stars did not. He thought the stars must be farther away than the planets. Stars are so distant that a new unit of length, the light-year, was invented to measure their distance. A **light-year** is the distance that light travels in 1 year. One light-year is equal to 9.46 trillion kilometers. The farthest objects we can observe are more than 10 billion light-years away! ☑

Many of the stars in the sky look the same. For example, the stars in Orion all seem to be about the same size in the sky. However, some stars are much closer than others. The figure below shows how stars that are very far apart can look the same to people on the Earth.

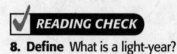

READING CHECK

8. Define What is a light-year?

_____

_____

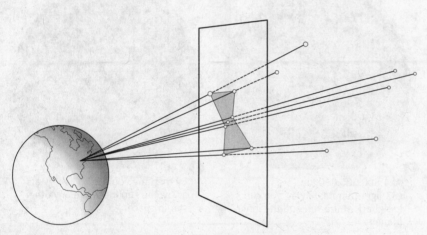

The stars in Orion seem to be very close together. However, they are actually very far apart.

THE SCALE OF THE UNIVERSE

When you think about the universe, it is important to think about scale. For example, stars appear to be very small when you see them in the sky. However, we know that most stars are much larger than the Earth. They look small in the sky because they are very far away. The figure on the next page shows how distance can affect the apparent size of objects. ☑

TAKE A LOOK

9. Infer How might the pattern of stars that we see in Orion change if the Earth were further away from the stars than it is?

_____

_____

READING CHECK

10. Explain Why do stars look very small, even though they are actually larger than the Earth?

_____

_____

SECTION 3 Mapping the Stars *continued*

1 Let's start with home plate in a baseball stadium. You are looking down from a distance of about 10 m.

2 At 100 km away, you see the city that contains the stadium and the countryside around the city.

3 At 1,500,000,000 km (83 light-minutes) away, you can look back at the sun and the inner planets.

4 By the time you are 10 light-years away, the sun looks like any other star in space.

TAKE A LOOK
11. Infer What is a light-minute?

_____

_____

TAKE A LOOK
12. Identify What is the local group?

_____

_____

_____

5 At 1 million light-years away, our galaxy looks like the Andromeda galaxy, a cloud of stars set in the blackness of space.

6 At 10 million light-years away, you can see a handful of galaxies called the *Local Group*.

SECTION 3 Mapping the Stars *continued*

How Do Scientists Know That the Universe Is Expanding?

We see stars and galaxies because they *emit*, or give off, visible light. The color of light that we see from stars can change if the stars are moving compared to the Earth. When stars or galaxies are moving away from the Earth, the light from them looks redder than normal. This effect is called *redshift*. When stars or galaxies are moving toward the Earth, the light from them looks bluer than normal. This effect is called *blueshift*. ☑

READING CHECK

13. Explain Why do we see stars and galaxies?

_____

_____

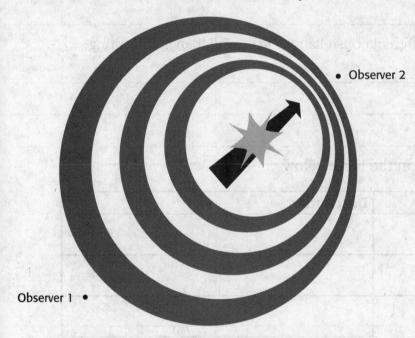

• Observer 2

Observer 1 •

The galaxy is moving away from Observer 1 and toward Observer 2. To Observer 1, the light from the galaxy looks redder than normal. The waves of light are more spread out, so the wavelength is longer. To Observer 2, the light from the galaxy looks bluer than normal. The waves of light are closer together, so the wavelength is shorter.

TAKE A LOOK

14. Identify Which observer is seeing light that is affected by redshift?

_____

Redshift and blueshift affect light only from objects that are moving very quickly. This is why cars and airplanes do not look redder or bluer when they pass by you. They are not traveling fast enough for you to see redshift or blueshift effects.

The astronomer Edwin Hubble studied the light from stars and galaxies in the universe. He discovered that light from all of these objects, except the ones closest to the Earth, is affected by redshift. This means that the stars and galaxies in the universe are moving away from each other and from the Earth. In other words, the universe is expanding. ☑

READING CHECK

15. Explain How did Edwin Hubble show that the universe is expanding?

_____

_____

_____

_____

Section 3 Review

SECTION VOCABULARY

altitude the angle between an object in the sky and the horizon	**horizon** the line where the sky and the Earth appear to meet
constellation a region of the sky that contains a recognizable star pattern and that is used to describe the location of objects in space	**light-year** the distance that light travels in one year; about 9.46 trillion kilometers
	zenith the point in the sky directly above an observer on Earth

1. Define Write your own definition for *constellation*.

_____

_____

2. Explain Why can we see different constellations in the fall than in the spring?

_____

_____

_____

3. Identify Fill in the spaces in the table below.

Term	Description
Declination	
Right ascension	
Celestial sphere	

4. Calculate About how many kilometers are in 0.5 light-years? Show your work.

_____

_____

5. Apply Concepts Suppose that Edwin Hubble had observed that light from most stars and galaxies was affected by blueshift. What conclusion about the universe could be drawn from this observation? Explain your answer.

_____

_____

_____

_____

CHAPTER 19 Stars, Galaxies, and the Universe

SECTION
1 **Stars**

BEFORE YOU READ

After you read this section, you should be able to answer these questions:

• Why are stars different colors?

• How can scientists learn what stars are made of?

• How can we measure the distance between stars?

• Why do stars seem to move across the sky?

Why Are Stars Different Colors?

Stars look like tiny points of light in the sky. However, they are actually huge, bright balls of burning gas. If you look closely at the night sky, you might see that stars are different colors. Scientists can tell how much heat a star gives off by studying its color.

Compare the yellow flame of a candle to the blue flame of a Bunsen burner. A blue flame is much hotter. Stars are similar: blue stars burn hotter than yellow ones. Red stars are coolest.

A blue flame is hotter than a yellow one.

What Are Stars Made Of?

Stars are made of gas. Hydrogen and helium are the two main elements that make up a star. Stars also contain small amounts of other elements, such as carbon, nitrogen, and oxygen. Each star is made up of a different mix of elements.

Most stars are trillions of miles away from Earth. Because scientists cannot visit the stars, they need to study stars from Earth. To find out what a star is made of, scientists study the light from the stars. ☑

STUDY TIP

Ask Questions Read this section quietly to yourself. Write down questions that you have about this section. Discuss your questions in a small group.

TAKE A LOOK

1. Color Use colored pencils to make these flames the correct color.

2. Identify Which of the flames is cooler?

_____

READING CHECK

3. Explain How do scientists learn about stars?

_____

_____

_____

How Can Scientists Learn About Stars from Their Light?

Light takes time to travel through space. Stars are so far away that their light takes millions of years to travel to Earth! When scientists look through telescopes, it is as if they are looking back in time. The light we see from stars today was made millions of years ago. Some stars that we see might have already burned out. However, we can still see them because their light is just reaching Earth.

What Can Scientists Learn from a Star's Light?

Scientists use the light from stars to find out what the stars are made of. When you look at white light through a glass prism, you can see a rainbow of colors. This rainbow is called a **spectrum** (plural, *spectra*). Millions of colors make up a spectrum, including red, orange, yellow, green, blue, indigo, and violet. Scientists use a machine called a *spectrograph* to break up a star's light into a spectrum.

Each element has a particular pattern of lines that appear in an *emission spectrum*. The emission spectrum shows scientists what elements are in the star.

These are the emission spectra for the elements hydrogen and helium. These two elements make up most stars. Each line represents a different color of visible light.

How Do Scientists Classify Stars?

Stars can be classified in several ways. Scientists classify stars most commonly by temperature and brightness.

TEMPERATURE

In the past, scientists classified stars by the elements they contained. Today, stars are classified by temperature. Each group of stars is named with a letter of the alphabet. The table on the next page shows the features of different groups of stars.

Critical Thinking

4. Apply Concepts When we look at the night sky, are we seeing the universe exactly as it is?

_____

_____

_____

_____

TAKE A LOOK
5. Compare Which emission spectrum contains more colors of visible light, hydrogen or helium?

_____

SECTION 1 Stars *continued*

Class	Color	Temperature (°C)	Elements detected
O	blue	above 30,000	helium
B	blue-white	10,000 to 30,000	hydrogen, helium
A	blue-white	7,500 to 10,000	hydrogen
F	yellow-white	6,000 to 7,500	hydrogen and heavier elements
G	yellow	5,000 to 6,000	calcium and heavier elements
K	orange	3,500 to 5,000	calcium and iron
M	red	less than 3,500	molecules, such as titanium dioxide

TAKE A LOOK

6. Identify A scientist discovers a star that is blue-white and is made of hydrogen. Which class should the scientist put the star in?

_____

7. Identify Which class has hotter stars—G or B?

_____

BRIGHTNESS

Before telescopes were invented, scientists judged the brightness of the stars with their naked eyes. They called the brightest stars they could see first-magnitude stars, and the dimmest stars, sixth-magnitude stars.

When telescopes were developed, scientists discovered this system had flaws. They could see more stars with the telescope than with the naked eye. They could also see the differences in brightness more clearly. The old system for classifying brightness was too general to include the dimmest stars that scientists were finding. A new system had to be created.

Today, scientists give each star a number to show its brightness, or *magnitude*. The dimmest stars have the largest numbers. The brightest stars have the smallest numbers. The magnitude of a very bright star can even be a negative number!

Critical Thinking

8. Apply Concepts Which star is brighter: one with a magnitude of 6.3 or one with a magnitude of −1.4?

_____

Magnitudes of Stars in the Big Dipper

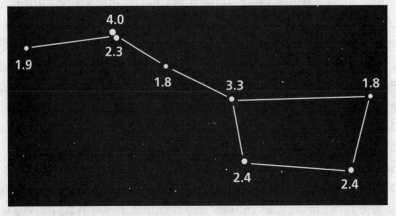

TAKE A LOOK

9. Identify Circle the brightest stars in the Big Dipper. What is their magnitude?

SECTION 1 Stars *continued*

Does Distance Change a Star's Brightness?

If you look at a row of street lights, do all of the lights look the same? The nearest lights look brightest, and the farthest ones look dimmest.

TAKE A LOOK
10. Identify Circle the dimmest light in the picture. Put a box around the brightest light.

11. Explain The street lights are all equally bright. Why do they appear different?

_____

_____

The closer a light is, the brighter it looks.

The brightness of a star as we see it from Earth is the star's **apparent magnitude**. A bright star can look very dim if it is very far away from Earth. A dim star can appear bright if it is closer to Earth.

A star's **absolute magnitude** is the actual brightness of the star. If all stars were the same distance away, their absolute magnitudes would equal their apparent magnitudes. For example, the sun's absolute magnitude is +4.8, but because it is close to Earth, its apparent magnitude is −26.8.

How Do Scientists Measure Distance to a Star?

Math Focus
12. Calculate What is the distance in kilometers from Earth to a star that is 30 light years away?

_____

The distance between Earth and the stars is too large to be measured in miles or kilometers. Instead, scientists use a unit called a **light-year**, which is the distance that light can travel in one year. One light year equals 9.46 trillion kilometers. How can scientists measure such a large distance?

As Earth revolves around the sun, stars close to Earth seem to move, but far-off stars do not. This is called **parallax**. Scientists use parallax and math to find the distance between Earth and stars. To understand parallax, think about riding in a car past a large mountain. As you drive past the mountain, it seems to move. However, the mountain is not actually moving. It is your motion compared to the mountain that makes the mountain seem to move.

Parallax

As the Earth revolves around the sun, a star's position seems to change.

Very distant stars

Apparent position in July

Apparent position in January

Nearer star

Parallax

Earth in January

Earth in July

Do Stars Move?

Stars move, but because they are so far away and move so slowly, we cannot see their movement easily. Every night stars seem to rise and set, but it is not the stars that are moving. It is the Earth.

The rotation of Earth causes daytime and nighttime. Because of Earth's rotation, the sun moves across the sky during the daytime. For this same reason, the stars seem to move across the sky at night. All of the stars that you see appear to rotate around Polaris, the North Star. The stars seem to make a full circle around Polaris every 24 hours.

Earth's tilt and revolution cause the seasons. During each season, any point on Earth faces a different part of the sky at night. That means that different stars appear in the night sky at different times of the year.

TAKE A LOOK
13. Explain What causes parallax?

_____

_____

Because of the Earth's rotation, the stars seem to move across the sky.

TAKE A LOOK
14. Compare Circle one star, other than Polaris, in the picture on the left. Then circle the same star in the picture on the right. Draw a curved arrow in the first picture that shows the direction that the star seemed to move.

In addition to their apparent motion, stars are moving through space. Because the stars are so far away, it is difficult for us to see their motions. Over thousands of years, however, the movements of the stars can cause the shapes of constellations to change.

Section 1 Review

SECTION VOCABULARY

absolute magnitude the brightness that a star would have at a distance of 32.6 light-years from Earth	**light-year** the distance that light travels in one year; about 9.46 trillion kilometers
apparent magnitude the brightness of a star as seen from Earth	**parallax** an apparent shift in the position of an object when viewed from different locations
	spectrum the band of colors produced when white light passes through a prism

1. Identify What are the two main elements that make up most stars?

_____

_____

2. Apply Concepts Put the following star classes in order from hottest to coolest: A, B, G, K, O.

_____

3. Analyze Why do scientists use light-years to measure the distances between stars and Earth?

_____

_____

_____

4. Explain Why do stars seem to move in the sky?

_____

_____

5. Compare What is the difference between apparent magnitude and absolute magnitude?

_____

_____

_____

_____

6. Explain Why is the actual movement of stars hard to see?

_____

_____

_____

CHAPTER 19 | Stars, Galaxies, and the Universe

SECTION
2 | # The Life Cycle of Stars

BEFORE YOU READ

After you read this section, you should be able to answer these questions:

• How do stars change over time?

• What is an H-R diagram?

• What may a star become after a supernova?

How Do Stars Age?

Stars do not remain the same forever. Like living things, stars go through a life cycle from birth to death. The actual life cycle of a star depends on its size. An average star, such as the sun, goes through four stages during its life.

A star enters the first stage of its life cycle as a ball of gas and dust called a *protostar*. Gravity pulls the gas and dust together. As the ball becomes denser, it gets hotter. Eventually, the gas becomes so hot that it begins to react. These reactions produce energy, which keeps the new star from collapsing more.

The second, and longest, stage of a star's life cycle is the *main sequence star*. During this stage, hydrogen in the center of the star reacts to form helium. This produces a great deal of energy. As long as a main-sequence star has enough hydrogen to react, its size will not change very much.

When a main-sequence star uses up all of its hydrogen, it can start to expand and cool. This forms a huge star called a **red giant**.

In the final stage of its life cycle, an average star is classified as a white dwarf. A **white dwarf** is the small, hot, leftover center of a red giant.

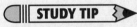

STUDY TIP

Compare Make a chart comparing the steps in the life cycles of average stars and massive stars.

Critical Thinking

1. Infer A star can live for billions of years. Therefore, scientists can't watch a star through its entire life. How do you think scientists figure out the life cycle of a star?

_____

_____

_____

TAKE A LOOK
2. Identify This average star is in the last stage of its life cycle. What is that stage?

_____

Life Cycle of an Average Star

> **1. Protostar** A *protostar* is a ball of gas and dust. Gravity pulls the gas and dust together, and its center gets hotter.

> **2. Main-Sequence Star** The *main sequence* is the longest stage of a star's life cycle. The fusion of hydrogen atoms makes energy in the star.

> **3. Red Giant** When a main sequence star uses up all its hydrogen, it can no longer give off energy. The star's center contracts and the outer layers expand and cool, forming a red giant.

> **4. White Dwarf** A white dwarf is the leftover center of a red giant. It is a small, hot, and dim star that can shine for billions of years.

TAKE A LOOK

3. Identify What causes a main sequence star to become a red giant?

_____

_____

READING CHECK

4. Identify Which two characteristics of a star must a scientist measure to make an H-R diagram?

_____

_____

What Is an H-R Diagram?

An **H-R diagram** is a graph that shows the relationship between a star's temperature and its brightness. The H-R diagram also shows how stars change over time. The diagram is named after Ejnar Hertzsprung and Henry Norris Russell, the scientists who invented it. ☑

Temperature is given along the bottom of the diagram. Hotter (bluer) stars are on the left, and cooler (redder) stars are on the right. Brightness, or absolute magnitude, is given along the left side of the diagram. Bright stars are near the top, and dim stars are near the bottom. The bright diagonal line on the H-R diagram is called the **main sequence**. A star spends most of its life on the main sequence.

Why Does a Star's Position on the H-R Diagram Change?

As a main-sequence star ages, it becomes a red giant. When this happens, the star moves to a new place on the H-R diagram. The star's position on the diagram changes again when it becomes a white dwarf. These changes happen because the brightness and temperature of a star change throughout its life. ☑

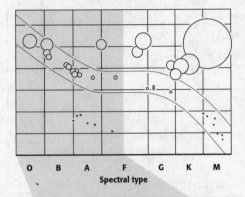

Spectral type
O B A F G K M

Main-sequence Stars
Stars on the main sequence form a band that runs accross the H-R diagram. The sun is a main-sequence star. The sun has been shining for about 5 billion years. Scientists think that the sun is in the middle of its life and will remain on the main sequence for another 5 billion years.

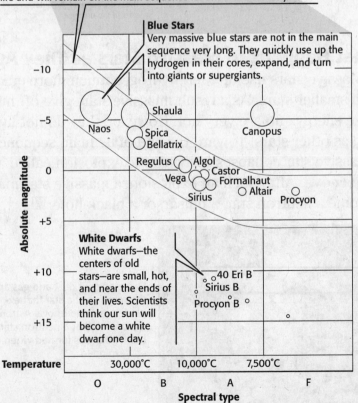

An H-R diagram can show the life cycle of a star.

READING CHECK

5. Explain Why does a star's position on the H-R diagram change at different stages of its life cycle?

_____

_____

_____

_____

_____

TAKE A LOOK
6. Identify Where in the H-R diagram are the brightest stars located?

_____

7. Identify Where in the diagram are the hottest stars located?

_____

A Continuation of the H-R Diagram

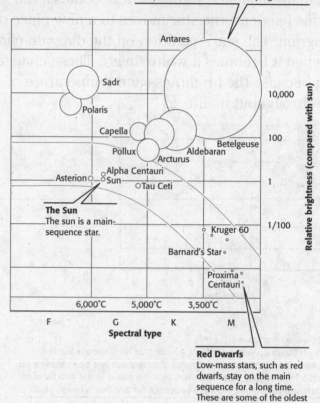

Giants and Supergiants
When a star runs out of hydrogen in its core, the center of the star contracts and the outer part expands. This forms a red giant. If the star is very massive, it becomes a supergiant.

The Sun
The sun is a main-sequence star.

Red Dwarfs
Low-mass stars, such as red dwarfs, stay on the main sequence for a long time. These are some of the oldest stars in the universe.

TAKE A LOOK

8. Compare Which star is hotter—Antares or Polaris?

_____

9. Read a Graph Is Betelgeuse on the main sequence?

_____

What Happens to Massive Stars as They Age?

Massive stars use up their hydrogen much more quickly than smaller stars. As a result, massive stars give off much more energy and are very hot. However, they do not live as long as other stars. Toward the end of its main sequence, a massive star collapses in a gigantic explosion called a **supernova**. After such an explosion, a massive star may become a neutron star, a pulsar, or a black hole. ☑

READING CHECK

10. Identify What can cause a main-sequence star to turn into a neutron star, a pulsar, or a black hole?

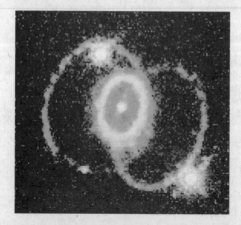

These rings of dust and gas are the remains of a star that exploded in a supernova. Astronomers think that a neutron star or black hole was formed when this star exploded.

NEUTRON STARS

After a supernova, the center of a collapsed star may contract into a tiny ball of neutrons. This ball, called a **neutron star**, is extremely dense. On Earth, a single teaspoon of matter from a neutron star would weigh 100 million metric tons!

PULSARS

If a neutron star is spinning, it is called a **pulsar**. Pulsars send out beams of radiation that sweep through space. A radio telescope, an instrument that can pick up radiation with long wavelengths, can detect pulsars. Every time a pulsar's beam sweeps by Earth, scientists hear rapid clicks, or pulses, in the radio telescope.

BLACK HOLES

If the collapsed star is extremely massive, the force of its gravity may cause it to contract even more. This contraction crushes the dense center of the star, creating a **black hole**. Even though they are called holes, black holes aren't really empty spaces. A black hole is an object so dense that even light cannot escape its gravity.

Because black holes do not give off light, it can be hard for scientists to locate them. Gas and dust from a nearby star may fall into the black hole and give off X rays. When scientists find these X rays, they can infer that a black hole is close by.

Critical Thinking

11. Infer Could an average star, such as our sun, become a neutron star? Explain your answer.

_____

_____

_____

_____

 **Say It**

Discuss In a small group, talk about other places you have heard about X rays. Where were they used? What were they used for?

After a supernova, a massive star may become

a neutron star	a pulsar	a black hole
which is	which is	which is

TAKE A LOOK

12. Describe Fill in the blank spaces to describe neutron stars, pulsars, and black holes.

Section 2 Review

SECTION VOCABULARY

black hole an object so massive and dense that even light cannot escape its gravity	**neutron star** a star that has collapsed under gravity to the point that the electrons and protons have smashed together to form neutrons
H-R diagram Hertzsprung-Russell diagram, a graph that shows the relationship between a star's surface temperature and absolute magnitude	**pulsar** a rapidly spinning neutron star that emits pulses of radio and optical energy
main sequence the location on the H-R diagram where most stars lie; it has a diagonal pattern from the lower right (low temperature and luminosity) to the upper left (high temperature and luminosity)	**red giant** a large, reddish star late in its life cycle
	supernova a gigantic explosion in which a massive star collapses and throws its outer layers into space
	white dwarf a small, hot, dim star that is the leftover center of an old star

1. List What are the four stages in the life cycle of an average star?

_____

_____

_____

2. Identify Label the axes on this H-R diagram.

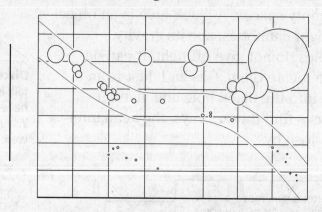

3. Explain How does a star's temperature change as the star ages from a main sequence star to a red giant and from a red giant to a white dwarf?

_____

_____

4. Compare How is the life cycle of a massive star different from the life cycle of an average star?

_____

_____

CHAPTER 19 Stars, Galaxies, and the Universe

SECTION 3 # Galaxies

> **BEFORE YOU READ**
>
> **After you read this section, you should be able to answer these questions:**
> - What are different shapes that a galaxy can have?
> - What are galaxies made of?
> - How do galaxies form?

What Is a Galaxy?

If you look out on a clear night far from city lights, you can see hundreds of stars. Many of these stars are part of our galaxy, which is called the *Milky Way*. Our galaxy actually contains many more stars than you can see.

A **galaxy** is a large group of gas, dust, and millions of stars. The biggest galaxies contain more than a trillion stars. Scientists can't actually count the stars, of course. They estimate how many stars are in a galaxy by measuring the size and brightness of the galaxy. The bigger and brighter the galaxy, the more stars it has. ☑

Galaxies come in different shapes and sizes. Scientists classify galaxies by shape. The three most common types of galaxies are spiral, elliptical, and irregular.

SPIRAL GALAXIES

A *spiral galaxy* has two parts: a central bulge and arms that form a spiral around the center. The bulge is a dense group of old stars. The arms are made of gas, dust, and much younger stars.

The Milky Way is a spiral galaxy. Our sun is one of the 200 billion stars in the Milky Way. From Earth, the edge of the Milky Way looks like a bright belt of stars that stretches across the night sky.

The Andromeda galaxy is a spiral galaxy. Our galaxy, the Milky Way, probably looks very much like Andromeda.

STUDY TIP

Compare As you read, make a table comparing the three different types of galaxies.

READING CHECK

1. Explain How do scientists estimate how many stars a galaxy has?

_____

_____

TAKE A LOOK

2. Identify Label the part of the galaxy that contains the oldest stars.

ELLIPTICAL GALAXIES

An *elliptical galaxy* is made of many stars and looks like a snowball. Elliptical galaxies are among the largest galaxies in the universe. Some may contain as many as 5 trillion stars! There is very little free gas in an elliptical galaxy. Therefore, few new stars form there.

TAKE A LOOK
3. Compare Name two ways that spiral galaxies differ from elliptical galaxies.

_____

_____

_____

_____

_____

Galaxy M87, an elliptical galaxy, has no spiral arms.

IRREGULAR GALAXIES

An *irregular galaxy* has no clear shape. It may have as few as 10 million or as many as several billion stars. Some irregular galaxies form when two other galaxies collide.

The Large Magellanic Cloud, an irregular galaxy, is close to our own.

READING CHECK
4. Identify What are galaxies made of?

_____

_____

_____

What Objects May Be Found in Galaxies?

Remember that galaxies are made of gas, dust, and billions of stars. Some of these stars form different features, such as nebulas, open clusters, and globular clusters. When scientists study the stars in galaxies, they look for these features. ☑

SECTION 3 Galaxies *continued*

NEBULAS

A **nebula** (plural, *nebulae* or *nebulas*) is a large cloud of gas and dust. Most stars are born in nebulas. Some nebulas glow or reflect starlight, but others absorb light and are too dark to see. Therefore, although nebulas can be found throughout a galaxy, they can be hard to see. ☑

This is part of a nebula. The tall, thin shape to the left of the bright star is wider than our solar system.

STAR CLUSTERS

An **open cluster** is a group of 100 to 1,000 stars. The stars in an open cluster are closer together than stars in other parts of space. Open clusters are usually found in the arms of a spiral galaxy. All of the stars in an open cluster are the same age. They formed at the same time from the same nebula. Newly formed open clusters have many bright blue stars.

A **globular cluster** is a group of up to 1 million stars that are packed closely together. A globular cluster looks like a ball. Some globular clusters orbit spiral galaxies, such as the Milky Way. Others can be found near giant elliptical galaxies.

What Are Quasars?

Remember that light from stars can take millions of years to reach Earth. Therefore, looking at distant stars is like looking back in time. Scientists study the early universe by studying objects that are very far away. Looking at distant galaxies shows what early galaxies looked like. By studying distant galaxies, scientists can learn how galaxies form and change.

Among the most distant objects are quasars. **Quasars** are starlike sources of light that are very far away. They are among the strongest energy sources in the universe. Some scientists think that quasars may be caused by black holes, but they are not sure how this happens.

READING CHECK

5. Explain Why are some nebulas hard to see?

_____

_____

Critical Thinking

6. Compare How is a nebula different from a star cluster?

_____

_____

_____

_____

_____

Section 3 Review

SECTION VOCABULARY

galaxy a collection of stars, dust, and gas bound together by gravity	**open cluster** a group of stars that are close together relative to surrounding stars
globular cluster a tight group of stars that looks like a ball and contains up to 1 million stars	**quasar** quasi-stellar radio source; a very luminous object that produces energy at a high rate; quasars are thought to be the most distant objects in the universe
nebula a large cloud of gas and dust in interstellar space; a region in space where stars are born	

1. Compare How is a nebula different from a galaxy?

_____

_____

_____

2. List What three shapes can galaxies be?

_____

_____

_____

3. Compare Complete the chart below to describe different features of galaxies.

Galaxy feature	What they are made of	Where they are found	Other characteristics
		throughout a galaxy	where stars form
	100 to 1,000 stars, relatively close together		may contain bright blue stars
Globular cluster		around a spiral galaxy or near a large elliptical galaxy	

4. Explain What do some scientists think causes quasars?

_____

_____

CHAPTER 19 Stars, Galaxies, and the Universe

SECTION 4 # Formation of the Universe

After you read this section, you should be able to answer these questions:

• What is the big bang theory?

• How is the universe structured?

• How old is the universe?

How Do Scientists Think the Universe Formed?

Like all scientific theories, theories about the beginning and end of the universe must be tested by observations or experiments. The study of how the universe started, what it is made of, and how it changes is called **cosmology**.

To understand how the universe formed, scientists study the movements of galaxies. Careful measurements have shown that most galaxies are moving away from each other. This indicates that the universe is expanding. Based on this observation, scientists have made inferences about how the universe may have formed. ☑

Imagine that the formation and evolution of the universe was recorded on a video tape. If you rewound the video, the universe would seem to contract. At the beginning of the universe, all matter and energy would be squeezed into one small space. Now imagine running that same video forward. All the matter and energy in the universe would explode and begin to expand in all directions.

STUDY TIP

Predict Before you read this section, write down your prediction of how scientists think the universe formed and what will happen to it in the future. As you read, take notes on these topics.

READING CHECK

1. Complete Scientists took careful measurements of galaxies and found that the universe is

_____

_____.

The Big Bang

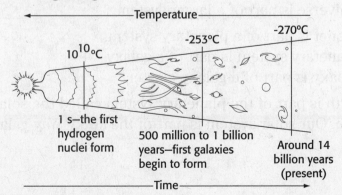

Most astronomers think that the big bang caused the universe to expand in all directions.

TAKE A LOOK

2. Identify After the big bang, how did the temperature of the universe change?

_____

_____

SECTION 4 Formation of the Universe *continued*

What Is the Big Bang Theory?

The theory that the universe began with a huge explosion is called the **big bang theory**. It is the scientific model that explains why the universe is expanding. According to this theory, all the contents of the universe were originally squeezed into a very small volume. These contents were at extremely high pressure and temperature. About 14 billion years ago, this small volume rapidly expanded and cooled. ☑

Just minutes after the big bang, the following things had already formed:

- the light elements, such as helium
- the forces of nature, such as gravity
- the beginnings of galaxies

EVIDENCE FOR THE BIG BANG

All scientific theories must have evidence to support them. The first piece of evidence for the big bang theory is the expansion of the universe. The second piece of evidence is called *cosmic background radiation*.

Scientists use radio telescopes to pick up radiation with long wavelengths. Several decades ago, some scientists noticed a background "noise" coming from all directions in space. They think this cosmic background radiation is energy left over from the big bang. ☑

What Is the Structure of the Universe?

The universe contains many different objects. However, these objects are not just scattered around the universe. They are grouped into systems. Every object in the universe is part of a larger system:

- A planet is part of a planetary system.
- A planetary system is part of a galaxy.
- A galaxy is part of a galaxy cluster.

Earth is part of the planetary system called the solar system. Our solar system is part of the Milky Way galaxy.

✓ **READING CHECK**

3. Identify Where were the contents of the universe before the big bang?

_____

_____

_____

✓ **READING CHECK**

4. List Give two pieces of evidence for the big bang theory.

_____

_____

_____

Critical Thinking

5. Apply Concepts Are there probably more planets or more galaxies in the universe? Explain your answer.

_____

_____

_____

_____

SECTION 4 Formation of the Universe *continued*

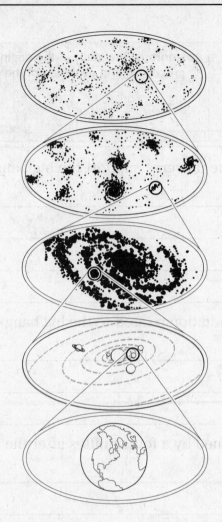

How Old Is the Universe?

Scientists can estimate the age of the universe by studying the oldest stars in the Milky Way galaxy. These stars are white dwarfs. The first stars that formed after the big bang became white dwarfs after about 1 billion years. The oldest white dwarfs are between 12 billion and 13 billion years old. Therefore, scientists think that the universe is about 14 billion years old. ☑

No one knows what will happen to the universe in the future. Some scientists think that the universe will continue to expand, faster and faster. Stars will age and die, and one day, the universe will become cold and dark. Even after the universe becomes cold and dark, it will continue to expand.

7. Identify How old do scientists think the universe is?

_____

_____

Section 4 Review

SECTION VOCABULARY

big bang theory the theory that all matter and energy in the universe was compressed into an extremely small volume that 13 billion to 15 billion years ago exploded and began expanding in all directions	**cosmology** the study of the origin, properties, processes, and evolution of the universe

1. Explain How does the expansion of the universe support the big bang theory?

_____

_____

_____

_____

2. Explain How is cosmic background radiation related to the big bang theory?

_____

_____

_____

3. Identify List three things that had formed by a few minutes after the big bang.

_____

_____

_____

4. Describe Explain how every object in the universe is part of a larger system.

_____

_____

_____

5. Explain Imagine you are a scientist studying the formation of the universe. How could you estimate the age of the universe?

_____

_____

_____

_____

CHAPTER 20 Formation of the Solar System
SECTION
1 **A Solar System Is Born**

National Science Education Standards
ES 3a, 3b, 3c

How Do Solar Systems Form?

You probably know that our solar system today includes the planets, moons, and other objects that orbit our sun. However, our solar system has existed for only about 5 billion years. It was not always the same as it is now. It began as a nebula

All solar systems start as clouds of gas and dust in space called **nebulas** (or *nebulae*). Our solar system probably formed from a nebula called the **solar nebula**. The gases in a nebula are mainly hydrogen and helium. The dust contains elements such as carbon and iron.

Solar systems form from nebulas. The gas and dust in nebulas become solar systems because of two forces: gravity and pressure. ☑

GRAVITY: PULLING MATTER TOGETHER

Remember that gravity pulls objects together. The particles of matter in a nebula are very small. There is a lot of space between them. Therefore, the force of gravity holding the particles together is very weak. It is just strong enough to keep the nebula from drifting apart.

STUDY TIP

Organize As you read this section, make a flowchart showing the steps in the formation of a solar system like ours.

READING CHECK

1. Identify Which two forces cause the gas and dust in nebulas to form solar systems?

_____

_____

Force of gravity
Force of gravity
Force of gravity

The force of gravity pulls the particles in a nebula together.

TAKE A LOOK
2. Identify What effect does gravity have on the particles in a nebula?

_____

_____

| SECTION 1 | A Solar System Is Born *continued* |

PRESSURE: PUSHING MATTER APART

Gravity pulls the particles in a nebula together. Why don't the particles collapse into a single point? The answer has to do with the pressure inside a nebula.

The particles in a nebula are always moving. As the particles move around, they sometimes bump into each other. When two particles bump into each other, they move apart. This produces *pressure* within the nebula. The closer the particles are, the more likely they are to bump into each other. Therefore, the pressure is high.

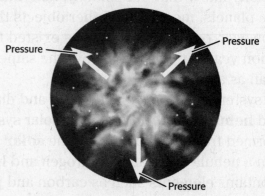

Pressure causes the particles in a nebula to move apart.

THE BALANCE BETWEEN GRAVITY AND PRESSURE

High pressure causes the nebula to *expand*, or get larger. The particles spread farther apart, and pressure decreases. However, gravity increases pressure by pulling the particles together. When the nebula is just the right size, the pressure inside it exactly balances the force of gravity. This balance keeps the nebula the same size. It does not expand or shrink.

UPSETTING THE BALANCE

The balance between gravity and pressure in a nebula can be *upset*, or changed. For example, a supernova can produce a force on the nebula. The force can cause small regions of the nebula to be *compressed*, or pushed together. These small regions are called *globules*. ☑

A globule can become very dense. Gravity can cause the globule to collapse. As it collapses, its temperature increases. The hot, dense globule can eventually become a star.

Critical Thinking

3. Apply Concepts How does high pressure probably affect the size of a nebula?

_____

_____

TAKE A LOOK

4. Identify What causes pressure inside a nebula?

_____

_____

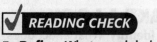 **READING CHECK**

5. Define What are globules?

_____

_____

_____

SECTION 1 A Solar System Is Born *continued*

How Did the Solar System Form?

It took about ten million years for our solar system to form from the solar nebula. The figures below show some of the important events in the formation of our solar system.

1. A globule formed near the center of the solar nebula. The nebula began to rotate, and the matter in it flattened into a disk.

2. In some parts of the rotating disk, bits of dust and rock collided and stuck together. The bodies they formed grew larger as more dust and rock collided with them. Some of these bodies got to be hundreds of kilometers wide. They were called *planetesimals*, or small planets.

TAKE A LOOK
6. Define What is a planetesimal?

_____

_____

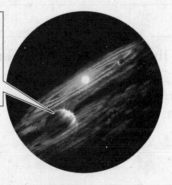

3. The largest planetesimals formed far from the center of the disk. Their gravity attracted some of the gases in the nebula. These planetesimals became the gas giant planets: Jupiter, Saturn, Uranus, and Neptune.

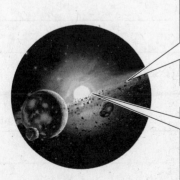

4. Smaller planetesimals formed near the center of the disk. These planetesimals attracted more dust and less gas. They became the inner, rocky planets: Mercury, Venus, Earth, and Mars.

5. Most of the gas in the solar nebula moved toward the center. The particles of gas became very densely packed at the center. They began to react with each other. The reactions released huge amounts of energy. Our sun was born.

STANDARDS CHECK

ES 3a The earth is the third planet from the sun in a system that includes the moon, the sun, eight other planets and their moons, and smaller objects, such as asteroids and comets. The sun, an average star, is the central and largest body in the solar system.

7. List What are the four gas giant planets?

_____

_____

_____

_____

Section 1 Review

SECTION VOCABULARY

nebula a large cloud of gas and dust in interstellar space; a region in space where stars are born	**solar nebula** a rotating cloud of gas and dust from which the sun and planets formed

1. Describe What happened to the solar nebula?

_____

2. Identify What force pulls the matter in a nebula together?

_____

3. Explain How did the gas giant planets form?

_____

_____

_____

_____

4. Explain How did the inner, rocky planets form?

_____

_____

_____

_____

5. Identify Where did most of the gas in the solar nebula end up?

_____

6. List What are the four inner, rocky planets?

_____

_____

_____

7. Apply Concepts What would happen to a nebula if the pressure inside it was greater than the force of gravity? Explain your answer.

_____

_____

_____

CHAPTER 20 Formation of the Solar System

SECTION
2 **The Sun: Our Very Own Star**

BEFORE YOU READ

After you read this section, you should be able to answer these questions:

• Where does the sun's energy come from?

• How do sunspots and solar flares affect Earth?

National Science Education Standards

ES 3a

What Is the Structure of the Sun?

The sun is the largest part of our solar system. Ninety-nine percent of the matter in our solar system is found in the sun. Although the sun may look like a solid ball in the sky, it is actually made of gas. The gas is held together by gravity. The figure below shows the structure of the inside of the sun.

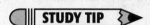
STUDY TIP

Ask Questions As you read, write down any questions you have. When you finish reading, discuss your questions with a partner or in a small group. Together, try to figure out the answers to your questions.

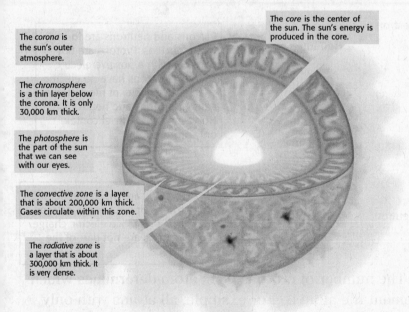

The *corona* is the sun's outer atmosphere.

The *chromosphere* is a thin layer below the corona. It is only 30,000 km thick.

The *photosphere* is the part of the sun that we can see with our eyes.

The *convective zone* is a layer that is about 200,000 km thick. Gases circulate within this zone.

The *radiative zone* is a layer that is about 300,000 km thick. It is very dense.

The *core* is the center of the sun. The sun's energy is produced in the core.

TAKE A LOOK
1. Identify What is the corona?

_____

_____

Energy is produced in the core of the sun. The energy produced in the core takes millions of years to move to the photosphere. First, the energy passes from the core into the radiative zone. Next, the energy reaches the convective zone. Within the convective zone, hot gases carry energy to the photosphere. Energy leaves the sun as light. It takes about 8.3 min for light to travel from the sun to Earth.

Where Does the Sun's Energy Come From?

Our sun has existed for about 4.6 billion years. Scientists have developed many theories about why the sun shines. For example, scientists used to think that the sun burns fuel, like a campfire. However, new observations about the age of the sun showed that this theory could not be correct. A sun that burns fuel could not last for more than about 10,000 years. ☑

Scientists now know that nuclear fusion is the process that powers our sun and most other stars. To understand nuclear fusion, you must know a little bit about the structure of atoms.

Remember that all matter is made of atoms. Atoms, in turn, are made of even smaller particles called electrons, protons, and neutrons. Protons and neutrons make up the *nucleus* (plural, *nuclei*) of the atom. The electrons move around the nucleus. The figure below shows a model of an atom of the element helium. ☑

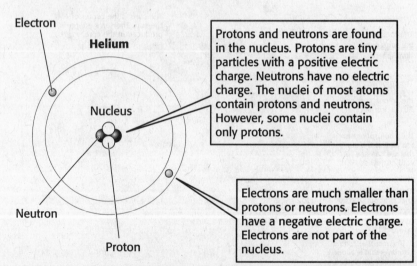

Electron

Helium

Nucleus

Neutron

Proton

Protons and neutrons are found in the nucleus. Protons are tiny particles with a positive electric charge. Neutrons have no electric charge. The nuclei of most atoms contain protons and neutrons. However, some nuclei contain only protons.

Electrons are much smaller than protons or neutrons. Electrons have a negative electric charge. Electrons are not part of the nucleus.

The number of protons in an atom determines which element the atom is. For example, all atoms with only one proton are atoms of the element hydrogen. However, atoms of an element can contain different numbers of neutrons. For example, most hydrogen atoms contain no neutrons, but some contain one neutron. An atom with one proton and one neutron is still hydrogen. It is simply a different form of hydrogen.

During **nuclear fusion**, two or more nuclei *fuse*, or join together, to form a new nucleus. This process releases a huge amount of energy. Within stars, nuclei of hydrogen fuse to form nuclei of helium. ☑

FUSION IN OUR SUN

Normally, hydrogen nuclei never get close enough to each other to fuse into helium. However, the pressure in the center of the sun is very high. This high pressure forces hydrogen nuclei together, so they can fuse. The figure below shows how hydrogen nuclei in the sun fuse to form helium. ☑

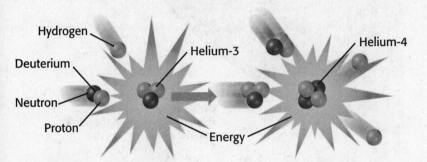

Deuterium is a form of hydrogen that contains one proton and one neutron in its nucleus. Within the sun, a nucleus of deuterium can collide with a nucleus of hydrogen, which contains only one proton. This releases a large amount of energy. It also forms a nucleus of a form of helium called *helium-3*. A nucleus of helium-3 contains two protons and one neutron.

Two helium-3 nuclei collide. This forms a nucleus of *helium-4*, which contains two protons and two neutrons. Two protons are released, along with a large amount of energy.

How Does Solar Activity Affect Earth?

The movement of energy in the photosphere causes the gases to churn. The circulation of gases and the sun's rotation produce magnetic fields. These magnetic fields reach far into space. They can cause changes in the photosphere. These changes can also affect the Earth.

SUNSPOTS

The sun's magnetic fields slow the movement of gases in the convective zone. This causes certain areas of the photosphere to become cooler than others. The cooler areas show up as sunspots. **Sunspots** are cooler, dark spots on the photosphere. They vary in size and shape. Some sunspots are as large as 80,000 km in diameter. ☑

✓ READING CHECK

6. Define What is nuclear fusion?

_____

_____

_____

TAKE A LOOK

7. Compare How is helium-3 different from helium-4?

_____

_____

_____

✓ READING CHECK

8. Define What are sunspots?

_____

_____

_____

391

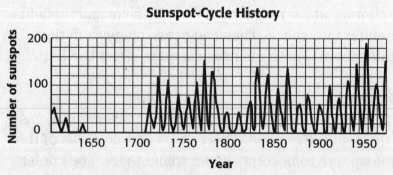

Sunspots are cooler areas of the photosphere. They are related to changes in the sun's magnetic field.

TAKE A LOOK

9. Identify What causes sunspots?

_____

_____

THE SUNSPOT CYCLE

The Italian scientist Galileo was one of the first to study sunspots. Using a telescope, he observed the numbers, sizes, and locations of sunspots over time. He found that the numbers and locations of sunspots change in a predictable pattern. This pattern is called the *sunspot cycle*.

Today, scientists know that the sunspot cycle is about 11 years long. Every 11 years, the number of sunspots reaches a peak. Then it declines. The graph below shows how the number of sunspots has changed over time.

Math Focus

10. Read a Graph In about what decade was the maximum number of sunspots observed?

_____

Sunspot-Cycle History

This graph shows the number of sunspots observed in different years. Notice that the number of sunspots changes in a regular way.

EFFECTS OF SUNSPOTS ON CLIMATE

Scientists think that sunspot activity may affect Earth's weather. For example, there were few sunspots between the years of 1645 and 1715. During this time, Europe's climate was much colder than usual. In fact, the climate was so cold that this period is sometimes called the "Little Ice Age." However, scientists do not understand how a small number of sunspots may change the Earth's climate.

Most scientists agree that sunspots may affect the Earth's climate. However, the connection between sunspots and the climate on Earth is not clear. More research is needed in order for us to fully understand how sunspots can affect our climate.

SOLAR FLARES

The magnetic fields that cause sunspots can also cause solar flares. *Solar flares* are extremely hot, bright regions on the sun's surface. They send huge streams of electrical particles throughout the solar system. Solar flares can extend up to several thousand kilometers within only a few minutes. ☑

Scientists do not know exactly what causes solar flares. However, they do know that most solar flares are associated with sunspots.

Solar flares can have significant effects on the Earth. The streams of charged particles solar flares emit can interact with the Earth's atmosphere. They can interfere with radio and television transmissions. Therefore, scientists are trying to find ways to predict solar flares.

Say It

Hypothesize What kinds of evidence could support the hypothesis that sunspots affect the Earth's climate? By yourself, think about some answers to this question. Then, talk about your answers with a partner or in a small group.

READING CHECK

11. Define What is a solar flare?

_____

_____

_____

Type of solar activity	Description	How can it affect the Earth?
Sunspots		may cause climate change, but the connection is not clear
Solar flares		

TAKE A LOOK

12. Describe Fill in the blank spaces in the table.

Name _____ Class _____ Date _____

Section 2 Review

SECTION VOCABULARY

nuclear fusion the process by which nuclei of small atoms combine to form a new, more massive nucleus; the process releases energy	**sunspot** a dark area of the photosphere of the sun that is cooler than the surrounding areas and that has a strong magnetic field

1. Identify What process powers most stars, including our sun?

_____

2. Describe Label the layers of the sun that are missing from this diagram.

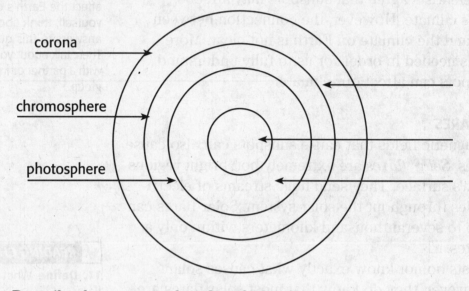

corona

chromosphere

photosphere

3. Explain Describe the process of nuclear fusion in our sun.

_____

_____

_____

_____

4. Identify What produces the sun's magnetic fields?

_____

_____

5. List What are two effects that are caused by changes in the sun's magnetic fields?

_____

_____

6. Describe How do solar flares affect the Earth?

_____

_____

SECTION 3 The Earth Takes Shape

BEFORE YOU READ

After you read this section, you should be able to answer these questions:

• How did the Earth form?

• How did the Earth's atmosphere form?

• How did the Earth's oceans form?

National Science Education Standards
ES 2b

How Did the Earth Form?

The Earth is made mostly of rock. Water covers nearly three-fourths of its surface. A protective atmosphere of nitrogen and oxygen surrounds it. However, the Earth has changed a lot since it formed 4.6 billion years ago.

THE EFFECTS OF GRAVITY

Earth formed when rocky planetesimals collided and combined. When Earth was a young planet, it was smaller and had an uneven shape, like a potato. As it attracted more matter, gravity increased. When Earth reached a diameter of about 350 km, the rock at Earth's center was crushed by gravity. As a result, the planet started to become rounder. ☑

THE EFFECTS OF HEAT

As Earth was changing shape because of gravity, it was also heating up. The energy from collisions with plan-etesimals, along with radiation from radioactive mate-rial, warmed the young planet. When the Earth got large enough, the temperature rose faster than the inside could cool. The rocky center began to melt.

Today, the Earth is still cooling from the energy produced when it formed. You can see evidence of Earth's internal heat in volcanoes, earthquakes, and hot springs.

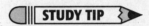

STUDY TIP

Summarize Before you read this section, make an outline using the headings from the section. As you read, fill in the main ideas of the section in your outline.

READING CHECK

1. Identify What caused the early Earth to change from an uneven shape to a round shape?

_____

_____

Force	Effect
	• attracted matter to the early Earth • crushed rock at the early Earth's center
	• caused the inside of the early Earth to melt • is a cause of volcanoes, earthquakes, and hot springs today

TAKE A LOOK

2. Identify Fill in the blank spaces in the table.

SECTION 3 The Earth Takes Shape *continued*

How Did the Earth's Layers Form?

Today, geologists divide the Earth into three main layers. Each layer has a different composition. The three layers are the crust, the mantle, and the core.

The **crust** is the thin, outermost layer. It is between 5 km and 100 km thick. The rock in the crust is rich in elements such as oxygen, silicon, and aluminum.

The **mantle** is the layer beneath the crust. It extends 2,900 km below the Earth's surface. The solid rock in the mantle is rich in elements such as magnesium and iron. It is more dense than the rock in the crust.

The **core** is the central part of Earth. The core is made mostly of iron and nickel. It is the densest layer of the Earth. The core has a radius of about 3,400 km.

How did these three layers form from the rocky material that made up the early Earth? Remember that heat within the Earth caused the rocks to melt. As the rocks melted, more dense materials, such as nickel and iron, sank to the center. They formed the core. Less dense materials floated to the surface. They became the crust and mantle. ☑

Critical Thinking

3. Compare Give three differences between the crust and the mantle.

_____

_____

_____

✓ **READING CHECK**

4. Explain How did the Earth's layers form?

_____

_____

_____

_____

_____

_____

_____

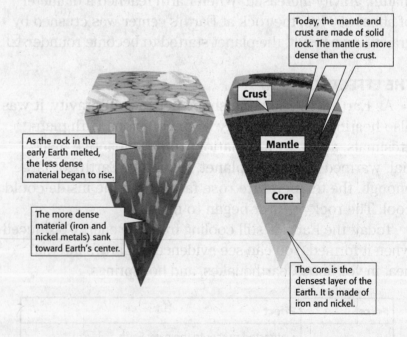

Today, the mantle and crust are made of solid rock. The mantle is more dense than the crust.

Crust

Mantle

Core

As the rock in the early Earth melted, the less dense material began to rise.

The more dense material (iron and nickel metals) sank toward Earth's center.

The core is the densest layer of the Earth. It is made of iron and nickel.

SECTION 3   The Earth Takes Shape *continued*

How Did the Earth's Atmosphere Form?

Today, the Earth's atmosphere contains 78% nitrogen, 21% oxygen and about 1% argon. It also contains tiny amounts of many other gases. However, like the inside of the Earth, the Earth's atmosphere has not always been the same. Early in Earth's history, the atmosphere was probably very different from the atmosphere today. ☑

EARTH'S EARLY ATMOSPHERE

Scientists think that Earth's early atmosphere was a mixture of gases that were given off as Earth cooled. Early in the Earth's history, its surface was very hot. In some places, it was *molten*, or melted. The molten rock gave off large amounts of carbon dioxide and water vapor. Therefore, the Earth's early atmosphere probably contained large amounts of these two gases. ☑

EARTH'S CHANGING ATMOSPHERE

As the Earth cooled and its layers formed, the atmosphere changed. Volcanoes released chlorine, nitrogen, and sulfur, as well as carbon dioxide and water vapor. These gases collected in the atmosphere.

Volcanoes on the early Earth released gases into the atmosphere.

Comets may also have helped to form the Earth's early atmosphere. *Comets* are planetesimals that are made mainly of ice. The material in comets contains many different elements. When the comets crashed into the Earth, these elements were released and became part of the atmosphere. Comets probably brought oxygen, nitrogen, and hydrogen gases to the Earth's atmosphere. They may also have brought some of the water that helped form the oceans.

READING CHECK

5. Identify Which two gases make up most of the Earth's atmosphere today?

_____

_____

READING CHECK

6. List Which two gases probably made up most of the Earth's earliest atmosphere?

_____

_____

TAKE A LOOK
7. Describe How did volcanoes affect the early Earth's atmosphere?

_____

_____

Where Did the Oxygen in Today's Atmosphere Come From?

Most living things on Earth today, including humans, need oxygen in order to survive. Comets brought only a small amount of oxygen to the atmosphere. Most of the oxygen in the atmosphere today is there because of life on Earth. ☑

ULTRAVIOLET RADIATION

You may know that ultraviolet (UV) radiation in sunlight can cause sunburns. Scientists think that this radiation may also have helped to produce the conditions necessary for life to form on Earth.

Ultraviolet radiation contains a lot of energy. Therefore, it can break apart molecules in the air and on the Earth's surface. This is probably what happened on the early Earth. The smaller molecules collected in water. In the water, these chemicals might have combined to form the complex molecules that made life possible.

THE SOURCE OF OXYGEN

The first life forms did not need oxygen to live. By 3.4 billion years ago, organisms that could carry out photosynthesis had evolved. *Photosynthesis* is a process in which a living thing uses sunlight, carbon dioxide, and water to produce food and oxygen. ☑

As these living things carried out photosynthesis, they added more and more oxygen to the atmosphere. At the same time, they removed carbon dioxide from the atmosphere.

Some of the oxygen reacted with sunlight to form ozone gas. This gas then formed the ozone layer. The ozone layer blocked much of the harmful UV radiation, making it possible for life to move onto land. The earliest forms of land life were simple plants, like algae today. They moved onto land about 2.2 billion years ago.

READING CHECK

8. Identify Where does most of the oxygen in today's atmosphere come from?

_____

_____

READING CHECK

9. Define What is photosynthesis?

_____

_____

_____

TAKE A LOOK

10. List Fill in the blank spaces in the table.

Source of atmospheric gases	Gases from this source
Molten rock on the early Earth's surface	
Volcanoes	
Comets	
Living things	

SECTION 3 The Earth Takes Shape *continued*

These *stromatolites* are mats of fossilized algae. Algae like these may have been some of the earliest forms of life on Earth.

How Did the Earth's Oceans and Continents Form?

At first, Earth was so hot that much of its water was in the form of water vapor in the atmosphere. Scientists think that the Earth's oceans formed once Earth had cooled enough for rain to fall. The rain collected on the Earth's surface. After millions of years of rainfall, water covered the planet. By about 4 billion years ago, the first global ocean covered the planet.

GROWTH OF CONTINENTS

There may not have been any dry land during the first few hundred million years of Earth's history. Scientists think that the rocks in the crust and mantle melted and cooled many times while the Earth formed. Each time the rocks melted, denser materials sank and less dense materials rose toward the surface.

After a while, some of the rocks were light enough to pile up on the surface. These rocks were the beginnings of the earliest continents. Over time, as the continents have moved over the surface, they have become even larger. The processes of plate tectonics, such as continental collisions, have produced new continental material. Most of the material in the continents today formed within the last 2.5 billion years. ☑

TAKE A LOOK
11. Define What is a stromatolite?

_____

_____

✓ READING CHECK
12. Explain What has caused the continents to become larger?

_____

_____

Section 3 Review

NSES ES 2b

SECTION VOCABULARY

core the central part of the Earth below the mantle	**crust** the thin and solid outermost layer of the Earth above the mantle
	mantle the layer of rock between the Earth's crust and core

1. List Which two forces caused the early Earth's size and structure to change?

_____

_____

2. Identify Label the three layers of the Earth in the figure below.

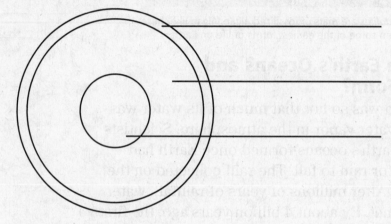

3. Describe How did photosynthesis change the Earth's atmosphere? Give two ways.

_____

_____

4. Explain How did the ozone layer form?

_____

_____

5. Identify When did the ocean on Earth form?

_____

_____

6. Explain How did the continents form?

_____

_____

_____

_____

CHAPTER 20 Formation of the Solar System

SECTION
4 **Planetary Motion**

BEFORE YOU READ

After you read this section, you should be able to answer these questions:

• What is the difference between rotation and revolution?

• What are Kepler's three laws of planetary motion?

• How does gravity affect the orbits of planets?

How Do Scientists Describe the Motions of the Planets?

Remember that the Earth, like all planets, spins on its axis. Scientists use the term **rotation** to describe the motion of a body spinning on its axis. As the Earth rotates, different parts of its surface face the sun. It is daytime in the part that faces the sun. It is night in the part that faces away from the sun. At any time, only one-half of the Earth faces the sun.

In addition to rotating, all planets move around the sun. The path that a planet follows around the sun is called its **orbit**. One complete trip around the sun is called a **revolution**.

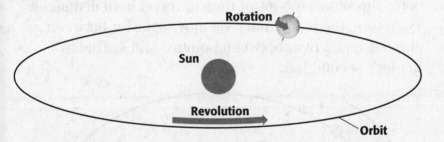

The amount of time it takes for a planet to complete one revolution is called its *period of revolution*. Each planet has a different period of revolution. For example, Earth's period of revolution is 365.24 days. Mercury's is only 88 days.

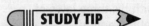
STUDY TIP

Summarize After you read this section, make a chart describing Kepler's laws of planetary motion.

Critical Thinking

1. Infer Venus rotates more slowly than the Earth. On which planet does daytime last longer?

_____

_____

TAKE A LOOK

2. Describe What are two ways that planets move?

_____

_____

SECTION 4 Planetary Motion *continued*

What Do We Know About How Planets Move?

Scientists have not always known how the planets move. Until the 1600s, scientists did not know the shapes of the planets' orbits or their periods of revolution accurately.

In the 1600s, a German scientist named Johannes Kepler made detailed observations of the motions of the planets. After analyzing his observations, he developed three laws of planetary motion. Kepler's observations and calculations were so accurate that scientists still use his laws today! ☑

KEPLER'S FIRST LAW

Kepler carefully observed the path that Mars takes through the sky. When he analyzed his observations, he found that Mars' orbit is not a perfect circle. Instead, it is shaped like an *ellipse*, or oval. Kepler's first law of planetary motion states that the orbits of all planets are ellipses.

KEPLER'S SECOND LAW

Kepler reasoned that the planets must move through their orbits faster in some places than in others. To understand why this is so, look at the figure below. The distance between point 1 and point 2 is longer than the distance between point 3 and point 4. The planet takes the same amount of time to travel both distances. Therefore, the planet must be moving faster between points 1 and 2 than between points 3 and 4. This is Kepler's second law.

READING CHECK

3. Describe How did Johannes Kepler come up with his three laws of planetary motion?

_____

_____

_____

TAKE A LOOK

4. Identify Label the place in the planet's orbit where it is moving the fastest.

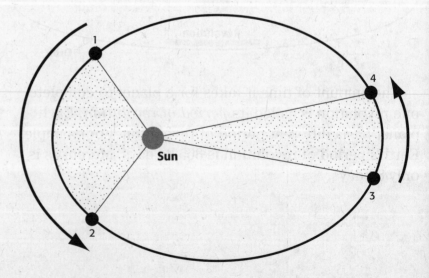

Interactive Textbook
Formation of the Solar System

KEPLER'S THIRD LAW

Kepler observed that planets that are far from the sun, such as Saturn, take longer to orbit the sun. This is Kepler's third law of planetary motion.

How Does Gravity Affect a Planet's Orbit?

Kepler never knew why planets orbit the sun. Another astronomer, Sir Isaac Newton, solved the puzzle. He combined the observations of earlier scientists with mathematical models to describe the force of gravity.

Newton observed that objects fall toward Earth. He reasoned that all objects are attracted to each other through the force of gravity. The strength of the force depends on two factors—the masses of the objects and the distance between them.

Newton's *law of universal gravitation* describes how the force of gravity is related to these two factors. When the masses are large, the force of gravity is strong. When the objects are far apart, the force of gravity is weak.

ORBITS AND GRAVITY

If gravity is pulling on the moon, why doesn't the moon fall to Earth? The answer has to do with the moon's inertia. *Inertia* is an object's resistance to changes in its speed or direction.

Gravity is like a string holding the moon in orbit around the Earth. Without gravity, the moon would move in a straight line away from the Earth. The moon's orbit is a balance between its inertia and the force of gravity. This balance is the reason that all bodies in orbit, including the Earth, travel along curved paths.

STANDARDS CHECK

ES 3b Most objects in the solar system are in regular and predictable motion. Those motions explain such phenomena as the day, the year, phases of the moon, and eclipses.

Word Help: predictable
able to be known ahead of time

Word Help: phenomenon
(plural *phenomena*) any fact or event that can be sensed or described scientifically

5. Identify Relationships How is the distance of a planet from the sun related to its period of revolution?

_____

_____

_____

_____

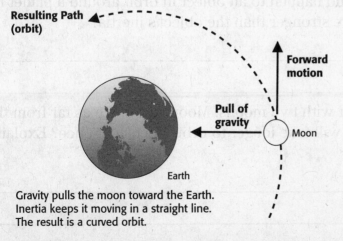

Gravity pulls the moon toward the Earth.
Inertia keeps it moving in a straight line.
The result is a curved orbit.

TAKE A LOOK
6. Explain Why doesn't the moon move away from the Earth in a straight line?

_____

_____

Name ____________________ Class ____________ Date ____________

Section 4 Review

NSES ES 3b

SECTION VOCABULARY

orbit the path that a body follows as it travels around another body in space	**revolution** the motion of a body that travels around another body in space; one complete trip along an orbit **rotation** the spin of a body on its axis

1. Compare How is rotation different from revolution?

____

2. Define Explain Kepler's second law of planetary motion in your own words.

____

3. Describe Fill in the blank spaces in the table.

Factor	How it affects the force of gravity
Mass of the objects	
Distance between the objects	

4. Identify What two factors must be balanced in order for an object to remain in orbit?

____

5. Predict Consequences What would happen to an object in orbit around a planet if the planet's force of gravity were stronger than the object's inertia?

____

6. Apply Concepts Imagine a planet with two moons. Moon A is twice as far from the planet as moon B. Which moon will take longer to orbit the planet once? Explain your answer.

____

CHAPTER 21 | A Family of Planets)

SECTION
1 **Our Solar System**

BEFORE YOU READ

After you read this section, you should be able to answer these questions:

• What are the parts of our solar system?

• When were the planets discovered?

• How do astronomers measure large distances?

National Science Education Standards
ES 1c, 3a, 3b, 3c

What Is Our Solar System?

Our *solar system* includes our sun, the planets, their moons, and many other objects. At the center of our solar system is a star that we call the sun. Planets and other smaller objects move around the sun. Most planets have one or more moons that move around them. In this way, our solar system is a combination of many smaller systems.

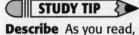

STUDY TIP

Describe As you read, make a chart showing the parts of our solar system that were discovered in the following time periods: before the 1600s; the 1700s; the 1800s; and the 1900s.

The planets and the sun are some of the objects in our solar system.

How Was Our Solar System Discovered?

Until the 1600s, people thought that there were only eight bodies in our solar system. These were the sun, Earth's moon, and the planets Earth, Mercury, Venus, Mars, Jupiter, and Saturn. These are the only objects in the solar system that we can see from Earth without a telescope.

Once the telescope was invented, however, scientists were able to see many more bodies in our solar system. In the 17th century, scientists discovered some of the moons of Jupiter and Saturn. Uranus and several other moons were discovered in the 1700s. Neptune was discovered in the 1800s. Pluto was not discovered until the 1900s.

STANDARDS CHECK

ES 3a The earth is the third planet from the sun in a system that includes the moon, the sun, eight other planets and their moons, and smaller objects, such as asteroids and comets. The sun, an average star, is the central and largest body in the solar system. **Note:** In 2006, the International Astronomical Union ruled that Pluto is no longer considered to be a planet.

1. Identify List three kinds of objects that make up our solar system.

_____

_____

_____

How Do Scientists Measure Long Distances?

Remember that astronomers use light-years to measure long distances in space. To measure distances within our solar system, astronomers use two other units: the astronomical unit and the light-minute.

One **astronomical unit** (AU) is the average distance between the sun and Earth. This distance is about 150,000,000 km. Earth is 1 AU from the sun. Neptune is about 30.1 AU from the sun. Therefore, Neptune is about 30.1 × 150,000,000 km = 4,500,000,000 km from the sun.

Another way to measure distances in space is by using the speed of light. Light travels at about 300,000 km/s in space. In one minute, light travels about 18,000,000 km. Therefore, one *light-minute* is equal to about 18,000,000 km. Light from the sun takes 8.3 minutes to reach Earth. Therefore, Earth is 8.3 light-minutes from the sun.

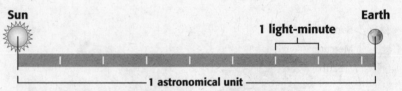

One astronomical unit equals about 8.3 light-minutes.

How Is Our Solar System Divided?

Astronomers divide our solar system into two main parts. These parts are called the *inner solar system* and the *outer solar system*.

THE INNER PLANETS

The inner solar system contains the four planets that are closest to the sun: Mercury, Venus, Earth, and Mars. The inner planets are also sometimes called the *terrestrial planets. Terrestrial* means "like Earth." Mercury, Venus, and Mars are like Earth because they have dense, rocky surfaces, as Earth does. The figure on the top of the next page shows the orbits of the inner planets.

Math Focus

2. Convert Scientists discover an asteroid that is 3 AU from the sun. How far, in kilometers, is the asteroid from the sun?

_____

Math Focus

3. Convert About how many light-minutes from the sun is Neptune?

_____

SECTION 1 Our Solar System *continued*

The Inner Planets

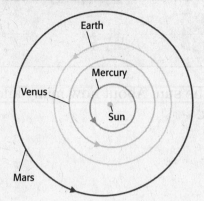

TAKE A LOOK
4. Identify Which of the inner planets is closest to the sun?

_____

5. Identify Which of the inner planets is furthest from the sun?

_____

THE OUTER PLANETS

The outer solar system contains four planets: Jupiter, Saturn, Uranus, and Neptune. The outer planets are very different from the inner planets.

The outer planets are very large and are made mostly of gases. Therefore, Jupiter, Saturn, Uranus, and Neptune are sometimes called the *gas giant* planets, or simply the "gas giants."

The distances between the outer planets are much larger than the distances between the inner planets. For example, the distance between Jupiter and Saturn is much larger than the distance between Mars and Earth. The figure below shows the orbits of the outer planets.

Critical Thinking
6. Compare Give two differences between the inner solar system and the outer solar system.

_____

_____

_____

_____

_____

_____

The Outer Planets

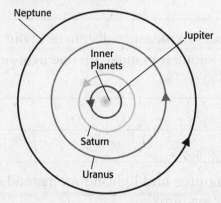

TAKE A LOOK
7. Identify What planet is farthest from the sun?

_____

Section 1 Review

NSES ES 1c, 3a, 3b, 3c

SECTION VOCABULARY

astronomical unit the average distance be-tween the Earth and the sun; approximately 150 million kilometers (symbol, AU)	

1. Calculate Mercury is about 0.39 AU from the sun. About how many kilometers from the sun is Mercury? Show your work.

2. Identify Name the four planets in the inner solar system.

_____

_____

_____

3. Identify Name the four gas giant planets.

_____

_____

_____

4. Infer Scientists sometimes use light-hours to measure distances in our solar system. What is a light-hour? About how many kilometers make up one light-hour?

_____

_____

_____

5. Explain Why do scientists use light-minutes and light-hours instead of light-years to measure distances within our solar system?

_____

_____

_____

CHAPTER 21 A Family of Planets

SECTION 2 The Inner Planets

National Science
Education Standards
ES 1c, 3a, 3b

BEFORE YOU READ

After you read this section, you should be able to answer
these questions:

• Which planets are known as the inner planets?

• What properties do the inner planets share?

Why Group the Inner Planets Together?

The inner solar system includes the only planet known
to support life, Earth, and three other planets. These four
inner planets are called **terrestrial planets** because
they all have a chemical makeup similar to that of Earth.
The terrestrial planets are much smaller, denser, and
more rocky than most of the outer planets. ☑

Which Planet Is Closest to the Sun?

Mercury is the planet closest to the sun. After Earth, it
is the second densest object in the solar system. This is
because, like Earth, Mercury has a large iron core in its
center. The surface of Mercury is covered with craters.

Mercury rotates on its axis much more slowly than
Earth. Remember that the amount of time that a planet
takes to rotate once is its *period of rotation*. It is the
length of a day on the planet. Mercury's period of rotation
is about 59 Earth days long. Therefore, a day on Mercury
is about 59 Earth days long.

On Mercury, a year is not much longer than a day.
Remember that the time it takes a planet to go around
the sun once is the planet's *period of revolution*. It is the
length of one year on the planet. A *Mercurian* year, or
a year on Mercury, is equal to 88 Earth days. Therefore,
each year on Mercury lasts only 1.5 Mercurian days.

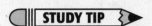

STUDY TIP

Compare In your notebook,
create a chart showing the
similarities and differences
among the inner planets.

READING CHECK

1. Explain Why are the
inner planets called
terrestrial planets?

_____

_____

Critical Thinking

2. Infer Which of the facts
on the table could scientists
use to infer that Mercury has
a core made of iron?

_____

Facts About Mercury

Distance from sun	0.38 AU
Period of rotation	58 Earth days, 19 hours
Period of revolution	88 Earth days
Diameter	4,879 km
Density	5.43 g/cm³
Surface gravity	38% of Earth's

Is Venus Earth's Twin?

The second planet from the sun is Venus. In some ways, Venus is more like Earth than any of the other planets. It is about the same size as Earth. However, Venus is slightly smaller, less dense, and less massive than Earth.

If you could observe the sun from the surface of Venus, you would see it rise in the west and set in the east. That is because Venus and Earth rotate on their axes in opposite directions.

If you looked down on Earth from above the North Pole, you would see Earth spinning counterclockwise. This is called **prograde rotation**. However, if you were to look down on Venus from above its north pole, you would see it spinning clockwise. This is called **retrograde rotation**. ☑

THE ATMOSPHERE OF VENUS

Venus has the densest atmosphere of the terrestrial planets. The atmospheric pressure on Venus's surface is 90 times that on Earth. This pressure would instantly crush a human on Venus. Venus's atmosphere is mostly made of carbon dioxide and thick clouds of sulfuric acid. The thick atmosphere causes a strong greenhouse effect. As a result, surface temperatures on Venus average about 464°C. This is hot enough to melt lead and some other metals.

Facts About Venus

Distance from sun	0.72 AU
Period of rotation	243 Earth days, 16 hours
Period of revolution	224 Earth days, 17 hours
Diameter	12,104 km
Density	5.24 g/cm³
Surface gravity	91% of Earth's

MAPPING THE SURFACE OF VENUS

Because of its thick atmosphere, we cannot observe the surface of Venus from Earth with telescopes. Between 1990 and 1992, the *Magellan* spacecraft made maps of Venus using radar waves. These waves can travel through the atmosphere and bounce off the surface. Maps made from the radar data showed that Venus has craters, mountains, lava plains, and volcanoes.

☑ **READING CHECK**

3. Compare How do prograde rotation and retrograde rotation differ?

_____

_____

_____

TAKE A LOOK
4. Compare Which is longer on Venus, one day or one year?

_____

_____

Critical Thinking

5. Analyzing Methods Why did scientists use *Magellan's* radar instead of telescopes to map the surface of Venus?

_____

_____

_____

_____

What Makes Earth Unique?

Until the mid-1900s, no one knew what Earth looked like from space. Today, satellites and spacecraft can take pictures of a sparkling blue planet. Light reflecting off ocean water makes Earth look blue from space. ☑

WATER ON EARTH

Earth is the only planet in the solar system that can support life as we know it. This is because Earth has a certain combination of factors that make life possible. These factors include abundant water and just the right amount of energy from the sun.

Liquid water is vital to life as we know it. Earth is not the only planet in the solar system to have water on its surface. However, Earth is the only planet that has large amounts of liquid water on its surface. Earth is close enough to the sun that all the water does not freeze. It is far enough away that the water does not boil away. If Earth were much closer to or farther from the sun, liquid water—and life—could not exist here.

Facts About Earth

Distance from sun	1.0 AU
Period of rotation	23 hours, 56 minutes
Period of revolution	365 Earth days, 6 hours
Diameter	12,756 km
Density	5.52 g/cm³
Surface gravity	100% of Earth's

STUDYING EARTH FROM SPACE

NASA's Earth Science Enterprise is a program to study Earth from space. Studying Earth from space lets scientists study the Earth as a whole system. It helps them understand changes in Earth's atmosphere, oceans, ice, landforms, and living things. It may also be able to help them understand how humans affect the global environment. By studying Earth from space, scientists can learn how different parts of the Earth interact.

☑ **READING CHECK**

6. Identify What feature of Earth causes it to appear blue from space?

_____

_____

Math Focus

7. Calculate Use the information on the table to explain why every fourth year on Earth is a leap year. Show your work.

(Hint: Compare Earth's period of revolution to the number of days in a calendar year.)

_____

_____

_____

_____

What Is the Red Planet?

Besides Earth, the most studied planet in the solar system is Mars. Mars looks red, so it is sometimes known as "the red planet." Some scientists think that there could be simple life on Mars.

Scientists have learned much about Mars by observing it from Earth. However, most of our knowledge of the planet has come from unmanned spacecraft. So far, these observations have found no evidence of life.

THE ATMOSPHERE OF MARS

Because it has a thinner atmosphere than Earth and is farther from the sun, Mars is colder than Earth. In the middle of the summer, the spacecraft *Mars Pathfinder* recorded a temperature range from $-13°C$ to $-77°C$. The Martian atmosphere is made mainly of carbon dioxide. ☑

The atmospheric pressure on Mars is very low. At the surface, it is about the same as the pressure 30 km above Earth's surface. Because of the low temperatures and air pressure, liquid water cannot exist on the surface of Mars. The only water on Mars's surface is in the form of ice.

READING CHECK

8. Explain What are two reasons that the surface of Mars is colder than that of Earth?

_____

_____

_____

_____

TAKE A LOOK

9. Compare How does the length of a day on Mars compare to the length of day on Earth?

_____

_____

Facts About Mars

Distance from sun	1.52 AU
Period of rotation	24 hours, 37 minutes
Period of revolution	687 Earth days
Diameter	6,794 km
Density	3.93 g/cm³
Surface gravity	38% of Earth's

WATER ON MARS

Even though water cannot exist on the surface of Mars today, it may have in the past. Evidence from spacecraft suggests that some of Mars's features were formed by liquid water. For example, some of Mars's features are similar to those caused by water erosion on Earth. Other features indicate that Mars's surface contains sediments that may have been deposited by the water from a large lake. ☑

Scientists cannot prove that these features were caused by liquid water. However, they indicate that at some time in the past, Mars may have had liquid water. If this is true, it would show that Mars was once warmer and had a thicker atmosphere than it does today.

READING CHECK

10. Identify What two features suggest that water once existed on the surface of Mars?

_____

_____

_____

WHERE THE WATER IS NOW

Mars has two polar icecaps made of a combination of frozen water and frozen carbon dioxide. Most of the water on Mars is trapped in this ice. There is some evidence from the *Mars Global Surveyor* that water could exist just beneath the surface. If so, it may be in liquid form. If Mars does have liquid water beneath its surface, there is a possibility that some form of life may exist on Mars. ☑

VOLCANOES ON MARS

The remains of giant volcanoes exist on the surface of Mars. They show that Mars has had active volcanoes in the past. Unlike on Earth, however, the volcanoes are not spread across the whole planet. There are two large volcanic systems on Mars. The largest one is about 8,000 km long.

The largest mountain in the solar system, Olympus Mons, is a Martian volcano. It is a shield volcano, similar to Mauna Kea on the island of Hawaii. However, Olympus Mons is much larger than Mauna Kea. The base of Olympus Mons is 600 kilometers —about 370 miles—across. It is nearly 24 kilometers tall. That is three times as tall as Mount Everest! It may have grown so tall because the volcano erupted for long periods of time. ☑

MISSIONS TO MARS

Scientists sent several vehicles to Mars in the early 21st century. The figure below shows *Mars Express Orbiter*, which reached Mars in December 2003. In January 2004, the exploration rovers *Spirit* and *Opportunity* landed on Mars. These solar-powered wheeled robots have found evidence that water once existed on the Martian surface. ☑

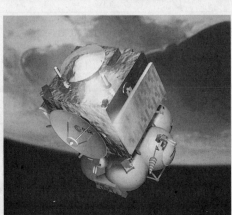

The *Mars Express Orbiter* helps scientists map Mars and study Mars's atmosphere.

☑ READING CHECK

11. Identify Where does water exist on Mars today?

_____

_____

☑ READING CHECK

12. Explain What may have allowed Olympus Mons to grow so large?

_____

_____

_____

☑ READING CHECK

13. Describe What have the rovers *Spirit* and *Opportunity* found?

_____

_____

Section 2 Review

SECTION VOCABULARY

prograde rotation the counterclockwise spin of a planet or moon as seen from above the planet's North Pole; rotation in the same direction as the sun's rotation	**retrograde rotation** the clockwise spin of a planet or moon as seen from above the planet's North Pole **terrestrial planet** one of the highly dense planets nearest to the sun; Mercury, Venus, Mars, and Earth

1. Identify Does Earth show prograde or retrograde rotation?

_____

2. Compare Fill in the blanks to complete the table.

Planet	Distance from sun	Period of revolution
	0.38 AU	58 Earth days, 19 hours
	0.72 AU	243 Earth days, 16 hours
	1.00 AU	365 Earth days, 6 hours
	1.52 AU	687 Earth days

3. Analyze Ideas Why do scientists think that Mars was once warmer and had a thicker atmosphere than it does today?

_____

_____

_____

4. Identify Relationships How is the period of revolution of a planet related to its distance from the sun? (Hint: examine the statistics tables.)

_____

_____

5. Explain Why is the surface temperature of Venus higher than the surface temperatures of the other inner planets?

_____

_____

6. Explain Why could life probably not have developed on Earth if Earth were closer to the sun?

_____

CHAPTER 21 | A Family of Planets

SECTION
3 **The Outer Planets**

BEFORE YOU READ

After you read this section, you should be able to answer these questions:

• Which planets are known as the outer planets?

• What properties do the outer planets share?

National Science Education Standards

ES 1c, 3a, 3b

How Are the Outer Planets Different from the Inner Planets?

The outer planets are very large and are made mostly of gases. These planets are called **gas giants**. Unlike the inner planets, they have very thick atmospheres and not very much hard, rocky material on their surfaces.

Which Planet Is the Biggest?

Jupiter is the largest planet in our solar system. Its mass is twice as large as the other eight planets combined. Even though it is large, Jupiter's rotation takes less than 10 hours.

Like the sun, Jupiter is made mostly of hydrogen and helium. Jupiter's atmosphere also contains small amounts of ammonia, methane, and water. These gases form clouds in the outer part of Jupiter's atmosphere. The outer atmosphere also contains storms, such as the Great Red Spot. This huge storm is about 3 times the diameter of Earth. It has lasted for over 400 years! ☑

Deeper into Jupiter's atmosphere, the pressure is so high that hydrogen turns to liquid. Deeper still, the pressure is even higher. Because of the high pressures, the inside of Jupiter is very hot. It is so hot that Jupiter produces more heat than it gets from the sun.

The information that scientists have about Jupiter has come from five space missions: *Pioneer 1*, *Pioneer 2*, *Voyager 1*, *Voyager 2*, and *Galileo*. The Voyager probes showed that Jupiter has a thin, faint ring.

STUDY TIP

Compare In your notebook, create a chart showing the similarities and differences among the outer planets.

READING CHECK

1. List Give four gases that are found in Jupiter's atmosphere.

_____

_____

_____

_____

Facts About Jupiter

Distance from sun	5.20 AU
Period of rotation	9 hours, 55.5 minutes
Period of revolution	11 Earth years, 313 days
Diameter	142,984 km
Density	1.33 g/cm³
Surface gravity	236% of Earth's

TAKE A LOOK

2. Identify Which of the facts in the table could you use to infer that Jupiter has a shorter day than Earth does?

_____

What Are Saturn's Rings Made Of?

Saturn is the second-largest planet in the solar system. Like Jupiter, Saturn is made up mostly of hydrogen with some helium and traces of other gases and water. Saturn has about 764 times more volume than Earth and about 95 times more mass. Therefore, it is much less dense than Earth.

Facts About Saturn

Distance from sun	9.54 AU
Period of rotation	10 hours, 42 minutes
Period of revolution	29 Earth years, 155 days
Diameter	120,536 km
Density	0.69 g/cm³
Surface gravity	92% of Earth's

Critical Thinking

3. Compare About how many times does Earth revolve around the sun in the time it takes Saturn to revolve once?

_____

_____

The inside of Saturn is probably similar to the inside of Jupiter. Also, like Jupiter, Saturn gives off more heat than it gets from the sun. Scientists think that Saturn's extra energy comes from helium condensing from the atmosphere and sinking toward the core. In other words, Saturn is still forming.

Saturn is probably best known for the rings that orbit the planet above its equator. They are about 250,000 km across, but less than 1 km thick. The rings are made of trillions of particles of ice and dust. These particles range from a centimeter to several kilometers across. ☑

READING CHECK

4. Identify What two materials make up the rings of Saturn?

_____

_____

This picture of Saturn was taken by the *Voyager 2* probe.

SECTION 3 The Outer Planets *continued*

How Is Uranus Unique?

Uranus is the third-largest planet in the solar system. It is so far from the sun that it does not reflect much sunlight. You cannot see it from Earth without using a telescope. ☑

Uranus is different from the other planets because it is "tipped" on its side. As shown in the figure below, the north and south poles of Uranus point almost directly at the sun. The north and south poles of most other planets, like Earth, are nearly at right angles to the sun.

For about half the Uranian year, one pole is constantly in sunlight, and for the other half of the year it is in darkness. Some scientists think that Uranus may have been tipped over by a collision with a massive object.

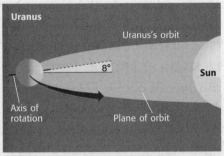

Uranus is tilted so that its poles point almost directly at the sun.

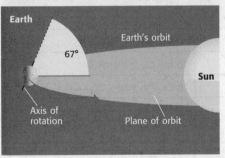

In contrast, Earth's poles, like those of most other planets, are nearly at right angles to the sun.

Like Jupiter and Saturn, Uranus is made mostly of hydrogen, helium, and small amounts of other gases. One of these gases, methane, filters sunlight and makes the planet look bluish-green. ☑

Facts About Uranus

Distance from sun	19.22 AU
Period of rotation	17 hours, 12 minutes
Period of revolution	83 Earth years, 273 days
Diameter	51,118 km
Density	1.27 g/cm³
Surface gravity	89% of Earth's

READING CHECK

5. Identify Why can't Uranus be seen from Earth without a telescope?

_____

_____

TAKE A LOOK

6. Explain Why do scientists say that Uranus is "tipped over"?

_____

_____

TAKE A LOOK

7. Compare How does the length of a year on Uranus compare to the length of a year on Earth?

_____

_____

What Is Neptune Like?

Some astronomers predicted that there was a planet beyond Uranus before the planet was observed. Uranus did not move in its orbit exactly as they expected. The force of gravity due to another large object was affecting it. Using predictions of its effect on Uranus, astronomers discovered Neptune in 1846. ☑

Neptune is the fourth-largest planet in the solar system. Like the other gas giants, Neptune is made up mostly of hydrogen, helium, and small amounts of other gases. It has a deep blue color, which is caused by methane in its atmosphere.

Clouds and weather changes are seen in the atmosphere of Neptune. The spacecraft *Voyager 2* flew past Neptune in 1989 and observed a Great Dark Spot in the southern hemisphere. This spot was a storm as large as Earth. It moved across the planet's surface at about 300 m/s. By 1994, the Great Dark Spot had disappeared. Another dark spot was then found in the northern hemisphere. *Voyager 2* images also showed that Neptune has very narrow rings.

Facts About Neptune

Distance from sun	30.06 AU
Period of rotation	16 hours, 6 minutes
Period of revolution	163 Earth years, 263 days
Diameter	49,528 km
Density	1.64 g/cm³
Surface gravity	112% of Earth's

Why Is Pluto Called a Dwarf Planet?

Pluto has been called the ninth planet since its discovery in 1930. However, in 2006, astronomers defined *planet* in a new way. Pluto does not fit the new definition of a planet. So, Pluto has been reclassified as a dwarf planet.

A *dwarf planet* is any object that orbits the sun and is round because of its own gravity, but has not cleared its orbital path. In addition to Pluto, Eris and Ceres have been classified as dwarf planets. Eris is larger than Pluto. Ceres was previously classified as an asteroid.

✓ READING CHECK

8. Explain What evidence did astronomers have that Neptune existed before they actually observed it?

_____

_____

TAKE A LOOK

9. Compare How does Neptune's average distance from the sun compare to Earth's?

_____

_____

_____

SECTION 3 The Outer Planets *continued*

A SMALL WORLD

Pluto is made of rock and ice and has a thin atmosphere made of methane and nitrogen. Scientists do not know if Pluto formed along with the planets.

AN UNUSUAL ORBIT

The shape of Pluto's orbit is different from the shapes of the orbits of the outer planets. As shown in the figure below, sometimes Pluto is closer to the sun than Neptune is. At other times, Neptune is closer to the sun.

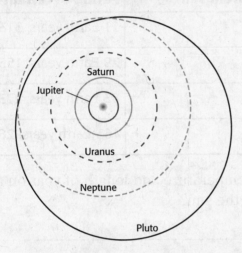

The shape of Pluto's orbit is very different from the orbits of the outer planets.

TAKE A LOOK
10. Compare How is Pluto's orbit different from the orbits of the outer planets?

_____
_____
_____
_____

Facts About Pluto

Distance from sun	39.5 AU
Period of rotation	6 days, 10 hours
Period of revolution	248 Earth years, 4 days
Diameter	2,390 km
Density	1.75 g/cm³
Surface gravity	6% of Earth's

A LARGE MOON

Pluto's moon, Charon, is more than half the size of Pluto. From Earth, it is hard to separate the images of Pluto and Charon because they are so far away. Charon may be covered by frozen water.

Critical Thinking

11. Infer How do you think scientists learned that Pluto has a moon if it is difficult to separate their images?

_____
_____
_____
_____

Section 3 Review

SECTION VOCABULARY

gas giant a planet that has a deep, massive atmosphere, such as Jupiter, Saturn, Uranus, or Neptune	

1. Identify What is the main element found in the atmosphere of a gas giant planet?

_____

2. Compare Fill in the blanks to complete the table.

Planet	Distance from sun	Period of revolution
	5.20 AU	11 Earth years, 313 days
	9.54 AU	29 Earth years, 155 days
	19.22 AU	83 Earth years, 273 days
	30.06 AU	163 Earth years, 263 days

3. Evaluate Data How are the surface temperature and length of year on a planet related to the planet's distance from the sun?

_____

_____

_____

4. Make Comparisons How do the gas giants differ from the inner planets of the solar system? In your answer, discuss composition, size, distance from the sun, length of a year, and solar energy received.

_____

_____

_____

5. Identify What gives Neptune and Uranus their blue to blue-green colors?

_____

6. Describe What did the Voyager probes discover about Jupiter?

_____

CHAPTER 21 A Family of Planets

SECTION 4 # Moons

National Science
Education Standards
ES 1a, 3a, 3b, 3c

BEFORE YOU READ

After you read this section, you should be able to answer
these questions:

- How did Earth's moon probably form?
- How does the moon's appearance change with time?
- What moons revolve around other planets?

What Are Moons?

Satellites are natural or artificial bodies that revolve
around larger bodies in space, such as planets. Except for
Mercury and Venus, all of the planets have natural satellites called *moons*. Moons come in a wide variety of sizes,
shapes, and compositions.

What Do We Know About Earth's Moon?

Scientists have learned a lot about Earth's moon,
which is also called *Luna*. Much of what we know comes
from observations from Earth, but other discoveries have
come from visiting the moon. Some lunar rocks brought
back by Apollo astronauts were found to be almost
4.6 billion years old. These rocks have not changed much
since they were formed. This tells scientists that the solar
system itself is at least 4.6 billion years old. ☑

THE MOON'S SURFACE

The moon is almost as old as Earth. It is covered with
craters, many of which can be seen from Earth on a clear
night. Because the moon has no atmosphere and no erosion, its surface shows where objects have collided with
it. Scientists think that many of these collisions happened
about 3.8 billion years ago. They were caused by matter
left over from the formation of the solar system.

Facts About Luna

Period of rotation	27 Earth days, 9 hours
Period of revolution	27 Earth days, 7 hours
Diameter	3,475 km
Density	3.34 g/cm3
Surface gravity	16% of Earth's

STUDY TIP

Describe In your notebook,
create a Concept Map about
Earth's moon, including
information about its origin,
why it shines, phases, and
eclipses.

✓ **READING CHECK**

1. Explain How do scientists
know what moon's crust is
made of?

_____

_____

_____

Math Focus

2. Identify What fraction of
Earth's gravity is the moon's
gravity?

_____

SECTION 4 Moons *continued*

THE ORIGIN OF THE MOON

When scientists studied the rock samples brought back from the moon by astronauts, they found some surprises. The composition of the moon is similar to that of Earth's mantle. This evidence led to a theory about the moon's formation. ☑

Scientists now think that the moon formed when a large object collided with the early Earth. The object was probably about the size of Mars. The collision was so violent that a large mass of material was thrown into orbit around Earth. Gravity pulled this material into a sphere. The sphere continued to revolve around the planet. Eventually, it became the moon.

READING CHECK

3. Identify What discovery caused scientists to revise their theory about the origin of the moon?

_____

_____

Formation of the Moon

❶ About 4.6 billion years ago, a large body collided with Earth. At this time, Earth was still mostly molten. The collision blasted part of Earth's mantle into space.

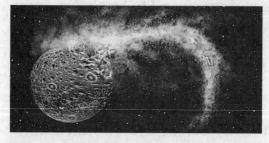

❷ Within a few hours of the collision, the debris began to orbit the Earth. The debris was made of mantle material from Earth and some iron core material from the colliding body.

❸ In time, the material began to clump together. Eventually, the moon formed. As it cooled, collisions with smaller objects produced cracks in the moon's crust. Lava flowed onto the moon's surface. This formed the dark patches, or *maria*, that we can see on the moon today.

TAKE A LOOK

4. Identify According to this theory, material was thrown from Earth in clumps. What caused the material to come together as a sphere?

_____

_____

SECTION 4 Moons *continued*

PHASES OF THE MOON

The moon revolves around the Earth once each month. It rotates on its axis in almost the same period. Therefore, we always see the same side of the moon. However, the moon does not always look the same. This is because we cannot always see all of the part that is reflecting light.

As the moon's position changes compared to the sun and Earth, it looks different to people on Earth. During a month, the face of the moon that we can see changes from a fully lit circle to a thin crescent and then back to a circle. The figure below shows how the moon's appearance changes as it moves around Earth.

Critical Thinking
5. Explain The moon does not produce its own light. How can the moon be seen from Earth?

_____

_____

_____

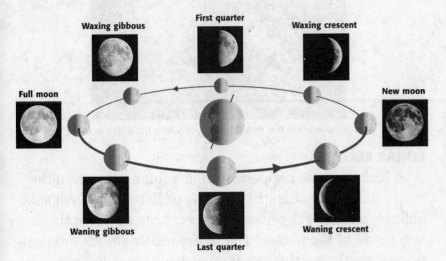

Waxing gibbous   First quarter   Waxing crescent

Full moon   New moon

Waning gibbous   Last quarter   Waning crescent

TAKE A LOOK
6. Explain Why does the moon look different on different nights?

_____

_____

_____

The different appearances of the moon are called **phases**. When the moon is *waxing*, the part of the sunlit side that we can see increases every day. The moon appears to get bigger. When the moon is *waning*, the part of the sunlit side that we can see decreases every day. The moon appears to get smaller.

What Is an Eclipse?

An **eclipse** happens when the shadow of one body in space falls on another. A *solar eclipse* happens when the moon comes between the sun and Earth. Then, the shadow of the moon falls on part of Earth's surface. A *lunar eclipse* happens when Earth comes between the sun and the moon. Then, the shadow of Earth falls on the moon. ☑

✓ **READING CHECK**
7. Explain What happens during a solar eclipse?

_____

_____

_____

_____

_____

SOLAR ECLIPSES

Because the moon's orbit is elliptical (oval-shaped) instead of circular, the distance between Earth and the moon changes. When the moon is close to Earth, the moon appears to be the same size as the sun. If the moon passes between the sun and Earth during that time, there is a *total solar eclipse*. If the moon is farther from earth, the eclipse is an annular eclipse. During an *annular eclipse*, a thin ring of the sun can be seen around the moon.

During a solar eclipse, the moon passes between the Earth and the sun.

LUNAR ECLIPSES

A lunar eclipse happens during a full moon when the moon passes through the shadow of Earth. Unlike a solar eclipse, a lunar eclipse can be seen from much of the night side of the planet. The figure below shows the position of Earth and the moon during a lunar eclipse.

During a lunar eclipse, the Earth passes between the sun and the moon.

Lunar eclipses are interesting to watch. At the beginning and end of a lunar eclipse, the moon is in the outer part of the shadow. In this part of the shadow, Earth's atmosphere filters out some of the blue light. As a result, the light that is reflected from the moon is red.

TAKE A LOOK
8. Explain Why can't a solar eclipse be seen from every point on Earth?

_____

_____

_____

TAKE A LOOK
9. Describe What happens during a lunar eclipse?

_____

_____

_____

THE MOON'S TILTED ORBIT

The moon rotates around Earth each month, so you might expect that there would be an eclipse each month. However, eclipses happen only about once a year.

Eclipses don't happen every month because the moon's orbit is slightly tilted compared to Earth's orbit. This tilt is enough to place the moon out of Earth's shadow during most full moons. It also causes the Earth to be out of the moon's shadow during most new moons. ☑

10. Explain Why don't solar eclipses occur each month?

_____

_____

_____

_____

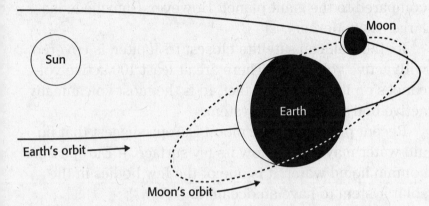

The moon's orbit is tilted compared to the Earth's. Therefore, eclipses do not happen every month.

Are Other Moons Like Earth's Moon?

All of the planets, except Mercury and Venus, have moons. Mars has two moons. All of the gas giants have many moons. Many of these moons were discovered fairly recently using spacecraft cameras or the Hubble Space Telescope. Some moons may not have been discovered yet. ☑

The solar system's moons vary widely. They range in size from very small bits of rock to objects as large as a terrestrial planet. Their orbits range from nearly circular to very elliptical. Most moons orbit in the same direction as the planets orbit the sun. However, some orbit in the opposite direction.

THE MOONS OF MARS

Mars has two moons, Phobos and Deimos. They are small, oddly shaped satellites. Both moons have dark surfaces and resemble *asteroids*, or rocky bodies in space. Phobos is about 22 km across at its largest dimension. Deimos is about 15 km across. Both moons may be asteroids that were captured by Mars's gravity. ☑

Say It

Discuss In a group, discuss why you can't look at the sun during a solar eclipse but you can look at the moon during a lunar eclipse.

11. Compare Which types of planets have the most moons—terrestrial planets or gas giants?

_____

12. Identify What are the names of Mars's moons?

_____

_____

SECTION 4 Moons *continued*

THE MOONS OF JUPITER

Jupiter has more than 60 moons. The four largest were discovered in 1610 by Galileo. When he observed Jupiter through a telescope, Galileo saw what looked like four dim stars that moved with Jupiter. He observed that they changed position compared to Jupiter and each other from night to night.

These moons—Ganymede, Callisto, Io, and Europa— are known as the *Galilean satellites*. They appear small compared to the giant planet. However, Ganymede is larger than Mercury. ☑

Io, the Galilean satellite closest to Jupiter, is covered with active volcanoes. There are at least 100 active volcanoes on its surface. In fact, Io is the most volcanically active body in the solar system.

Recent pictures of the moon Europa suggest that liquid water may exist below its icy surface. If Europa does contain liquid water, it is one of the few bodies in the solar system to have an ocean. ☑

✓ **READING CHECK**

13. Identify What are the names of the Galilean satellites?

_____

_____

_____

✓ **READING CHECK**

14. Identify What may lie below the icy surface of Europa?

_____

_____

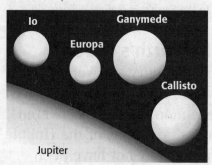

This figure shows the sizes of the four Galilean satellites compared to Jupiter.

THE MOONS OF SATURN

Saturn has more than 50 moons. Saturn's largest moon, Titan, is slightly smaller than Ganymede. Unlike most moons, Titan has an atmosphere. Its atmosphere is composed mostly of nitrogen, with small amounts of other gases, such as methane. Scientists think that Titan's atmosphere is similar to Earth's early atmosphere.

None of Saturn's other moons are as large as the Galilean moons of Jupiter. Most of them are from several kilometers to several hundred kilometers across. They are made mostly of frozen water and rocks.

Critical Thinking

15. Make Inferences Would humans be able to live unprotected on the surface of Titan? Explain your answer.

_____

_____

_____

_____

SECTION 4 Moons *continued*

THE MOONS OF URANUS

Uranus has at least 27 moons. Most of them are small. They were discovered by space probes or orbiting observatories, such as the Hubble Space Telescope. Like the moons of Saturn, Uranus's largest moons are made of ice and rock. ☑

THE MOONS OF NEPTUNE

Neptune has 13 known moons. The largest, Triton, revolves in a *retrograde*, or "backward," orbit. Triton's unusual orbit suggests that it was captured by Neptune's gravity after forming somewhere else in the solar system. Triton has a thin nitrogen atmosphere. Its surface is mostly frozen nitrogen and methane. It has active "ice volcanoes" that send gas high into its atmosphere. Neptune's other moons are small and are made of ice and rock.

THE MOONS OF PLUTO

Although Pluto is not considered a planet, it does have at least three moons. The diameter of Charon, the largest moon, is about half that of Pluto. Charon revolves around Pluto in 6.4 days, the same period as Pluto's rotation. That means that Charon is always located at the same place in Pluto's sky. Two additional moons of Pluto, discovered by the Hubble telescope in 2005, are much smaller than Charon. These moons are called Hydra and Nix.

✓ READING CHECK

16. Describe What are Uranus's largest moons made of?

_____

_____

Some of the Moons of the Solar System

Planet	Moon	Diameter	Period of revolution
Earth	Luna	3475 km	27.3 Earth days
Mars	Phobos	26 km	0.3 Earth days
Mars	Deimos	15 km	1.3 Earth days
Jupiter	Io	3636 km	1.8 Earth days
Jupiter	Europa	3120 km	3.6 Earth days
Jupiter	Ganymede	5270 km	7.1 Earth days
Jupiter	Callisto	4820 km	16.7 Earth days
Saturn	Titan	5150 km	15.9 Earth days
Uranus	Titania	1580 km	8.7 Earth days
Neptune	Triton	2700 km	5.9 Earth days
Pluto	Charon	1180 km	6.4 Earth days

Critical Thinking

17. Identify Relationships Some of the moons of the gas giants are larger than Mercury. Why are they not considered to be planets?

_____

_____

_____

Section 4 Review

NSES ES 1a, 3a, 3b, 3c

SECTION VOCABULARY

eclipse an event in which the shadow of one celestial body falls on another **phase** the change in the sunlit area of one celestial body as seen from another celestial body	**satellite** a natural or artificial body that revolves around a planet

1. Compare How is a solar eclipse different from a lunar eclipse?

_____

_____

2. Identify Fill in the blanks to complete the chart.

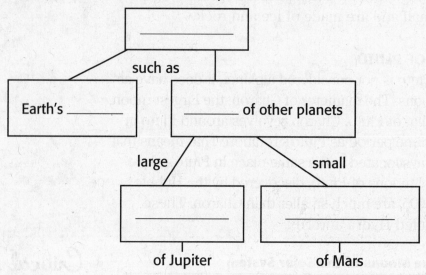

An object that revolves around a planet is called a

_____

such as

Earth's _____   _____ of other planets

large   small

_____   _____

of Jupiter   of Mars

3. Analyze Methods How can astronomers use rocks from the moon to estimate the age of the solar system?

_____

_____

_____

4. Explain Why don't eclipses happen every month?

_____

_____

_____

CHAPTER 21 | A Family of Planets
SECTION 5

Small Bodies in the Solar System

After you read this section, you should be able to answer these questions:

- What are comets?
- What are asteroids?
- What are meteoroids?

What Are Comets?

The sun, the planets, and their moons are not the only objects in our solar system. There are also a large number of smaller bodies, including comets, asteroids, and meteoroids. Scientists study these objects to learn about the formation and composition of the solar system.

A **comet** is a small, loosely packed body of ice, rock, and dust. The *nucleus*, or core, of a comet is made of rock, metal, and ice. A comet's nucleus can range from 1 km to 100 km in diameter. A spherical cloud of gas and dust, called a *coma*, surrounds the nucleus. The coma may extend as far as 1 million kilometers from the nucleus. ☑

COMET TAILS

A comet's tail is its most spectacular feature. Sunlight changes some of the comet's ice to gas, which streams away from the nucleus. Part of the tail is made of *ions*, or charged particles. The *ion tail*, pushed by the solar wind, always points away from the sun, no matter which way the comet is moving. A second tail, the *dust tail*, follows the comet in its orbit. Some comet tails are more than 80 million kilometers long, glowing brightly with reflected sunlight.

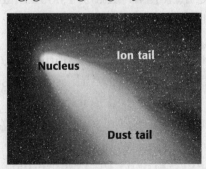

This image shows the physical features of a comet when the comet comes close to the sun. The nucleus of the comet is hidden by the brightly lit gases and dust of the coma.

Say It

Compare In your notebook, create a table that compares comets, asteroids, and meteoroids.

READING CHECK

1. Describe What are comets made of?

_____

_____

TAKE A LOOK

2. Identify Draw an arrow from the nucleus label showing the direction the comet is moving.

SECTION 5 Small Bodies in the Solar System *continued*

COMET ORBITS AND ORIGINS

Remember that the planets move in *elliptical*, or oval-shaped, orbits. Comets also move in elliptical orbits. However, the orbits of comets are much more stretched out than the orbits of planets.

Scientists think that many comets come from the Oort cloud. The *Oort cloud* is a spherical cloud of dust and ice. It surrounds the solar system, far beyond the orbit of Pluto. Pieces of the Oort cloud may fall into orbits around our sun and become comets. Some comets may also come from the *Kuiper belt*, a flat ring of objects just beyond Neptune's orbit. ☑

✓ **READING CHECK**

3. Identify Where is the Oort cloud located?

_____

_____

TAKE A LOOK

4. Explain Why does the ion tail extend in different directions during most of the comet's orbit?

_____

_____

_____

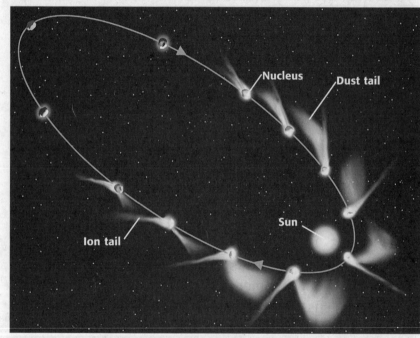

Comets have very long orbits that take them close to the sun and well beyond Pluto.

What Are Asteroids?

Asteroids are small, rocky bodies that revolve around the sun. They range from a few meters to almost 1,000 km in diameter. More than 50,000 asteroids have been discovered. None of them can be seen from Earth without a telescope. In fact, scientists didn't know that asteroids exist until 1801.

Most asteroids orbit the sun in the **asteroid belt**. This is a region that is 300 million km wide and is located between the orbits of Mars and Jupiter. Astronomers think that asteroids are made of material from the early solar system. The pull of Jupiter's gravity prevented this material from coming together to form a planet. ☑

✓ **READING CHECK**

5. Identify Where is the asteroid belt?

_____

SECTION 5 Small Bodies in the Solar System *continued*

COMPOSITION OF ASTEROIDS

It is hard to determine what asteroids are made of. This is because they are small and usually far away from Earth. Mostly, they are composed of either rock or metal. Some asteroids may contain carbon and carbon compounds.

In general, asteroids do not have a spherical shape because of their small size. Gravity must be very strong to pull matter together into a spherical shape. Only the largest asteroids are spherical.

What Are Meteoroids?

Pieces of dust and debris from asteroids and comets, called **meteoroids**, are scattered throughout the solar system. Most meteoroids are about the size of a grain of sand. When a meteoroid enters Earth's atmosphere, it can reach a speed of up to 250,000 km/h.

Friction with the atmosphere heats meteoroids and the air around them, causing them to glow brightly. The glowing trails that form when meteoroids burn up in the atmosphere are called **meteors**. A meteor can be a few hundred meters in diameter and tens of kilometers long before it fades.

Sometimes, a larger meteoroid enters the atmosphere. Some of these meteoroids pass through the atmosphere without burning up completely. When they reach Earth's surface, they are called **meteorites**. ☑

Critical Thinking

6. Make Inferences How do you think scientists know what asteroids are made of?

_____

_____

_____

TAKE A LOOK

7. Compare How do asteroid sizes compare to planet sizes?

_____

_____

✓ **READING CHECK**

8. Compare What is the difference between a meteoroid and a meteorite?

_____

_____

_____

_____

_____

SECTION 5 Small Bodies in the Solar System *continued*

TYPES OF METEORITES

Scientists classify meteorites based on composition. There are three main types of meteorites: stony, metallic, and stony-iron. *Stony meteorites* are similar to rocks on Earth. Some of them contain carbon compounds similar to those found in living organisms. Stony meteorites probably come from carbon-rich asteroids. ☑

Metallic meteorites have a distinctive metallic appearance and do not look like terrestrial rocks. They are made mainly of iron and nickel. *Stony-iron meteorites* are made of a combination of rocky material, iron, and nickel.

Three Major Types of Meteorites

| Stony Meteorite: rocky material | Metallic Meteorite: iron and nickel | Stony-iron Meteorite: rocky material, iron, and nickel |

Scientists study meteorites to learn about the early solar system.

Remember that asteroids and comets are probably made of debris from the formation of our solar system. Meteorites are easier for scientists to study than asteroids and comets. Because meteorites are pieces of asteroids and comets, scientists study meteorites to learn about the early solar system.

METEOR SHOWERS

Meteors can be seen on most clear nights. When many small meteoroids enter the atmosphere in a short period, it is called a *meteor shower*. During some meteor showers, several meteors are visible every minute. Meteor showers happen at the same time each year. These showers happen when Earth passes through orbits of comets that have left a dust trail.

☑ **READING CHECK**
9. List What are the three types of meteorites?

_____

_____

_____

TAKE A LOOK
10. Identify What two metals are found in metallic meteorites?

_____

_____

Critical Thinking
11. Infer Why are meteorites easier to study than asteroids or comets?

_____

_____

SECTION 5 Small Bodies in the Solar System *continued*

IMPACTS IN OUR SOLAR SYSTEM

Impacts are common in our solar system. An *impact* happens when an object in space collides with another object. In many cases, impacts produce impact craters. Many of the planets and moons in our solar system, including Earth, have visible impact craters.

Planets and moons with atmospheres have fewer impact craters than those without atmospheres. For example, there are only a few visible impact craters on Earth. However, the surface of our moon is covered with impact craters. Earth has fewer craters because the atmosphere acts as a shield. Most objects that enter Earth's atmosphere burn up before they reach the surface. ☑

Another reason that there are few visible impact craters on Earth is that Earth has a very active surface. Plate tectonics, weathering, erosion, and deposition act to smooth out and change Earth's surface. These processes are less common on other planets and moons.

Most objects that enter Earth's atmosphere are small and burn up completely before reaching the surface. However, scientists think that impacts powerful enough to cause a natural disaster happen every few thousand years. An impact large enough to cause a global catastrophe may happen once every 50 to 100 million years.

✔ **READING CHECK**

12. Identify Why do fewer meteorites hit Earth's surface than the surface of the moon?

_____

_____

_____

_____

THE TORINO SCALE

Scientists can track objects that are close to Earth to learn whether they might hit Earth. Scientists use the *Torino scale* to rate the chance than an object will hit the Earth. The Torino scale ranges from 0 to 10. Zero indicates that an object has a very small chance of hitting the Earth. Ten indicates that the object will definitely hit the Earth. The Torino scale is also color coded, as shown in the table below.

Color	Number	Hazard level
White	0	very low; almost certainly will not hit the Earth
Green	1	low
Yellow	2, 3, or 4	moderate
Orange	5, 6, or 7	high
Red	8, 9, or 10	very high; almost certainly will hit the Earth

TAKE A LOOK

13. Identify Which color on the Torino scale is used to describe an object that will probably hit the Earth?

_____

Section 5 Review

SECTION VOCABULARY

asteroid a small, rocky object that orbits the sun; most asteroids are located in a band between the orbits of Mars and Jupiter	**meteor** a bright streak of light that results when a meteoroid burns up in Earth's atmosphere
asteroid belt the region of the solar system that is between the orbits of Mars and Jupiter and in which most asteroids orbit	**meteorite** a meteoroid that reaches the Earth's surface without burning up completely
comet a small body of ice, rock, and cosmic dust that follows an elliptical orbit around the sun and that gives off gas and dust in the form of a tail as it passes close to the sun	**meteoroid** a relatively small, rocky body that travels through space

1. Describe How can a comet become the source of meteoroids and meteors?

_____

_____

2. Classify Fill in the blanks to complete the table.

Object	Composition	Main Location
	Large chunk of rock or metal— much smaller than planets	
		Oort cloud and Kuiper belt
	small chunk of rock or metal	throughout the solar system

3. Identify Connections Why is information about comets, asteroids, and meteoroids important for understanding the development of the solar system?

_____

_____

_____

4. Apply Concepts Why would scientists want to know if an asteroid is on a course to collide with Earth in 20 years?

_____

_____

CHAPTER 22 Exploring Space
SECTION 1 **Rocket Science**

After you read this section, you should be able to answer these questions:

• How were rockets developed?

• How do rockets work?

How Did Rocket Science Begin?

Years ago, people could only dream of traveling into space. The problem was that no machine could generate enough force to overcome Earth's gravity.

In the early 1900s, a Russian teacher named Konstantin Tsiolkovsky proposed that rockets could take people into space. A **rocket** is a machine that moves because of the force produced by escaping gas.

Tsiolkovsky proved that rockets could produce enough force to reach outer space. However, he never built any rockets himself. The first rockets were built by Robert Goddard, an American physicist and inventor. Over time, Goddard tested more than 150 rocket engines. He is sometimes called the "father of modern rocketry." ☑

During World War II, the United States military became interested in Goddard's work on rocket engines. They were interested in rockets because of a new German weapon.

Germany had developed the V-2 rocket. It could carry explosives over long distances. The German scientists who developed this rocket surrendered to the Americans near the end of the war. The United States gained 127 of the best German rocket scientists. They helped improve rocket science in the United States during the 1950s.

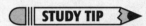

STUDY TIP

Compare After you read this section, make a chart comparing orbital velocity and escape velocity. In your chart, define each velocity and give the speed an object has to travel in order to reach each velocity.

READING CHECK

1. Identify Who was Robert Goddard?

_____

_____

_____

_____

The German V-2 rocket is the ancestor of all modern rockets.

TAKE A LOOK
2. Describe What was the V-2 rocket?

_____

_____

_____

SECTION 1 Rocket Science *continued*

Say It

Investigate The U.S. government formed NASA as part of a "space race" with the U.S.S.R. during the Cold War. Find out more about the Cold War. Share what you learn with a small group.

Math Focus

3. Calculate The space shuttle is about what fraction of the height of the *Saturn V* rocket?

_____

4. Explain Why do rockets move?

_____

_____

5. Define What is thrust?

_____

_____

THE FORMATION OF NASA

In 1958, the U.S. government combined all of the rocket-development teams in the country into one new group. This group was called the National Aeronautics and Space Administration, or **NASA**. The scientists at NASA have developed many rockets to explore space. Two of these rockets are shown in the figure below.

Saturn V rocket
First successful launch: 1967
Height: 111 m

Space shuttle and boosters
First successful launch: 1981
Height: 56 m

How Do Rockets Work?

Many people think that rockets move because gases from the rocket push down on the launch pad. However, this is not the case. If it were, rockets could not move in space, where there is nothing for the gases to push against. Instead, rockets move because the forces on them are not balanced. ☑

To understand how this works, imagine blowing up a balloon and holding its end closed. The air in the balloon pushes on all parts of the balloon with the same force. The forces on the balloon are balanced, so it does not move. Now, imagine letting go of the end of the balloon. The end of the balloon is open, so the air cannot push on it. The air still pushes on the front of the balloon. The forces on the balloon are not balanced, so it moves forward.

This is similar to how a rocket works. The fuel inside the rocket produces gases as it burns. These gases push on the inside of the rocket. The rocket is like the balloon with the open end. The gases can't push on the bottom of the rocket, but they can push on the top. Therefore, the rocket moves upward. The force that makes the rocket move is called **thrust**. The figure on the top of the next page shows how unbalanced forces cause rockets to move. ☑

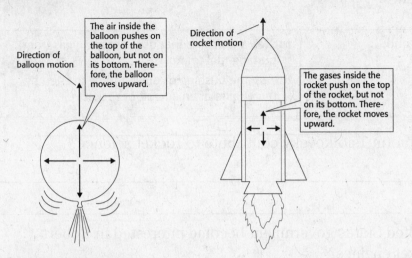

Direction of balloon motion

The air inside the balloon pushes on the top of the balloon, but not on its bottom. Therefore, the balloon moves upward.

Direction of rocket motion

The gases inside the rocket push on the top of the rocket, but not on its bottom. Therefore, the rocket moves upward.

TAKE A LOOK
6. Explain Why do the balloon and the rocket move upward?

_____

_____

_____

CARRYING OXYGEN

Rockets burn fuel to produce the gas that causes them to move. Fuel requires oxygen to burn, but there is almost no oxygen in space. Therefore, rockets that go into outer space must carry oxygen with them.

ORBITAL VELOCITY AND ESCAPE VELOCITY

Remember that gravity attracts objects toward the Earth. A rocket must reach a certain *velocity*, or speed and direction, to orbit the Earth or escape its gravity.

Orbital velocity is the velocity a rocket must travel in order to orbit a planet or moon. The orbital velocity for the Earth is about 8 km/s. If a rocket goes any slower, it will fall back to Earth.

Escape velocity is the velocity a rocket must travel to break away from a planet's gravitational pull. The escape velocity for the Earth is about 11 km/s.

Critical Thinking

7. Compare How is orbital velocity different from escape velocity?

_____

_____

_____

_____

_____

_____

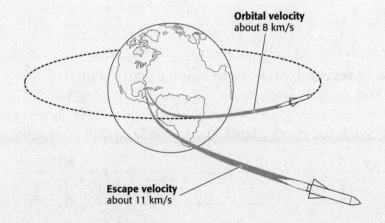

Orbital velocity
about 8 km/s

Escape velocity
about 11 km/s

Section 1 Review

SECTION VOCABULARY

NASA the National Aeronautics and Space Administration	**rocket** a machine that uses escaping gas from burning fuel to move
	thrust the pushing or pulling force exerted by the engine of an aircraft or rocket

1. Describe How did Konstantin Tsiolkovsky contribute to rocket science?

_____

_____

2. Explain Why did the United States government become interested in Robert Goddard's work on rocket engines?

_____

_____

3. Identify How did NASA form?

_____

_____

4. Apply Concepts Draw an arrow on the figure below to show which direction the rocket will move.

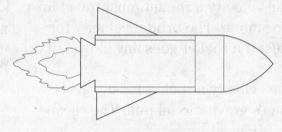

5. Explain Why do rockets carry oxygen in addition to fuel?

_____

_____

6. Identify What must a rocket overcome in order to reach escape velocity?

_____

7. Infer Is escape velocity the same for every planet and moon? Explain your answer.

_____

_____

_____

CHAPTER 22 | Exploring Space
SECTION
2 **Artificial Satellites**

BEFORE YOU READ

After you read this section, you should be able to answer these questions:

- What is an artificial satellite?
- What are LEO and GEO?
- How do satellites benefit people?

How Do Satellites Affect Our Lives?

Imagine that you are watching TV. An emergency weather bulletin interrupts your favorite show. There is a hurricane warning. You grab your cell phone to call your friend. She lives in the hurricane's path.

The TV show, the weather bulletin, and probably even the phone call were made possible by artificial satellites. An **artificial satellite** is a human-made object that orbits a body in space. ☑

There are many kinds of artificial satellites. *Weather satellites* track the movements of gases in the atmosphere. This allows us to predict weather on Earth's surface. *Communications satellites* relay TV programs, phone calls, and computer data. *Remote-sensing satellites* monitor changes in the environment.

THE FIRST ARTIFICIAL SATELLITES

The Soviet Union launched the first artificial satellite, *Sputnik 1*, in 1957. It orbited for 57 days before it fell back to Earth. Two months later, *Sputnik 2* carried the first living thing, a dog, into space. The United States launched its first satellite, *Explorer 1*, in 1958. By 1964, communications satellite networks were sending messages around the world.

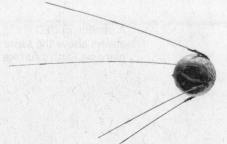

Sputnik 1 was the first artificial satellite to orbit the Earth.

STUDY TIP

Compare As you read this section, make a table comparing the advantages and disadvantages of low Earth orbits and geostationary orbits.

✔ READING CHECK

1. Define What is an artificial satellite?

_____

_____

_____

TAKE A LOOK

2. Identify What was *Sputnik 1*?

_____

_____

How Do Artificial Satellites Orbit?

Artificial satellites orbit the Earth in two main ways: low Earth orbit and geostationary orbit. Early satellites were placed in **low Earth orbit** (LEO). Low Earth orbit is a few hundred kilometers above the Earth's surface. A satellite in LEO moves around the Earth very quickly. Because it is close to the Earth's surface, it can provide clear images of Earth. ☑

Satellites transmit their information back to the Earth's surface. In order for a place on Earth to receive information from a satellite, the satellite must be nearby. Satellites in LEO can be above different places on Earth at different times. They cannot always transmit their information back to receivers on Earth, because they are not always above their recievers. Therefore, they are out of contact with the surface for much of the time.

Most communications satellites and weather satellites move in geostationary orbits. In **geostationary orbit** (GEO), a satellite travels in an orbit that matches Earth's rotation. Therefore, the satellite is always above the same spot on Earth. Satellites in GEO are farther from the Earth than satellites in LEO. Therefore, satellites in GEO cannot produce such detailed images of the Earth's surface.

A satellite in GEO is always above its receivers on Earth. Therefore, these satellites are not out of contact with the surface. This allows the satellite transmissions, such as TV programs and phone calls, to be uninterrupted.

☑ **READING CHECK**

3. Describe About how far above the Earth's surface are satellites in LEO?

_____

_____

Critical Thinking

4. Infer Why is it important for communications satellites to be in GEO?

_____

_____

_____

_____

_____

TAKE A LOOK

5. Identify Which kind of orbit allows a satellite to stay over the same place on Earth all the time?

_____

_____

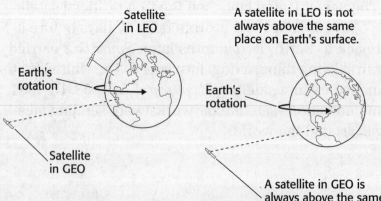

Satellite in LEO

Earth's rotation

Satellite in GEO

A satellite in LEO is not always above the same place on Earth's surface.

Earth's rotation

A satellite in GEO is always above the same place on Earth's surface.

MILITARY SATELLITES

Cameras on satellites in LEO can take very detailed photos. The United States and the former Soviet Union used photographs from satellites to spy on each other. The United States and many other countries continue to use spy satellites for defense purposes. Scientists can also use photographs from military satellites to study changes on the Earth's surface. These changes include flooding, erosion, forest fires, and animal movements. ☑

READING CHECK

6. Explain How do scientists use the information from military satellites?

_____

_____

This photograph of San Francisco was taken by a Soviet spy satellite in LEO in 1989. The satellite was about 220 km above San Francisco when it took this picture.

Math Focus

7. Convert About how many miles above San Francisco was the satellite that took this picture?
1 km = 0.62 mi

_____

THE GLOBAL POSITIONING SYSTEM

The *Global Positioning System* (GPS) is an example of military technology that has become part of everyday life. The GPS is made of 27 satellites that constantly send radio signals to the Earth. Receivers on Earth pick up these signals. The receivers can determine how far they are from each satellite. By combining information from four satellites, a GPS receiver can calculate its exact location on the Earth's surface. ☑

People use GPS receivers for many different reasons. Some are placed in cars to help prevent people from getting lost. Hikers and boaters use them as guides when they travel.

READING CHECK

8. Identify What is the GPS?

_____

_____

_____

SECTION 2 Artificial Satellites *continued*

How Do Weather Satellites Work?

Weather satellites provide important information about conditions in Earth's atmosphere. Some weather satellites are in GEO. These satellites provide a "big-picture" view of the Earth's atmosphere. They monitor the atmosphere and look for places where severe weather may happen.

Some weather satellites are in LEO. These satellites are usually placed in polar orbits. *Polar orbits* are orbits that are nearly at right angles to Earth's direction of rotation. The figure below shows a satellite in polar LEO. ☑

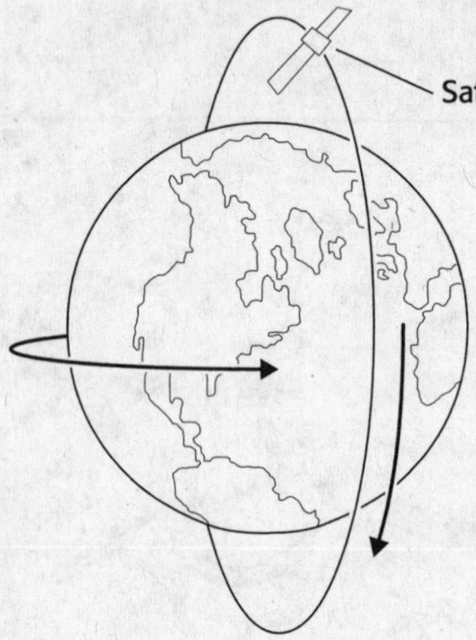

Weather satellites in polar LEO can provide detailed information about the weather in an area.

How Do Communications Satellites Work?

Many types of communications use radio waves or microwaves to carry messages. These waves are useful for communications because they can travel over long distances. However, the waves travel in straight lines, and they cannot travel through the Earth. This means that it is impossible to send a message directly to someone on the other side of the Earth. ☑

Communications satellites in GEO have helped to solve this problem. They *relay*, or send, information from one point on Earth's surface to another. For example, people in the United States can watch television programs from China. The television signals travel from China to a communications satellite in GEO. The satellite transmits the signal to other communications satellites. These satellites transmit the signal to receivers in the United States.

READING CHECK

9. Define What is a polar orbit?

_____

_____

_____

TAKE A LOOK

10. Infer How is the information from a weather satellite in LEO different from that of a satellite in GEO?

_____

_____

_____

READING CHECK

11. Explain Why is it impossible to send a microwave or radio wave message directly to someone on the other side of the Earth?

_____

_____

_____

SECTION 2 Artificial Satellites *continued*

What Are Remote-Sensing Satellites?

Scientists can use satellites to study the Earth in ways that were not possible before. Satellites gather information by remote sensing. *Remote sensing* is the gathering of images and data from a distance. Remote-sensing satellites measure light and other forms of energy that are reflected from the Earth's surface. They use the measurements to make detailed maps of the Earth's surface. ☑

THE LANDSAT PROGRAM

The Landsat program is one of the most successful remote-sensing projects. It began in 1972 and continues today.

Landsat images have helped scientists to identify and track environmental changes. For example, the figures below show Landsat images of part of the Mississippi Delta. One image was taken in 1973. The other was produced in 2003. The images show that the delta is shrinking because less silt is reaching it. They also show that wetlands in the area are disappearing. The information from these satellite images can help scientists reduce these problems.

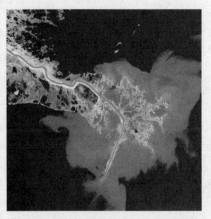 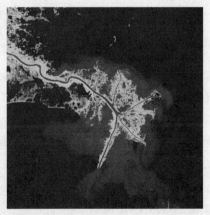

These Landsat images show how the Mississippi Delta changed over 30 years. Because of human activities, the delta is shrinking and wetlands are disappearing.

NEWER REMOTE-SENSING SATELLITES

One of the newest remote-sensing satellites is *Terra 1*. This satellite is part of NASA's Earth Observing System (EOS). Satellites in the EOS program are designed to work together to combine many kinds of data. For example, these satellites can track changes in the land, the atmosphere, the oceans, and the icecaps.

☑ **READING CHECK**

12. Explain How do remote-sensing satellites make maps of the Earth's surface?

_____

_____

_____

TAKE A LOOK

13. Describe What do the Landsat images of the Mississippi Delta show?

_____

_____

Section 2 Review

SECTION VOCABULARY

artificial satellite any human-made object placed in orbit around a body in space	**low Earth orbit** an orbit that is less than 1,500 km above Earth's surface
geostationary orbit an orbit that is about 36,000 km above Earth's surface and in which a satellite is above a fixed spot on the equator	

1. Identify What was the first artificial satellite that the United States placed in orbit around the Earth?

_____

2. Compare Give two differences between a satellite in GEO and one in LEO.

_____

_____

_____

3. Describe How does the Global Positioning System work?

_____

_____

_____

_____

_____

4. Identify What do weather satellites do?

_____

_____

5. Explain How do communications satellites allow people in the United States to watch television programs from China?

_____

_____

_____

6. Define What is remote sensing?

_____

CHAPTER 22 Exploring Space

SECTION 3 Space Probes

BEFORE YOU READ

After you read this section, you should be able to answer these questions:

• Why do we use space probes to visit other planets?

• What kinds of information can space probes gather?

What Are Space Probes?

What does the surface of Mars look like? Does life exist anywhere else in the solar system? To answer questions like these, scientists send space probes through the solar system. A **space probe** is a vehicle that carries scientific instruments into outer space, but has no people on board. Space probes visit planets or other bodies in space. They can complete missions that would be too dangerous or expensive for humans to carry out.

LUNA AND *CLEMENTINE*

Luna 1, the first space probe, was launched by the Soviet Union in 1959. It flew past the moon. In 1966, *Luna 9* made the first soft landing on the moon's surface. In all, space probes from the United States and the Soviet Union have completed more than 30 lunar missions.

In 1994, the United States probe *Clementine* discovered that craters on the moon may contain water. The water may have been left from comet impacts. In 1998, the *Lunar Prospector* confirmed that frozen water exists on the moon. This ice would be valuable to a human colony on the moon.

STUDY TIP

Summarize Make a timeline showing when the space probes in this section were launched. On the timeline, describe the destination of each probe.

Critical Thinking

1. Infer Why would frozen water on the moon be useful for a human colony there?

_____

_____

_____

_____

Luna 9 (U.S.S.R)
Launched: January 1966
Purpose: to land the first spacecraft on the moon

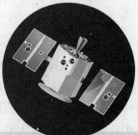

Clementine (U.S.)
Launched: January 1994
Purpose: to map the composition of the moon's surface

TAKE A LOOK

2. Identify What was the first spacecraft to land softly on the moon?

_____

SECTION 3 Space Probes *continued*

VENERA 9: THE FIRST PROBE TO LAND ON VENUS

The Soviet probe *Venera 9* was the first probe to land on Venus. It parachuted into Venus's atmosphere and transmitted images of the surface to Earth. *Venera 9* found that the surface temperature and atmospheric pressure on Venus are much higher than on Earth. ☑

Venera 9 and earlier missions to Venus showed that Venus has a severe greenhouse effect. Today, scientists study Venus's atmosphere to learn about the effects of increased greenhouse gases in Earth's atmosphere.

THE MAGELLAN MISSION: MAPPING VENUS

The United States launched the *Magellan* probe in 1989. This probe mapped 98% of the surface of Venus. The data were transmitted back to Earth. Computers used the data to produce three-dimensional images of the surface of Venus. The Magellan mission showed that Venus has surface features that are similar to Earth's. Some of these features suggest that plate tectonics occurs there. Venus also has volcanoes.

✓ **READING CHECK**

3. Identify What was the first space probe to land on Venus?

_____

_____

Math Focus

4. Reduce Fractions What fraction of Venus' surface did *Magellan* map? Give your answer as a reduced fraction.

_____

_____

TAKE A LOOK

5. Identify What was the purpose of the *Venera 9* probe?

_____

_____

Venera 9 (U.S.S.R.)
Launched: June 1975
Purpose: to record the surface conditions of Venus

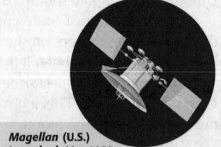

Magellan (U.S.)
Launched: May 1989
Purpose: to make a global map of the surface of Venus

THE VIKING MISSIONS: EXPLORING MARS

The United States sent a pair of probes called *Viking 1* and *Viking 2* to Mars in 1975. The surface of Mars is similar to the Earth's surface. Therefore, one of the main goals of the Viking mission was to look for signs of life on Mars. The probes gathered soil and tested it for evidence of life. They did not find signs of life. However, they did discover that Mars was once much warmer and wetter than it is now.

SECTION 3 Space Probes *continued*

THE MARS PATHFINDER MISSION: REVISITING MARS

A NASA space probe, the *Mars Pathfinder*, visited the surface of Mars again in 1997. The goal of the Mars Pathfinder mission was to explore Mars more cheaply than the Viking missions. The probe sent back images of channels on the planet's surface. The channels look like dry river valleys on Earth. These images suggest that running water may once have flowed on Mars.

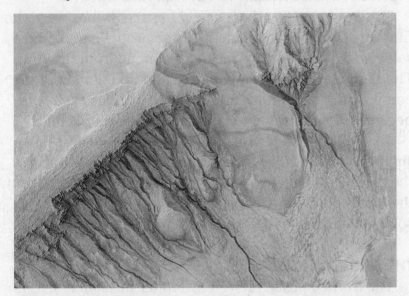

The *Mars Pathfinder* probe took many photographs of the surface of Mars. These photographs showed many features on the surface of Mars. Some of the features, like those shown here, indicate that liquid water may once have flowed over Mars' surface.

The *Mars Pathfinder* probe landed on Mars and sent out the *Sojourner* rover. The *Sojourner* traveled across the surface of Mars for almost three months, collecting data and recording images. The European Space Agency and NASA have more Mars missions planned in the near future. ☑

TAKE A LOOK
6. Explain How do scientists know that liquid water may once have flowed on the surface of Mars?

_____

_____

_____

_____

✓ **READING CHECK**
7. Identify What did the Sojourner rover do?

_____

_____

_____

Viking 2 (U.S.)
Launched: September 1975
Purpose: to search for life on the surface of Mars

Mars Pathfinder (U.S.)
Launched: December 1996
Purpose: to use inexpensive technology to study the surface of Mars

SECTION 3 Space Probes *continued*

How Can Space Probes Help Us Learn About the Outer Solar System?

Jupiter, Saturn, Uranus, and Neptune make up the outer solar system. These planets are very far away. Probes to these planets may take 10 years or more to complete their missions. ☑

PIONEER AND VOYAGER: TO JUPITER AND BEYOND

The *Pioneer 10* and *Pioneer 11* space probes were the first to visit the outer planets. These probes sampled the *solar wind*—the flow of particles coming from the sun. The Pioneer probes also gathered a lot of data about the composition of Jupiter's atmosphere. In 1983, *Pioneer 10* became the first probe to travel past the orbit of Pluto.

The Voyager space probes were the first to detect Jupiter's faint rings. *Voyager 2* was the first probe to fly by the gas giants—Jupiter, Saturn, Uranus, and Neptune. Today, the Pioneer and Voyager probes are near the edge of the solar system. Some of them are still sending back data! ☑

THE GALILEO MISSION

The *Galileo* probe arrived at Jupiter in 1995. It sent a smaller probe to measure the composition, density, temperature, and cloud structure in Jupiter's atmosphere. *Galileo* gathered data about Jupiter's magnetic properties and the geology of some of Jupiter's moons. *Galileo* discovered that some of the moons have magnetic fields. It also found that one of the moons, Europa, may have an ocean of liquid water under its icy surface.

READING CHECK

8. Explain Why do probes to the outer solar system take so long to complete their missions?

_____

_____

READING CHECK

9. Identify What was the first probe to fly by all four of the gas giants?

_____

TAKE A LOOK

10. Infer How long did it take Pioneer 10 to travel from Earth to Pluto?

_____

_____

Pioneer 10 (U.S.)
Launched: March 1972
Purpose: to study Jupiter and the outer solar system

Galileo (U.S.)
Launched: October 1989
Purpose: to study Jupiter and its moons

SECTION 3 Space Probes *continued*

THE CASSINI MISSION: EXPLORING SATURN'S MOONS

The *Cassini* space probe was launched in 1997 on a seven-year journey to Saturn. In 2005, it launched a small probe to examine the atmosphere of Titan, one of Saturn's moons. Scientists think that Titan's atmosphere is similar to the Earth's early atmosphere. Therefore, studying Titan's atmosphere may help scientists learn how Earth's atmosphere formed and changed.

What Are Some More Recent Space Probes?

The early space probe missions were large and expensive. Probes such as *Voyager 2* and *Galileo* took many years to develop. Today, NASA is trying to develop missions that are "faster, cheaper, and better."

STARDUST: COMET DETECTIVE

The *Stardust* space probe was the first probe to focus only on a comet. The probe was launched in 1999 and arrived at the comet in 2004. It gathered samples of the comet's dust tail and returned the samples to Earth in 2006.

DEEP SPACE 1: TESTING ION PROPULSION

Another new space probe project is the New Millennium program. The purpose of this program is to test new technologies that can be used in the future. *Deep Space 1*, launched in 1998, is the first mission of this program. ☑

The purpose of the *Deep Space 1* mission is to test a new type of propulsion system. *Deep Space 1* uses an ion-propulsion system. Ion-propulsion systems may help future space probes travel more quickly with less fuel.

Deep Space 1 uses an ion rocket to move.

Critical Thinking

11. Explain Why are scientists interested in the atmosphere of Titan?

_____

_____

_____

_____

Critical Thinking

12. Infer What was the purpose of the *Stardust* probe?

_____

_____

_____

 **READING CHECK**

13. Describe What is the purpose of the New Millennium program?

_____

_____

Section 3 Review

SECTION VOCABULARY

space probe an uncrewed vehicle that carries scientific instruments into space to collect scientific data

1. Identify What was the first space probe to fly past the moon?

_____

2. Describe Complete the table below.

Probe	Year of launch	Purpose
Venera 9		
Viking 2		
Clementine		
Pioneer 10		
Galileo		
Cassini		
Deep Space 1		

3. Describe What did the Magellan probe discover?

_____

_____

4. Explain What was unique about the mission of the *Stardust* probe?

_____

_____

5. Identify Relationships How can missions to Venus and Titan help scientists learn about the Earth?

_____

_____

_____

_____

_____

CHAPTER 22 | Exploring Space

SECTION
4 **People in Space**

BEFORE YOU READ

After you read this section, you should be able to answer
these questions:

• Who was the first person in space?

• How are space shuttles different from other space
vehicles?

• Why are space stations important?

Who Were the First People in Space?

On April 12, 1961, a rocket stood on a launch pad
in the Soviet Union. A cosmonaut named Yuri Gagarin
waited inside. He was about to do what no human had
done before. He was about to travel to outer space. No
one knew if his brain would function in space or if he
would be instantly killed by radiation.

Later that day, Yuri Gagarin became the first human
to orbit Earth. His flight lasted 108 minutes. News of his
success was quickly broadcast around the world. ☑

In February 1962, John Glenn became the first
American to orbit the Earth. Seven years later, the world
watched on television as the *Apollo 11* landing module
landed on the moon. During the *Apollo 11* mission, Neil
Armstrong became the first person to stand on a world
other than Earth. *Apollo 11* carried moon rocks back to
Earth for scientists to study. Its crew also put devices on
the moon to study moonquakes and the solar wind.

Person	Accomplishment
Yuri Gagarin	
John Glenn	
	first person to stand on a world other than Earth

REUSABLE SPACE VEHICLES

The Saturn V rockets, which carried the Apollo astronauts to the moon, were large and expensive. To save
money and natural resources, NASA began to develop the
space shuttle program in 1972. A **space shuttle** is a reus-
able space vehicle that takes off like a rocket and lands
like an airplane. Since 1981, NASA has completed more
than 100 space shuttle missions.

STUDY TIP

Outline Before you read this
section, make an outline of
it using the headings from
the section. As you read, fill
in the outline with the main
ideas of the section.

READING CHECK

1. Identify Who was the first
person to orbit the Earth?

_____

_____

TAKE A LOOK

2. Describe Fill in the blank
spaces in the table.

SECTION 4 People in Space *continued*

PARTS OF A SPACE SHUTTLE

A space shuttle has three main parts. The *orbiter* is the part that looks like an airplane. The orbiter carries the astronauts and equipment into space. The *liquid-fuel tank* carries some of the fuel for the rockets. Two white *solid-fuel booster* rockets help the shuttle reach orbit. Then, they fall back to Earth, along with the liquid-fuel tank. The booster rockets are reused, but the fuel tank is not.

Critical Thinking

3. Apply Concepts Why doesn't the orbiter need booster rockets to keep it in orbit?

(Hint: Why doesn't the moon need booster rockets to stay in orbit?)

_____

_____

_____

TAKE A LOOK
4. Identify What part of the space shuttle carries the crew?

_____

_____

Solid-fuel booster rockets

Liquid-fuel tank

Orbiter

SHUTTLE TRAGEDIES

On January 28, 1986, a booster rocket on the space shuttle *Challenger* exploded just after takeoff. In 2003, the space shuttle *Columbia* exploded as it reentered the atmosphere. In both instances, all the astronauts on board each shuttle were killed. Tragedies such as these show some of the dangers of space travel. However, scientists have learned from these accidents. They try to make sure that future space flights are safer.

SPACESHIPONE: PRIVATE SPACECRAFT

In 2004, *SpaceShipOne* won the $10 million X Prize. *SpaceShipOne* was the first privately owned, reusable vehicle to carry a crew to an altitude of 100 km. The X Prize was offered to encourage private companies to develop space technology. Some day, ships like *SpaceShipOne* may offer flights to space, much like airplanes carry people to places on Earth today. ☑

Math Focus

5. Convert About how high, in miles, did *SpaceShipOne* carry its crew?

1 km = 0.62 mi

_____

What Is a Space Station?

A **space station** is a large artificial satellite where people can live and work. In 1971, the Soviet Union became the first to successfully place a space station in orbit. By 1982, the Soviets had put up seven space stations, including the space station *Mir*. Cosmonauts on these space stations studied the effects of weightlessness on humans and carried out many other experiments. ☑

Skylab was the United States' first space station. It was a science and engineering laboratory. Scientists on *Skylab* carried out experiments in biology, astronomy, and manufacturing.

THE *INTERNATIONAL SPACE STATION*

The *International Space Station* (*ISS*) is being constructed in LEO by Russia, the United States, and 14 other countries. Construction began in 1998. The first crew boarded in 2000. The *ISS* will provide a unique laboratory for research and experiments. It is scheduled to be completed by 2010.

What Are the Benefits of the Space Program?

Space offers resources beyond those on Earth. For example, a rare form of helium is found on the moon. Some day, this helium might be a fuel in nuclear fusion reactors. These reactors could produce a lot of energy with little harmful waste.

A base on the moon could be used to manufacture materials in low gravity or in a vacuum. A colony on the moon or Mars could help bring space resources to Earth. It could also be a base to explore the rest of the solar system.

Space exploration can be expensive. Why do we continue to explore space? Space exploration serves our quest for new knowledge. It challenges our courage. It also has expanded our technology. Many technologies that were developed for the space program are now used in everyday life. A few of these technologies are listed below.

• smoke detectors	• cordless power tools
• pacemakers	• artificial heart pumps
• land mine removal devices	• medical lasers
• fire fighting equipment	• invisible dental braces
• video game joysticks	• ear thermometers

READING CHECK

6. Identify What country was the first to successfully place a space station in orbit?

_____

_____

Critical Thinking

7. Apply Ideas Is the International Space Station always above the same point on Earth's surface? Explain your answer.

_____

_____

_____

_____

Section 4 Review

SECTION VOCABULARY

space shuttle a reusable space vehicle that takes off like a rocket and lands like an airplane	**space station** a long-term orbiting platform from which other vehicles can be launched or scientific research can be carried out

1. Compare What is the main difference between space shuttles and Saturn V rockets?

_____

_____

2. Identify Who was the first person on the moon?

_____

_____

3. List What are the three main parts of the space shuttle?

_____

_____

_____

_____

4. Identify Which two parts of the space shuttle are reused?

_____

_____

_____

5. List Give five technologies that were developed for the space program that are now used in everyday life.

_____

_____

_____

_____

_____

_____

6. Identify What is *SpaceShipOne*?

_____

_____

7. Describe Give two ways that space exploration may benefit people on Earth.

_____

_____

_____

Photo Credits

Abbreviations used: c-center, b-bottom, t-top, l-left, r-right, bkgd-background.

1 (b) Woods Hole Oceanographic Institution; 2 (c) Dr. Howard B. Bluestein; 3 (bl) Andy Christiansen/HRW; 3 (br) Andy Christiansen/HRW; 11 (b) PhotoDisc/Getty Images; 12 (b) Peter Essick/Aurora; 19 (b) Guy Grenier/Masterfile; 20 (b) Sam Dudgeon/HRW; 25 (bl) Andy Christiansen/HRW; 25 (br) Andy Christiansen/HRW; 29 (c) Texas Department of Transportation; 30 (c) Spaceimaging.com/Getty Images/NewsCom; 35 (c) U. S. Geological Survey; 37 (c) Sam Dudgeon/HRW; 38 (br) Dr. Rainer Bode/Bode-Verlag Gmb; 39 (l) Victoria Smith/HRW; 39 (r) Sam Dudgeon/HRW; 39 (c) Sam Dudgeon/HRW; 39 (copper) Dr. E. R. Degginger/Color-Pic, Inc.; 39 (calcite) Dr. E. R. Degginger/Color-Pic, Inc.; 39 (fluorite) Dr. E. R. Degginger/Color-Pic, Inc.; 39 (corundum) Dr. E. R. Degginger/Color-Pic, Inc.; 39 (gypsum) SuperStock; 39 (galena) Ken Lucas/Visuals Unlimited; 42 (bl) Sam Dudgeon/HRW; 42 (br) Sam Dudgeon/HRW; 42 (bc) Tom Pantages Photography; 43 (tl) in table Mark A. Schneider/Photo Researchers, Inc.; 43 (bl) in table Sam Dudgeon/HRW; 43 (tc) in table Sam Dudgeon/HRW; 43 (bc) in table Sam Dudgeon/HRW; 43 (tr) in table Sam Dudgeon/HRW, Courtesy Science Stuff, Austin, TX; 43 (br) in table Tom Pantages Photography; 48 (t) PhotoDisc/Getty Images; 49 (tr) ©SuperStock; 51 (bl) Michael Melford/Getty Images; 51 (br) Joseph Sohm/Visions of America/CORBIS; 52 (c) © Royalty Free/CORBIS 54 (limestone) Breck P. Kent; 54 (calcite) Mark Schneider/Visuals Unlimited; 54 (aragonite) Breck P. Kent; 54 (granite) Pat Lanza/Bruce Coleman, Inc.; 54 (biotite) Dr. E. R. Degginger/Color-Pic, Inc.; 54 (feldspar) Mark Schneider/Visuals Unlimited; 54 (quartz) PhotoDisc/Getty Images; 55 (cl) Sam Dudgeon/HRW; 55 (c) Copyright Dorling Kindersley; 55 (cr) Breck P. Kent; 55 (bl) Dr. E. R. Degginger/Color-Pic, Inc.; 55 (br) Pat Lanza/Bruce Coleman, Inc.; 58 (cl) Breck P. Kent; 58 (cr) Breck P. Kent; 58 (bl) Breck P. Kent; 58 (br) Victoria Smith/HRW; 61 (b) © Royalty Free/CORBIS; 62 (l) Breck P. Kent; 62 (cl) Joyce Photographics/Photo Researchers, Inc.; 62 (cr) Sam Dudgeon/HRW; 62 (r) Sam Dudgeon/HRW; 63 (t) Breck P. Kent; 63 (b) Franklin P. OSFAnimals Animals/Earth Scenes; 65 (b) George Wuerthner; 67 (c) Jim Wark/Airphoto; 67 (bl) Dane S. Johnson/Visuals Unlimited; 67 (bl) Carlyn Iverson/Absolute Science and Photography; 67 (bl) Breck P. Kent; 67 (br) Breck P. Kent/Animals Animals/Earth Scenes; 69 (shale) Ken Karp/HRW; 69 (slate) Sam Dudgeon/HRW; 69 (phyllite) Sam Dudgeon/HRW; 69 (schist) Courtesy Stan Celestian; 69 (gneiss) Breck P. Kent; 71 (bl) Russell Illiq/Photodisc/gettyimages; 71 (bc) Andy Christiansen/HRW; 71 (br) Mark Lewis/Getty Images; 72 (bl) James Randklev/Getty Images; 72 (br) Myrleen Furgusson Cate/PhotoEdit; 77 (b) Martin Harvey; 79 (cl) NYC Parks Photo Archive/Fundamental Photographs; 79 (cr) Kristen Brochmann/Fundamental Photographs; 83 (c) Bob Rowan/Progressive Image/CORBIS; 84 (b) © Mark Gibson Photography; 101 (b) Layne Kennedy/CORBIS; 102 (b) © Louie Psihoyos/psihoyos.com; 103 (c) G.R. Roberts Photo Library; 104 (cl) Jonathan Blair/CORBIS; 104 (cr) Chip Clark/Smithsonian; 105 (bl) Thomas R. Taylor/Photo Researchers, Inc.; 107 (c) James L. Amos/CORBIS; 109 (c) Chip Clark/Smithsonian; 110 (c) Neg. no. 5793 Courtesy Dept. of Library Svcs., American Museum of Natural History; 111 (t) Neg. no. 5799 Courtesy Dept. of Library Svcs., American Museum of Natural History; 111 (b) Neg. no. 5801 Courtesy Dept. of Library Svcs., American Museum of Natural History; 127 (cl) Peter Van Steen/HRW; 127 (cr) Peter Van Steen/HRW; 131 (t) Tom

Bean; 136 (c) Roger Ressmeyer/CORBIS; 149 (c) Paul Chesley/Getty Images/Stone; 152 (b) Breck P. Kent/Animals Animals/Earth Scenes; 154 (cl) Tui De Roy/Minden Pictures; 154 (cr) Martin Miller/Visuals Unlimited; 154 (bl) B. Murton/Southampton Oceanography Centre/Science Photo Library/Photo Researchers, Inc.; 154 (br) Tom Bean/DRK Photo; 155 (tl) Francois Gohier/Photo Researchers, Inc.; 155 (tr) Glenn Oliver/Visuals Unlimited; 155 (cl) Dr. E. R. Degginger/Color-Pic, Inc.; 155 (cr) Tom Bean/DRK Photo; 155 (b) Alberto Garcia/CORBIS; 168 (cl) Ron Niebrugge/Niebrugge Images; 168 (c) Visuals Unlimited; 168 (cr) Grant Heilman/Grant Heilman Photography, Inc.; 170 (b) Laurence Parent; 184 (b) Paul Chesley/Getty Images; 185 (t) Mark Lewis/ImageState; 185 (c) AgStockUsa; 190 (cl) Pat O'Hara/CORBIS; 190 (cr) Roger Ressmeyer/CORBIS; 192 (t) Laurence Parent; 192 (b) Frans Lanting/Minden Pictures; 193 (t) Galen Rowell/Peter Arnold, Inc.; 193 (b) G.R. Roberts Photo Library; 195 (c) Glenn M. Oliver/Visuals Unlimited; 196 (cl) Earth Satellite Corporation/Photo Researchers, Inc.; 196 (cr) Martin G. Miller/Visuals Unlimited; 197 (c) Jerry Laizure/AP/Wide World Photos; 203 (t) Rich Reid/Animals Animals/Earth Scenes; 203 (b) Leif Skoogfers/Woodfin Camp & Associates, Inc.; 205 (b) Tony Freeman/Photo Edit; 206 (b) Morton Beebe/CORBIS; 211 (b) Tom Bean; 214 (tl) © Royalty Free/CORBIS; 214 (tr) James Randklev/Getty Images; 214 (bl) Jonathan Weston/Imagestate; 214 (br) Rich Reid/Getty Images; 215 (c) InterNetwork Media/Getty Images; 215 (b) Aerial by Caudell; 221 (b) Tom Bean/CORBIS; 223 (c) Tom Bean; 229 (b) Tom Van Sant, Geosphere Project/Planetary Visions/Science Photo Library; 237 (c) James Wilson/Woodfin Camp & Associates; 244 (b) Stuart Westmoreland/CORBIS; 245 (t) Mike Bacon/Tom Stack & Associates; 245 (b) James B. Wood; 246 (t) Al Giddings Images, Inc.; 246 (b) JAMESTEC; 247 (t) Mike Hall/Getty Images; 247 (b) Norbert Wu; 249 (b) Fred Bavendam/Peter Arnold, Inc.; 251 (b) Institute of Oceanographic Sciences/NERC/Science Photo Library/Photo Researchers, Inc.; 253 (b) Dr. E. R. Degginger/Color-Pic, Inc.; 254 (t) Fred Bavendam/Peter Arnold, Inc.; 254 (b) Greenpeace; 255 (t) Courtesy Mobil; 256 (t) Andy Christiansen/HRW; 256 (c) Richard Hamilton Smith/CORBIS; 256 (b) Tony Freeman/PhotoEdit, Inc.; 257 (b) David Young-Wolff/PhotoEdit, Inc.; 296 (cl) © Royalty Free/CORBIS; 296 (cr) Steve Starr/CORBIS; 299 (b) Francis Dean/The Image Works; 303 (b) Sam Dudgeon/HRW; 305 (c) John Morrison/Morrison Photography; 316 (cl) Howard B. Bluestein/Photo Researchers, Inc.; 316 (cr) Howard B. Bluestein/Photo Researchers Inc.; 316 (bl) Howard B. Bluestein/Photo Researchers Inc.; 316 (br) Howard B. Bluestein/Photo Researchers Inc.; 317 (b) NASA; 321 (b) Graham Neden/Ecoscene/CORBIS; 322 (c) G.R. Roberts Photo Library; 333 (b) © Royalty Free/CORBIS; 336 (c) © Royalty Free/CORBIS; 336 (b) Grant Heilman Photography, Inc.; 337 (t) Andrew Brown/Ecoscene/CORBIS; 338 (b) Kathy Bushue/Getty Images; 339 (c) SuperStock; 344 (t) Roger Werth/Woodfin Camp & Associates; 353 (c) NASA; 355 (c) NASA; 355 (c) NASA; 355 (c) NASA; 355 (c) NASA; 362 (tl) Jim Cummings/Getty Images; 362 (tr) NASA; 362 (bl) Jerry Lodriguss/Photo Researchers, Inc.; 362 (br) Tony & Daphne Hallas/Science Photo Library/Photo Researchers, Inc.; 365 (cl) Dr. E. R. Degginger/Color-Pic, Inc.; 365 (cr) Scott Van Osdol/HRW; 368 (t) Andre Gallant/Getty Images; 371 (b) V. Bujarrabal (OAN, Spain), WFPC2, HST, ESA/NASA; 374 (b) Dr. Christopher Burrows, ESA/STScl/NASA; 377 (b) Billy & Sally Fletcher/Tom Stack & Associates; 378 (t) David Malin/Anglo-Australian Observatory; 378 (b) Dennis Di Cicco/Peter Arnold, Inc.; 392 (c) NASA/Mark Marten/Photo Researchers, Inc.;

397 (b) SuperStock; 399 (t) Breck P. Kent/Animals Animals/Earth Scenes; 413 (b) ESA; 416 (b) ESA; 423 (c) John Bova/Photo Researchers, Inc.; 423 (c) John Bova/Photo Researchers, Inc.; 423 (c) John Bova/Photo Researchers, Inc.; 423 (c) John Bova/ Photo Researchers, Inc.; 423 (c) John Bova/Photo Researchers, Inc.; 423 (c) John Bova/Photo Researchers, Inc.; 423 (c) John Bova/Photo Researchers, Inc.; 429 (cr) Bill & Sally Fletcher/Tom Stack & Associates; 432 (l) Breck P. Kent/Animals Animals/Earth Scenes; 432 (c) Dr. E. R. Degginger/Bruce Coleman, Inc.; 432 (r) Ken Nichols/Institute of Meteorites/University of New Mexico; 435 (b) Hulton Archive/Getty Images; 439 (b) Brian Parker/ Tom Stack & Associates; 441 (c) Aerial Images, Inc. and SOVINFORMSPUTNIK; 443 (cl) U. S. Geological Survey; 443 (cr) U. S. Geological Survey; 447 (c) NASA; 449 (b) NASA; 452 (t) Royalty Free/CORBIS